Nanoscience and Nanoengineering

Advances and Applications

T0225549

Nanoscience and Nanoengineering

Advances and Applications

Edited by

Ajit D. Kelkar · Daniel J.C. Herr · James G. Ryan

CRC Press
Taylor & Francis Group
Boca Raton London New York

CRC Press is an imprint of the
Taylor & Francis Group, an **Informa** business

CRC Press
Taylor & Francis Group
6000 Broken Sound Parkway NW, Suite 300
Boca Raton, FL 33487-2742

First issued in paperback 2017

Version Date: 20140422

ISBN 13: 978-1-138-07656-3 (pbk)
ISBN 13: 978-1-4822-3119-9 (hbk)

Library of Congress Cataloging-in-Publication Data

Nanoscience and nanoengineering : advances and applications / editors, Ajit D. Kelkar, Daniel J.C. Herr, James G. Ryan.
　　　pages cm
　　Summary: "Preface The scientific prefix "nano" means one billionth. Therefore, a nanometer is one billionth of a meter, a nanosecond is one billionth of a second and so on. Clusters of atoms and molecules have dimensions in the order a a few nanometers. For example, the diameter of a carbon nanotube is approximately two nanometers and a typical DNA molecule is a little over two nanometers wide. Nanotechnology is often defined as the scientific and engineering know-how to control the arrangement of atoms and molecules enabling novel applications with customized properties. Most formal definitions of nanotechnology usually cites a size upper bound of one hundred nanometers (100 nm). Particles, features, structures, devices, etc., that have dimensions less than 100 nm are referred to as "nano", but in many technologies, this "cutoff" is arbitrary and it is often useful to view structures larger than 100 nm as nanotechnology as well. In order to provide perspective to the reader, it is good to think of the dimensions that nanotechnologists work with compared to objects in the macroscopic world. The two comparisons that I often use to explain relative sizes are that 100 nm is roughly 1000 times smaller than the diameter of a human hair. I also explain that approximately one million carbon nanotubes could be lined up side to side across the diameter of the head of a pin. People have used nanotechnology for hundreds of years but it is only in the last fifty years or so that the drive for miniaturization and the ability to manipulate nanoscale particles, fibers, films and structures has created a technology revolution. Early use of nanoparticles can be seen in the stained glass windows of gothic cathedrals, dichroic glass and in photography"-- Provided by publisher.
　　Includes bibliographical references and index.
　　ISBN 978-1-4822-3119-9 (hardback)
　　1. Nanotechnology. I. Kelkar, A. (Ajit) editor of compilation.

T174.7.N35825 2014
620'.5--dc23   2014002765

Visit the Taylor & Francis Web site at
http://www.taylorandfrancis.com

and the CRC Press Web site at
http://www.crcpress.com

In the loving memory of my parents
Leela D. Kelkar and Dhundiraj D. Kelkar

Ajit D. Kelkar

Contents

SECTION I   Nanoelectronics

SECTION II   Nanobio

SECTION III Nano Medicine

SECTION IV Nanomodeling

SECTION V Nanolithography and Nanofabrication

SECTION VI Nanosafety

Preface

The scientific prefix "nano" means 1 billionth. Therefore, a nanometer is 1 billionth of a meter, a nanosecond is 1 billionth of a second, and so on. Clusters of atoms and molecules have dimensions in the order of a few nanometers. For example, the diameter of a carbon nanotube is approximately 2 nanometers and a typical DNA molecule is a little over 2 nanometers wide. Nanotechnology is often defined as the scientific and engineering know-how to control the arrangement of atoms and molecules enabling novel applications with customized properties. Most formal definitions of nanotechnology usually cite a size upper bound of one hundred nanometers (100 nm). Particles, features, structures, devices, and so on, that have dimensions less than 100 nm are referred to as "nano," but in many technologies, this "cutoff" is arbitrary and it is often useful to view structures larger than 100 nm as nanotechnology as well. In order to provide perspective to the reader, it is good to think of the dimensions that nanotechnologists work with compared to objects in the macroscopic world. The two comparisons that I often use to explain relative sizes are that 100 nm is roughly 1000 times smaller than the diameter of a human hair. I also explain that approximately 1 million carbon nanotubes could be lined up side to side across the diameter of the head of a pin.

People have used nanotechnology for hundreds of years, but it is only over the past 50 years or so that the drive for miniaturization and the ability to manipulate nanoscale particles, fibers, films, and structures has created a technology revolution. Early use of nanoparticles can be seen in the stained glass windows of gothic cathedrals, dichroic glass, and in photography. Although the first users did not realize that they were using "nanotechnology," their work illustrated the amazing capabilities attainable by manipulating matter at nanoscale dimensions.

Nanotechnology's impact to the world's economy is measured in trillions of U.S. dollars, and it is expected to continue to grow as new applications are developed. In 2014, there are more than 1000 products available on the market from companies that identify the products as "nano" products, but there are many others that are "nano-enabled." Nanotechnology is part of the manufacturing processes for automobiles, computers, telecommunication devices, gaming, aerospace and defense applications, sporting equipment, cosmetics, sunscreen, clothing, and many others. There is also limited use of nanotechnology in medicine (e.g., drug delivery technologies, medical devices) and food (e.g., packaging, food safety sensors). Scientists and engineers are continuing to work to expand applications in medical and food applications while paying careful attention to health concerns. The economic impact of nanotechnology as well as the relation to aerospace and defense applications has caused many governments to focus on nanotechnology in order to promote economic development in areas.

Nanotechnology is normally used to improve the performance characteristics of "high tech" products. For example, nanolithography and thin-film technologies have enabled continuous improvements in microprocessor speeds and reduced power consumption. This book will focus on the nanoscience and nanoengineering behind the nanotechnologies that appear in high technology products. Nanoscience and nanoengineering are interdisciplinary in nature and are somewhat intertwined, because clearly there is a considerable amount of engineering involved with nanoscience and a substantial amount of science involved with engineering at the nanoscale.

Nanoscience and nanoengineering are used together to make the building blocks that eventually become part of nanotechnology-based products. For example, fundamental understanding developed for nanofiber synthesis by electrospinning enables the creation of products used in applications from nanomedicine to aircraft. Polymer nanofibers are used in antibacterial filtration applications as well as in more advanced medical applications such as tissue scaffolding. Devices such as sensors and structures based on nanocomposites (e.g., aircraft fuselages, windmill blades) can be fabricated by incorporating electrospun inorganic nanofibers to reduce their weight and improve their strength. Understanding the physics and chemistry of other nanotechniques such as thin-film deposition also enables the fabrication of a similar variety of products. Products that use antireflective films (e.g., sunglasses, cameras) or hard films (e.g., cutting tools) are common, and devices that use thin films as building blocks such as computer chips, MEMS and NEMS, and photovoltaic cells play a critical role in modern life. This book explains the mechanisms underpinning the nanotechnology-based products and technologies that we currently enjoy and those that will change our society in the not-too-distant future.

There are many new nanotechnology texts available to the reader, so the authors were seeking to make *Nanoscience and Nanoengineering: Advances and Applications* focus on emerging "nano" areas and provide researchers and students with a reference that provides depth and breadth in the subject area. Nanotechnology is widely used, but books on the subject often focus on narrow sub-fields. In order to show the true interdisciplinary nature of nanotechnology, the authors wanted to address the breadth of the field, from research to manufacturing, while also providing sufficient depth that the reader would gain understanding of some of the most important discoveries. For example, although general information on topics like nanoelectronics is available elsewhere, the nanoelectronics portion of the book provides an in-depth description of three new device types.

Conventionally, nanoscience involves the development of fundamental understanding applied to nanoscale phenomena such as physical properties of materials and objects such as clusters of atoms, molecules, and biological structures. Nanoscience is sometimes further broken down into sub-disciplines such as nanochemistry, nanophysics, and nanobiology. Our primary nanoscience focus areas in this book will involve nanobiology and its extensions. A field such as nanobiology is an extremely broad sub-discipline with many different sub-categories including nanoenvironmental, nanomedicine, nanotechnologies used in food and agriculture, nanotechnologies used in analysis and characterization, and many others. This book will focus on interdisciplinary areas such as the interface between nanobiology and nanophysics where microscopy and characterization methods are producing discoveries every day. The authors also focus on nanomedicine issues related to drug development and delivery as well as applications in diagnostics. The authors discuss emerging areas that are envisioned to provide fundamental understanding for the subjects that are most likely to lead to the great increase in economic impact associated with the nanotechnology.

Nanoengineering involves the use of fundamental understanding to design and fabricate new materials and devices at nanoscale dimensions. Although, as in nanoscience, there are many sub-disciplines, for the purposes of this book we will focus our discussions in the areas of nanomaterials, nanoelectronics, nanoelectromechanical systems (NEMS) and nanofluidic devices, and computational nanotechnology (nanomodeling) and nanomanufacturing issues including nanosafety.

In the nanomaterials field, we will discuss nanocomposite materials and their advantages over conventional structural materials as well as self-assembly and its potential for nanomanufacturing applications. Compound semiconductors and their applications in communications, display technology and infrared optics will be discussed in the nanoelectronics section. The design and fabrication of NEMS and their applications in nanomedicine will also be discussed. Computational nanotechnology is used in both nanoscience and nanoengineering and is critical to better understanding and design for process materials and nanobiotechnologies.

Nanomanufacturing is vital to the high technology sector but has not been extensively described in existing nanotechnology-oriented books. Due to concerns regarding the use of nanomaterials, understanding the topic of nanosafety has become a critical enabler for building nanomanufacturing capabilities. The authors deal with these two topics in series so that that the reader would be able to see how the technologies involved interact.

The facilities, tools, and techniques used in these areas of nanoscience and nanoengineering are critical to carrying out leading-edge research in these fields. For example, charged particle microscopes must have the ability to resolve objects the size of molecules. The fabrication of structures so tiny that small specks of dirt or a human hair would look like giant objects, must be carried out in a "cleanroom," where room air is continuously filtered and technologists wear special suits to prevent hair and skin flakes from falling on the surface where the structures are to be built. Certain types of nanobiological research must be carried out in "negative pressure" laboratories to assure that biological entities do not escape the confines of the laboratory. Also, simulation facilities that are capable of showing models in three dimensions are critical to understanding molecular structure and issues like drug interaction with target molecules. The authors explain how the combination of facility and instrumental capabilities are brought together with scientific and engineering insights to enable discovery.

This book is intended to be used by students and professionals alike with a goal of sparking their interest to investigate more deeply into the technological advances achieved through manipulation of atomic building blocks.

Most of the authors of this book are faculty, staff, and students at the Joint School of Nanoscience and Nanoengineering of North Carolina A&T State University and the University of North Carolina at Greensboro. The editors wish to thank the publisher, their universities, and their families for their support during the writing of this book.

Editors

Dr. Ajit D. Kelkar is a professor and chair of Nanoengineering Department at Joint School of Nanoscience and Nanoengineering (JSNN) of North Carolina A&T State University and the University of North Carolina at Greensboro. He earned his PhD degree from Old Dominion University, Virginia in mechanical engineering. Dr. Kelkar also serves as an associate director for the Center for Advanced Materials and Smart Structures (CAMSS) at North Carolina A&T State University and is a member of the National Institute of Aerospace (NIA). For the past 32 years he worked in the area of performance evaluation and modeling of polymeric composites and ceramic matrix composites. His expertise are in the area of low-cost fabrication and processing of woven composites using VARTM process, fatigue and impact testing of composites, analytical modeling of woven composites. Currently, he is involved in the development of nanoengineered multifunctional materials using carbon nanotubes (CNNTs), boron nitride nanotubes (BNNTs), electrospun nanofiber materials, and alumina nanoparticles. In the modeling area, he is working on atomistic modeling of polymers embedded with CNTs, BNNTs and alumina nanoparticles. He is also involved in high-velocity impact modeling of ceramic matrix composites and polymeric matrix composites embedded with electrospun nanofibers. He has published more than 200 papers in these areas. In addition, he has edited a book in the area of nanoengineered materials. He is member of several professional societies, including ASME, SAMPE, AIAA, ASM, and ASEE.

Dr. Daniel J.C. Herr is a professor and chair of the Nanoscience Department at the Joint School of Nanoscience and Nanoengineering (JSNN) of North Carolina A&T State University and the University of North Carolina at Greensboro. He earned his PhD degree from University of California Santa Cruz in chemistry. Prior to joining JSNN, Dr. Herr served as Semiconductor Research Corporation's (SRC's) director of Nanomanufacturing Sciences. For the past 28 years with the semiconductor industry, his research focused on designed nanoengineered materials, advanced patterning and directed self-assembly, nanomanufacturing, formulation, process qualification and optimization, sustainable technologies, and advanced device design. Recently, he was elected to serve as the AAAS Industrial Science and Technology section's Member-at-Large, and Fellow of the International Society for Optical Engineering for the design, development, and commercialization of two early families of chemically amplified resists. He is the inventor of several foundational patents and disclosures on defect-tolerant patterning, controlled nanotube synthesis and placement, deterministic semiconductor doping, and ultimate CMOS (complementary metal–oxide–semiconductor) devices. He serves as senior editor for *IEEE Transactions in Nanotechnology, coordinating editor for the Journal of Nanoparticle Research*, and reviewer for the *Journal of Vacuum Science and Technology*. Dr. Herr was a key member of the SRC (Semiconductor Research Corporation) team that was awarded the 2005 National Medal of Technology. He has strong research interests in advanced functional nanomaterials by design, nanoenergy, semiconductor manufacturing, bio-electronics, biomimetics, and sustainable hydroponic composites.

Dr. James G. Ryan is a founding dean of the Joint School of Nanoscience and Nanoengineering (JSNN) of North Carolina A&T State University and the University of North Carolina at Greensboro. He earned his PhD degree from Rensselaer Polytechnic Institute (RPI) in chemistry. Dr. Ryan joined JSNN after working at the College of Nanoscale Science and Engineering (CNSE) of the University at Albany as associate vice president of technology and professor of nano-science from 2005 to 2008. At CNSE, he managed the cleanrooms and numerous consortia involving CNSE and its industrial partners such as IBM, TEL, AMAT, ASML, and others. Dr. Ryan joined CNSE after a 25-year career with IBM. From 2003 to 2005, he was a distinguished engineer and director of Advanced Materials and Process Technology Development and served as the site executive for IBM at Albany Nanotech. Prior to that assignment Dr. Ryan managed interconnect technology groups in research, development and manufacturing engineering areas at IBM. He is the author of more than 100 publications, has 52 U.S. patents, and is the recipient of numerous awards including 17 IBM invention plateaus, an IBM Corporate Patent Portfolio award, an IBM Division Patent Portfolio Award, IBM Outstanding Technical Achievement Awards for Dual Damascene and for Copper technologies, and the 1999 SRC Mahboob Khan Mentor Award. Dr. Ryan's research interests include thin film deposition, interconnect technology, semiconductor manufacturing technology, and radiation hardened nanoelectronics.

Contributors

Shyam Aravamudhan
Department of Nanoengineering
Joint School of Nanoscience and
  Nanoengineering
North Carolina Agricultural
  and Technical State University
Greensboro, North Carolina

Adam Boseman
Department of Nanoscience
Joint School of Nanoscience and
  Nanoengineering
The University of North Carolina
  at Greensboro
Greensboro, North Carolina

Adam Bowen
Department of Electrical and
  Computer Engineering
North Carolina Agricultural and
  Technical State University
Greensboro, North Carolina

Autumn T. Carlsen
Department of Nanoscience
Joint School of Nanoscience and
  Nanoengineering
The University of North Carolina
  at Greensboro
Greensboro, North Carolina

Alan Covell
Department of Nanoscience
Joint School of Nanoscience and
  Nanoengineering
The University of North Carolina
  at Greensboro
Greensboro, North Carolina

Anthony Dellinger
Department of Nanoscience
Joint School of Nanoscience and
  Nanoengineering
The University of North Carolina
  at Greensboro
Greensboro, North Carolina

Hao Fong
Department of Chemistry and
  Applied Biological Sciences
South Dakota School of Mines
  and Technology
Rapid City, South Dakota

Komal Garde
Department of Nanoengineering
Joint School of Nanoscience and
  Nanoengineering
North Carolina Agricultural
  and Technical State University
Greensboro, North Carolina

Syed Gilani
Department of Nanoscience
Joint School of Nanoscience and
  Nanoengineering
The University of North Carolina
  at Greensboro
Greensboro, North Carolina

Joseph L. Graves Jr.
Department of Nanoengineering
Joint School of Nanoscience and
  Nanoengineering
North Carolina Agricultural
  and State University
Greensboro, North Carolina

Adam R. Hall
Department of Nanoscience
Joint School of Nanoscience and
    Nanoengineering
The University of North Carolina
    at Greensboro
Greensboro, North Carolina

Daniel J.C. Herr
Department of Nanoscience
Joint School of Nanoscience and
    Nanoengineering
The University of North Carolina
    at Greensboro
Greensboro, North Carolina

Albert Hung
Department of Nanoengineering
Joint School of Nanoscience and
    Nanoengineering
North Carolina Agricultural and
    Technical State University
Greensboro, North Carolina

Shanthi Iyer
Department of Nanoengineering
Joint School of Nanoscience and
    Nanoengineering
North Carolina Agricultural and
    Technical State University
Greensboro, North Carolina

Pavan Kumar Kasanaboina
Department of Electrical and
    Computer Engineering
North Carolina Agricultural and
    Technical State University
Greensboro, North Carolina

Ajit D. Kelkar
Department of Nanoengineering
Joint School of Nanoscience and
    Nanoengineering
North Carolina Agricultural and
    Technical State University
Greensboro, North Carolina

Christopher Kepley
Department of Nanoscience
Joint School of Nanoscience and
    Nanoengineering
The University of North Carolina
    at Greensboro
Greensboro, North Carolina

Karshak Kosaraju
Department of Nanoscience
Joint School of Nanoscience and
    Nanoengineering
The University of North Carolina
    at Greensboro
Greensboro, North Carolina

Dennis LaJeunesse
Department of Nanoscience
Joint School of Nanoscience and
    Nanoengineering
The University of North Carolina
    at Greensboro
Greensboro, North Carolina

Ram V. Mohan
Department of Nanoengineering
Joint School of Nanoscience and
    Nanoengineering
North Carolina Agricultural and
    Technical State University
Greensboro, North Carolina

Kyle Nowlin
Department of Nanoscience
Joint School of Nanoscience and
    Nanoengineering
The University of North Carolina
    at Greensboro
Greensboro, North Carolina

Sai Krishna Ojha
Department of Electrical and
    Computer Engineering
North Carolina Agricultural and
    Technical State University
Greensboro, North Carolina

Soodeh B. Ravari
Department of Nanoscience
Joint School of Nanoscience and
  Nanoengineering
The University of North Carolina
  at Greensboro
Greensboro, North Carolina

Thomas Rawdanowicz
Department of Materials Science
  and Engineering
North Carolina State University
Raleigh, North Carolina

Lew Reynolds
Department of Materials Science
  and Engineering
North Carolina State University
Raleigh, North Carolina

James G. Ryan
Joint School of Nanoscience and
  Nanoengineering
North Carolina Agricultural and
  Technical State University
and
The University of North Carolina
  at Greensboro
Greensboro, North Carolina

Marinella G. Sandros
Department of Nanoscience
Joint School of Nanoscience and
  Nanoengineering
The University of North Carolina
  at Greensboro
Greensboro, North Carolina

Furat Sawafta
Department of Nanoscience
Joint School of Nanoscience and
  Nanoengineering
The University of North Carolina
  at Greensboro
Greensboro, North Carolina

Goundla Srinivas
Department of Nanoengineering
Joint School of Nanoscience and
  Nanoengineering
North Carolina Agricultural and
  Technical State University
Greensboro, North Carolina

Joseph M. Starobin
Department of Nanoscience
Joint School of Nanoscience and
  Nanoengineering
The University of North Carolina
  at Greensboro
Greensboro, North Carolina

Ethan Will Taylor
Department of Nanoscience
Joint School of Nanoscience and
  Nanoengineering
The University of North Carolina
  at Greensboro
Greensboro, North Carolina

Stephen Vance
Department of Nanoscience
Joint School of Nanoscience and
  Nanoengineering
The University of North Carolina
  at Greensboro
Greensboro, North Carolina

Smith Woosley
Department of Nanoengineering
Joint School of Nanoscience and
  Nanoengineering
North Carolina Agricultural and
  Technical State University
Greensboro, North Carolina

Jun Yan
Department of Nanoengineering
Joint School of Nanoscience and
  Nanoengineering
North Carolina Agricultural
  and State University
Greensboro, North Carolina

Osama K. Zahid
Department of Nanoscience
Joint School of Nanoscience and
　Nanoengineering
The University of North Carolina
　at Greensboro
Greensboro, North Carolina

Effat Zeidan
Department of Nanoscience
Joint School of Nanoscience and
　Nanoengineering
The University of North Carolina
　at Greensboro
Greensboro, North Carolina

Lifeng Zhang
Department of Nanoengineering
Joint School of Nanoscience and
　Nanoengineering
North Carolina Agricultural
　and State University
Greensboro, North Carolina

1 Introduction
The Mechanical and Biological Paradigms

Ethan Will Taylor, Daniel J.C. Herr, and James G. Ryan

CONTENTS

1.1 BACKGROUND

One of the most unifying concepts in nanotechnology is that of "molecular manufacturing," ultimately to be achieved at the nanoscale. This concept, or at least the study of technologies that may contribute to its realization, is central to much of the theory and practice of nanoscience and nanoengineering. This focus is manifest not only in the development and application of instrumentation-based nanotechnologies such as the scanning tunneling microscope (STM) and the atomic force microscope (AFM), that have made it possible to manipulate and organize individual atoms, but also in what are often cited as foundational documents of nanotechnology, such as Feynman's "There's plenty of room at the bottom" (based on a 1959 lecture [1]) and "Infinitesimal machinery" (a 1983 lecture published a decade later [2]), and Drexler's 1981 "Molecular engineering" paper [3]. Despite Feynman's tipping of his hat to "the marvelous biological system" and the power of biological information systems, or Drexler's early focus on protein engineering and biological equivalents to mechanical devices, and the later emergence of a robust field of biomimetic nanotechnology, the concepts of what came to be called the "Feynman machine" (able to manipulate individual atoms) and the construction of ever-smaller machines by larger machines, or nanoscale mimics of macroscale devices such as gears, and so on, has had a huge influence in the popular conception of what defines nanotechnology. For the purposes of discussion, for lack of a better term, this approach will be referred to as "mechanical nanotechnology," or the *mechanical paradigm*, as opposed to "bio-mimetic nanotechnology," or the *biological paradigm*, with the understanding at the outset that there may be some overlap between the two.

Certainly, the structure and function of biomolecules and the machinery of life has been and can be viewed as molecular machinery, and therefore mechanical, but in this chapter the term *mechanical paradigm* is used to imply a *model for nano-manufacturing based upon analogies to macroscopic machines and manufacturing methods*, as succinctly articulated by Feynman: "I want to build a billion tiny facto-ries, models of each other, which are manufacturing simultaneously, drilling holes, stamping parts, and so on" [1, p. 34].

The actual introduction of the word "nanotechnology" by Norio Taniguchi 15 years later [4, pp. 18–23] was also specifically referring to technologies necessary for the extension of traditional approaches of manufacturing down to the limit of atomic size: "In the processing of materials, the smallest bit size of stock removal, accretion or flow of materials is probably of one atom or one molecule namely 0.1–0.2 nm in length. Therefore, the expected size limit of fineness would be of the order of 1 nm. Accordingly, 'Nano-technology' mainly consists of the processing of separa-tion, consolidation, and deformation of materials by one atom or one molecule" [4]. While the modern concept of nanotechnology encompasses materials well beyond atoms in size (up to 100 nm), the legacy and pervasiveness of this *atom-by-atom* concept of nanomanufacturing is apparent at the highest and most public levels of discussion in the United States, as evidenced by the very title of the Nanotechnology brochure put forth by the Interagency Working Group on Nanoscience, Engineering, and Technology (IWGN), under the White House National Science and Technology Council (NSTC) Committee on Technology [5].

The *mechanical paradigm* is largely distinct from the manner in which living systems are structured, function, and replicate, that is, the *biological paradigm*, and furthermore is simply not appropriate or even technically feasible for construction of the materials of which living systems are composed. An analysis of the clear distinc-tions that can be made between the mechanical and biological paradigms of molecu-lar assembly permits a realistic assessment of the strengths and limitations of each. Insight into these differences is essential for the future development of nanotechnol-ogy in which these two approaches can ultimately be merged, so that the strengths of both approaches can be exploited. To fully understand these issues, we will start with a brief review of the origins of the fundamental concepts of nanotechnology.

1.2 HISTORICAL ORIGINS OF THE NANOTECHNOLOGY CONCEPT

It has become part of the mythology of nanotechnology that Richard Feynman's 1959 American Physical Society lecture [1] laid the conceptual foundations for what became nanotechnology, although when that word was first introduced by Taniguchi in 1974, there was no citation or evidence of a Feynman influence [4]. In contrast, Feynman's thinking apparently had an inspiring effect on Eric Drexler, who was one of the first to cite this little known fragment of Feynman's oeuvre, in his seminal 1981 paper [3]. So, what did Feynman really talk about in this piece? As summarized by Toumey, Feynman "describes multiple possibilities, including the nano-etching of texts; the storing and retrieving of data in an atom-size code; the need to improve electron microscopes; the wonders of biological information systems; the miniaturization of computers; the difficulties of miniaturization; a mechanical surgeon that could be swallowed; a system of Waldos; a system of 'a billion tiny factories' working together; Van der Waals attractions; superconductivity; and simplified synthetic chemistry, to name only twelve ideas in that paper" [6, p. 161].

The question of the actual influence of Feynman's lecture, published in 1960 [1], on the future development of nanoscience and nanoengineering has been retrospectively dissected by a number of authors. Two of the most insightful and useful reviews in this regard are "From an idea to a vision: There's plenty of room at the bottom" by Junk and Riess from 2006 [7], and a 2008 article by Toumey, "Reading Feynman into nanotechnology: a text for a new science" [6].

Analyses of subsequent citations of the published versions of Feynman's key lectures viewed as the foundation of nanotechnology, and interviews with scientists who made key advances in the field, suggest that Feynman's direct influence, especially on *early* (pre-1985) nanotechnology development, has been exaggerated, and that it is perhaps more the case that his towering stature as a scientist has been invoked retroactively to legitimize the field [6,7]. Some of the things he envisioned *were* later achieved, but without prior knowledge of his ideas. With the notable exception of the realization in the mid-1980s by the inventors and early proponents of the STM (particularly Conrad Schneiker [8]) that it could be used as a "Feynman machine" for repositioning individual atoms, Toumey concludes that "much of the important scientific work that happened in the early years of nanotech, especially the big three breakthroughs in instrumentation, occurred without being influenced by Feynman or Drexler" [6, p. 159]. Similarly, Junk and Riess conclude that the attribution of Feynman via "Plenty of room" as the progenitor/instigator of nanotechnology is "misleading because there is no direct link from Feynman's talk to today's micromachines" [7, p. 825]. Nonetheless, many still view him as a prophet or visionary who foresaw the advent of nanotechnology, even if others developed much of it independently. In this view, perhaps the technology first had to catch up with Feynman's vision, whose writings were then discovered by others retroactively.

One idea that Feynman chose to highlight in his 1959 talk was inspired by a science fiction story that he had heard about from a colleague, Albert Hibbs, perhaps only weeks earlier [9]. This was Robert Heinlein's concept of "Waldoes," a term that is used to the present day to describe a "master-slave" arrangement of a remote set of mechanical hands controlled directly by a human user's hands (e.g., to manipulate radioactive materials). In the Heinlein story, titled *Waldo* (originally published in 1942 [10]), the hero was a disabled genius who devised exactly such instruments, at successively decreasing size dimensions, so that he could perform microsurgery and treat his own medical condition. This visionary piece of science fiction gave rise to Feynman's idea of "swallowing the doctor," and his detailed exposition of how to use a set of tools to create a nearly identical set of tools a quarter of the size, and so on until a set of microscopic tools was obtained. While Feynman acknowledged certain difficulties that could arise at extremely small size, such as issues of lubrication and surface tension, and the increasing role of van der Waals forces, he seemed fairly optimistic about the prospects for success in this endeavor, saying at various points: "It is rather a difficult program, but it is a possibility... If you thought of it very carefully, you could probably arrive at a much better system for doing such things" [1, p. 30]. Feynman later stated (regarding van der Waals forces), "There will be several problems of this nature that we will have to be ready to design for" [1, p. 34].

However, the reality is that over 40 years later, no one has yet created vastly reduced yet *functional* copies of common machines by following a successive step-wise scale-down procedure as prescribed by both Heinlein and Feynman (the "Waldo protocol"). The reasons why this is doomed to failure are now well understood, and will be discussed in the following section.

1.3 WHY NANOSCIENCE AND NANOENGINEERING ARE NECESSARY

In biology, examples abound of creatures that would not be able to function if their size was modified by one or two orders of magnitude; for example, insects' diffusive "breathing" would fail to provide adequate oxygenation in a much larger but similarly proportioned and constructed creature. Similarly, significant *decreases* in the size and weight of flying creatures require concomitant *increases* in wing beat frequency as size is diminished, as the air gets proportionally "thinner" relative to the size of the wings. Typical migrating birds can have wing beat frequencies between 3 and 10 Hz or somewhat higher if they are small [11], whereas the much smaller insects have higher frequencies, ranging up to as much as 1000 Hz. To attain wing beat frequencies above about 100–200 Hz, insects have had to evolve a special "asynchronous" mode of flight, because the "synchronous" mode of flight used by vertebrates and larger insects is inadequate for body sizes smaller than that of a bumblebee [12]. The synchronous flight mode requires a nerve impulse for each beat. Nerves cannot fire fast enough to attain the high frequencies required to support the flight of the smallest insects, because wing beat frequency in insects is more or less inversely proportional to the wing area [13]. The bottom line is: if we could somehow perform even just a 2-step Waldo protocol to shrink a bird by a factor of only 16 times, it would probably be unable to fly.

As Feynman himself pointed out, a reduction of a mechanical system to near-molecular size would lead to some unprecedented problems; for example, "the problem that materials stick together by the molecular (Van der Waals) attractions. It would be like this: After you have made a part and you unscrew the nut from a bolt, it isn't going to fall down because the gravity isn't appreciable; it would even be hard to get it off the bolt. It would be like those old movies of a man with his hands full of molasses..." [1, p. 34].

Another fundamental problem with executing the Waldo protocol for the design of miniature machines involves the issue of scaling when trying to make a tool at decreasing size. Because the power of a machine is proportional to its volume (e.g., for electric motors), but the friction in its moving parts is proportional to the surface area, at smaller scales frictional forces will ultimately predominate, because volume decreases faster than area, as size is decreased. So an extremely small but accurate replica of a tool will have inadequate power to overcome frictional forces.

It is precisely for such reasons, as well as the increasing importance of things like quantum effects and van der Waals forces at the atomic scale, and the loss of ensemble properties, that the fields of nanoscience and nanoengineering have been developed. We can't just proportionally extrapolate structures to a much smaller size and expect them to work in about the same way.

However, it is no surprise that biological evolution has found solutions for many of these problems, producing nanoscale motors, pumps, conveyor belts, and elegant machines of various types. This has been accomplished not only by exploitation of the types of forces that predominate at the molecular scale, but by the use of materials and construction methods that are entirely different from those used for building macroscopic machines. So, if we want to fully realize the potential of our endeavors at the nanoscale, it is of paramount importance that we understand how the machinery of life—the *biological paradigm*, differs from the *mechanical paradigm*. And a big part of that difference is in the materials themselves—the types of molecules that comprise what we have traditionally used to build our technology, as compared to the carbon-based molecules that comprise living systems.

1.4 FUNDAMENTAL DIFFERENCES IN THE MATERIALS, COMPOSITION, DESIGN, AND MANUFACTURING OF MACROSCOPIC MACHINES VS. THE NANOMACHINERY OF LIFE

A good place to start this analysis is with Feynman's discussion, near the end of "Plenty of room," about the possibility that chemical synthesis might be achieved by a physical approach in which, if atoms could be laid down in just the right spatial positions, one might be able to make any desired compound, without resorting to the arcane methodologies used by synthetic chemists. This fantasy of chemical synthesis by simple atomic assembly, along with Feynman's "billion tiny factories" comment cited earlier, seems to have been a significant influence in the proliferation of the "atom by atom" concept of nanotechnology fabrication over the last few decades. Although the atom-by-atom assembly technique at the atomic scale is rarely used in practical applications, it has been demonstrated using a scanning tunnelling microscope to inscribe IBM's logo using individual xenon atoms [14] and applications such as the fabrication of nanopores [1] and other nanoscale devices. However, there is every reason to believe that such techniques would fail miserably in an attempt to make even a relatively simple organic molecule, to say nothing of things like enzymes or DNA.

Semiconductor device fabrication is the closest large-scale nanomanufacturing technology to the atom-by-atom construction referred to above. Processes like ion implantation and certain thin film fabrication processes put atoms in close proximity and chemical reactions driven by relatively high temperatures are used to promote bond formation. Molecular-level self-assembly techniques have also been used to produce structures that cannot be formed by other means.

The chemical reactions used for semiconductor fabrication are driven by conditions that are too extreme for living organisms. Many elements do not covalently combine on their own, and depending upon the type of bonding necessary, conditions must be "right" to form certain compounds. Molecules having the appropriate steric and electronic properties to induce what chemists call a "reaction," involving the rearrangement of electron distributions, must be brought together in order to form new covalent bonds and break existing ones. Successful synthesis depends hugely upon how the electrons flow around the nuclei, and simply placing different atom types close to the position they are desired in a target molecule—even if that were attainable—is in itself not going to be sufficient to create an intended pattern of chemical bonding at temperatures and pressure compatible with life. The idea that one could make even some of the simpler building blocks of biomolecules, for example, an amino acid or a nucleotide, via physical assembly of the constituent atoms, without creating the appropriate microenvironment for the necessary bond-making electron flows, is far-fetched indeed. For living systems, biochemical reactions usually require an active site of an enzyme in an aqueous environment presenting the required combination of functional groups (i.e., electronic charges and atoms in specific states of hybridization) in a specific spatial orientation, with just the right amount of flexibility, capable of binding the required substrate and facilitating the required flow of electrons, all within the narrow range of "physiological" pH and temperature! A single enzyme may require thousands of atoms to attain the required structure to enable a single biochemical transformation that might be one of many steps in the formation of a desired product. None of this is intended to belittle the immense benefits and astonishing achievements of the mechanical paradigm, at the macro scale and, potentially, at the nanoscale. So we have much to learn and much to gain by mastering both the mechanical and biological paradigms.

At this point, it would be instructive to tabulate the characteristics of machines that we use in everyday life versus the workings of living systems. In Table 1.1, the mechanical paradigm as practiced on the macro scale is compared to the biological paradigm in terms of various factors ranging from the fundamental characteristics of the materials typically used, the type of chemical bonding in those materials, their chemical reactivity, typical fabrication methods used, and how they are scaled up for mass production.

The mechanical paradigm is characterized by materials that exhibit a variety of bonding in materials (ionic, covalent, metallic, van der Waals), typically with extensive regions of homogeneity (e.g., structural metal). In the final assembled products, these materials tend to be used largely in a structural manner, with fabrication methods involving shape transformations (which may be additive or subtractive) of the raw materials, with typically linear piece-by-piece assembly into a final product.

In almost complete contrast, the biological paradigm is characterized by extensive heterogeneity of materials, with largely covalent bonding in materials which tend to be inherently chemically reactive in performing their functions. Fabrication is fundamentally via enzymatic chemical synthesis as described above, so the making and breaking of covalent bonds is a critical characteristic. Higher-level structures are formed by self-assembly according to structural properties and a genetic program. All of this is performed in a massively parallel manner due to biochemical amplification and cascade-type mechanisms, and the potential for exponential cell division.

TABLE 1.1
A Comparison of the Mechanical and Biological Paradigms

	Mechanical Paradigm	**Biological Paradigm**
Characteristics of materials	*Broad variety of materials*, for example, <ul><li>metals, nonmetals, compounds</li><li>homogenous organic polymers (e.g., plastics)</li><li>natural materials (e.g., wood, adhesives) mostly used structurally in older and larger machines</li></ul>	*Carbon-based*, for example, <ul><li>Cellular contents</li><li>The most important biomolecules, for example, proteins, contain multiple chemical elements structured in a nonhomogenous manner, although they have repeating elements (monomers)</li></ul>
Type of chemical bonding in materials	*Varied chemical bonding* <ul><li>Ordered or semi-ordered arrays of molecules, for example, in ionic, metallic and van der Waals structures</li><li>Covalent bonding in ceramics, glasses, metals, bulk organic polymers used for structural elements</li></ul>	*Mostly covalent bonding* <ul><li>Covalent bonding predominates in all classes of biomolecules</li><li>Ionized or polar functional groups (e.g., biopolymer side chains) engage in nonbonded (e.g., H-bonding, ionic) interactions for structural purposes and in the induction (catalysis) of chemical reactions</li></ul>
Chemical reactivity of materials in final manufactured products	*May be inert or active* <ul><li>Chemically inert in mechanical parts (goal)</li><li>Chemically active chemistry for power generation, (combustion, batteries, etc.), sensing/detection, imaging, bonding, etc.</li><li>Mechanical paradigm parts do not regenerate</li></ul>	*Chemically active* at almost all levels <ul><li>Constant metabolic activity, supporting biosynthetic and degradative processes in balance</li><li>Even structural components (e.g., bone, cell membranes, microtubule structures), are typically being actively degraded and rebuilt in an ongoing process</li></ul>
Fabrication methods (most predominant)	*Shape transformations* of materials: <ul><li>Molding, casting, cutting, milling, joining, assembly, deposition, etching, etc.</li><li>Typically *piece-by-piece assembly* via manual or automated methods, often according to a computer-assisted design</li></ul>	<ul><li>*Chemical synthesis* of classes of bioorganic compounds (nucleic acids, proteins, etc.) via enzymatic methods of covalent bond making and breaking. Requires networks of highly evolved (i.e., optimized) stereospecific scaffolds (enzymes)</li><li>*Self-assembly* based on molecular complementarity and specificity, and the temporal- and site-specific expression of specific sets of genes according to a DNA-encoded program</li></ul>
Scalability for mass production	*Inherently linear* <ul><li>Amenable to assembly line type processes for mass production</li></ul>	*Inherently parallel*, due to: <ul><li>Constitutive amplification mechanisms (one gene gives multiple mRNA molecules, each of which gives multiple protein molecules, etc.)</li><li>Potential for exponential cell division (e.g., bacterial growth) with each cell as a "factory"</li></ul>

TABLE 1.2

Speed and Energy Required for Semiconductor Manufacturing and a Biological System

Comparative Metrics	Semiconductor Factory (Using EUV Lithography & Subtractive Etching)	Growth of a Baby (Biological Self-Assembly)
Bits patterned per second	<1.3 E9 bits/s	>1.5E+17 amino acids/s
Energy required per bit	>2.1E–8 J/bit	<6.6E–17 J/amino acid

Source: Updated from Herr, D.J.C., *Proceedings of SPIE* 2007. 6519: 651903.

If reduced to the nanoscale, the mechanical paradigm would essentially reach the limit of atomic size. At that point, fabrication (fundamentally via shape transformations with addition or removal of existing material) would indeed become an atom-by-atom process, as pointed out by Taniguchi in 1974 [4]. The point of the comparison in Table 1.1 is that, in the light of the previous discussion in this section, such processes are unlikely to enable us to create even the simplest building blocks of a living system, let alone a complex biomolecular network.

Table 1.2 shows a comparison of the nanomanufacturing capabilities of a semiconductor factory using extreme ultraviolet lithography making 10 Gb memory chips and a biological system building amino acids into a protein in a growing baby. The manufacture of one bit in the computer chip is analogous to fabrication of one amino acid in the growing baby [15]. As can be clearly seen in Table 1.2, biological self-assembly is faster and more energy efficient than our current semiconductor fabrication techniques by a substantial margin.

1.5   CONCLUSIONS

Particularly in electronics and semiconductor manufacturing, the advancing technology of miniaturization—the central theme of Feynman's 1959 talk—has forced us to confront the limitations of the mechanical paradigm and master a new set of rules and physical forces at the nanoscale, which are increasingly well understood by nanotechnologists. The expanding use of biological materials and biomimetic approaches in information technology (e.g., the use of DNA as an ultra-high density data storage medium [16]) and other areas of application reflect a growing appreciation for the power and potential of a future merging of the two paradigms, for example, between semiconductor and synthetic biology technologies [17].

Biological structures are formed through self-assembly controlled by a genetic program. Self-assembly in nonbiological systems is limited because there is no "controlling program" like the genetic code, but more and more self-assembly applications are being developed using fundamental principles of chemistry, physics, and materials science.

At some point in the future, perhaps we may be able to increase our capabilities so that techniques like self-assembly will offer the same advantages to the mechanical paradigm as they currently provide to the biological paradigm. Both paradigms are important to our lives today and through the discoveries of nanoscientists and nano-engineers, new technologies will be enabled and new industries will be founded. This book offers a vision into the opportunities that lie ahead.

REFERENCES

1. Feynman, R.P., There's plenty of room at the bottom. *Engineering & Science*, 1960. 23: 22–36.
2. Feynman, R.P., Infinitesimal machinery. *J. Microelectromech. Syst.*, 1993. 2: 4–14.
3. Drexler, K.E., Molecular engineering: An approach to the development of general capabilities for molecular manipulation. *Proc. Natl. Acad. Sci. USA*, 1981. 78: 5275–5278.
4. Taniguchi, N., On the basic concept of "nano-technology". in *Proc. Intl. Conf. Prod. Eng. Tokyo, Part II, Japan Society of Precision Engineering*. 1974. 18–23.
5. The Interagency Working Group on Nanoscience, Engineering and Technology, National Science and Technology Council Committee on Technology, *Nanotechnology: Shaping The World Atom by Atom*, September 1999. Available at: http://www.whitehouse.gov/sites/default/files/microsites/ostp/IWGN.Nanotechnology.Brochure.pdf
6. Toumey, C., Reading Feynman into nanotechnology. *Techné*, 2008. 12: 133–168.
7. Junk, A. and Riess, F. From an idea to a vision: There's plenty of room at the bottom. *Am. J. Phys.*, 2006. 74: 825–830.
8. Schnciker, C. et al., Scanning tunneling engineering. *J. Microsc.*, 1988. 152: 585–596.
9. Regis, E., *Nano: The Emerging Science of Nanotechnology*. 1995, Boston: Little, Brown.
10. Heinlein, R.A., *Waldo*, in *Waldo & Magic Inc.* 1986, Ballantine: New York.
11. Pennycuick, C.J., Speeds and wingbeat frequencies of migrating birds compared with calculated benchmarks. *J. Exp. Biol.*, 2001. 204: 3283–3294.
12. Hunter, P., The nature of flight. The molecules and mechanics of flight in animals. *EMBO R.*, 2007. 8: 811–813.
13. Deakin, M.A.B., Formulae for insect wingbeat frequency. *J. Insect Sci.*, 2010. 10: 96.
14. D.M. Eigler and E.K. Schweizer, Positioning single atoms with a scanning tunnelling microscope. *Nature*, 1990. 344: 524–526.
15. Herr, D.J.C., Emerging patterning technologies. *Proceedings of SPIE* 2007. 6519: 651903.
16. Church, G.M., Gao, Y., and Kosuri, S., Next-generation digital information storage in DNA. *Science*, 2012. 337: 1628.
17. Zhirnov, V.V. and Cavin, R.K., Future microsystems for information processing: Limits and lessons from the living systems. *IEEE J. Electron Dev. Soc.* 2013. 1: 29–42.

Section I

Nanoelectronics

2 Nano-Electro-Mechanical Systems

Processes and Devices

Jun Yan and Shyam Aravamudhan

CONTENTS

2.1 INTRODUCTION

Nano-electro-mechanical systems (NEMS) integrate mechanical structures and electrical circuitry into one system. In other words, NEMS devices typically include transistor-like nanoelectronics with mechanical components such as cantilever, free-standing pillar, suspended beam, suspended thin film, nanochannel, nanopore, valves, actuators, pumps, or motors, and may thereby to be used for chemical, biological, and physical devices. At least one dimension of the mechanical device is below 100 nm (300 nm in some cases), and this nanoscale dimension enables unique and critical attributes for these systems. In terms of active functions, NEMS devices can perform a broad range of tasks such as oscillation, resonance, switching, filtering, transportation, and manipulation. Examples of NEMS devices include but are not limited to timers, chemical and biological sensors, medical devices, consumer electronics, and precision scientific instruments. The total market for NEMS is expected to reach \$108.88 million by 2022, at a compound annual growth rate (CAGR) of 29.69% from 2012 to 2022 [1].

NEMS fabrication technologies can be divided into two categories: top-down and bottom-up methodologies. The top-down method utilizes the traditional semiconductor microfabrication processes. In the top-down method, first, a blank thin layer of material is deposited, and then the unwanted areas in the layer are removed. The processes often used to deposit the film are thermal oxidation of silicon, atomic layer deposition (ALD), chemical vapor deposition (CVD), epitaxy, electrochemical deposition (plating), or physical vapor deposition (PVD). Lithographic methods are used to create the etching mask on top of the thin layer. Then, etching or milling methods are used to remove the nonmasked thin layer. After stripping the etching mask, the desired patterns are formed on the substrate. In some cases, the lithography step is avoided and the unwanted areas of the layer are directly cut away using a focused ion beam (FIB). Even though, top-down methods are complicated and expensive, they provide precise location control. On the other side, the bottom-up methods physically assemble or chemically synthesize materials (molecules in some cases) only at desired locations on substrates to form a pattern. The process flow is simple, but has no inherent location control available on a substrate. It is quite possible to fabricate the entire NEMS device using only top-down methodology. However, the bottom-up methods become necessary when natural or synthetic nanomaterials, such as DNA or carbon nanotubes (CNTs), need to be integrated on a NEMS device. For instance, precisely placing nanotubes at the desired locations is extremely challenging. In the past, various micro- and nanomanipulations methods, such as electrical, magnetic, shear force, and fluidic assembly, have been investigated [2–12]. The manipulation process is slow and untenable for large-volume manufacturing. Other processes still need to show repeatability and reliability. Therefore, it becomes important to consider design for manufacturability, particularly in the case of NEMS devices. It is expected that NEMS devices in large-volume production may come from either top-down methods alone or from a hybrid process involving top-down and bottom-up methods.

2.2 MINIATURIZATION

NEMS is a logical miniaturization of micro-electro-mechanical systems (MEMS) with one critical dimension in the nanoscale. This miniaturization can (a) enable new applications that utilize unique nanoscale attributes, for example, solid-state nanopores that can explore DNA or proteins [13] and nanoprobe medical devices that can penetrate into tissues or even cells [14]; and (b) improve the performance of existing structures or devices. For example, microscale cantilevers have been used in MEMS area for a long time. With one critical dimension in the nanoscale, nanocantilevers can provide higher resonance frequency, higher quality factor (Q factor), and smaller mass. The cantilever resonance frequency ω_0 is inversely proportional to the linear size of the cantilever. This inverse proportion means that ω_0 goes to infinity when size approaches zero. A high-ω_0 device can operate at GHz compared to MHz for low-ω_0 device, which in turn leads to faster response time and high resolution in timing circuits. Nanocantilevers also exhibit high Q factor at ambient atmosphere [15]. The Q factor measures energy lost in the vibration. Less energy lost leads to a high Q factor, which means low power consumption. Lastly, nano-cantilevers also have low mass and high surface-to-volume ratio. This is an important characteristic for chemical and biological sensing applications. An ultrasmall amount of absorbed species or molecules can now significantly change the mass of the cantilever. This small mass change can then be detected by frequency shift, ω_0 at high resolution.

2.3 NEMS PROCESSES

NEMS fabrication typically involves processes that form patterns or define structures on a substrate. These processes can be the top-down method, where a blank thin film is deposited and then selectively removed from unwanted areas or the bottom-up method that directly assembles pattern or structure from atoms and/or molecules. The top-down processes are rooted from traditional semiconductor manufacturing techniques. Bulk micromachining, surface micromachining, and LIGA (a German acronym for x-ray lithography, electrodeposition, and molding) are some of the widely used MEMS top-down fabrication processes. The bottom-up methods can include growing nanowires from catalyst predefined on a substrate; self-assembly of molecules on pretreated substrate; and spreading or manipulating presynthesized nanomaterials on a substrate. NEMS devices are typically fabricated from just top-down methods or through hybrid processes that include top-down and bottom-up methods.

2.3.1 TOP-DOWN SEMICONDUCTOR PROCESSES

Top-down semiconductor processes include surface preparation, oxidation, PVD, CVD, chemical mechanical polishing (CMP), electroplating, doping, lithography, wet etch, and dry etching techniques [16]. Before a thin film is deposited on a substrate, it is important to clean the surface. Organic contamination on the surface can lead to adhesion issues; the metal contamination can change electrical properties of the device or particles can be incorporated in the thin film. If the deposited film is photoresist, particles incorporated in the photoresist can lead to lithography failure, which can fail the device by open/short circuit. Surface preparation includes wet clean (piranha clean, RCA clean, dilute hydrofluoric acid) and dry clean (plasma clean). Piranha clean uses a sulfuric acid (96%) and hydrogen peroxide solution (30%) at 3:1 to 7:1 ratio, along with heating to 100–130°C. It will remove most organics and add OH group to the surface, which will make the surface hydrophilic. RCA cleans, also called standard cleans (SC-1 and SC-2), is a two-step procedure. The first step (RCA1 or SC-1) adds ammonium hydroxide, hydrogen peroxide, and water at 75°C to remove organics and a thin layer of oxide. The second step is hydrochloric acid, hydrogen peroxide, and water mixture heated at 75°C to remove metallic contaminants. Dilute hydrofluoric acid is used to clean oxide off the silicon and create H-terminated surface (hydrophobic). Dry cleaning methods use plasma or sputtering to clean organics and oxides off the surface. Thin film deposition (Figure 2.1) can be done by spin-on, electroplating, PVD, epitaxy, and CVD (including ALD). Photoresist is spun on substrate using a spinner. Glass and some low-k dielectrics can also be spun on. Electroplating supplies electrons to the ions in a solution. Those ions that have gained electrons deposit on the substrate. PVD and CVD methods deposit thin films from gas phase. In PVD, atoms are either evaporated or sputtered from a source; they travel through a vacuum and condensate on a substrate. In CVD, the gases are broken into ions and radicals; they travel through a vacuum and react at the substrate surface to form a thin film.

After the deposition of a thin film, lithography is generally carried out, with a layer of photosensitive polymer (photoresist or just "resist") spun on the thin film. Lithography includes two steps: exposure and development of the photoresist. Figure 2.2 illustrates the lithography process using two types of exposure tools: projection lithography (stepper) and contact lithography (mask aligner). The projection tool will shrink the image on the reticle at a defined ratio (such as 5:1), while contact tools provide 1:1 patterning.

Next, wet etch or dry etch is used to remove areas, which are not protected by the photoresist. Wet etch is often isotropic, which will lead to undercutting. Reactive ion

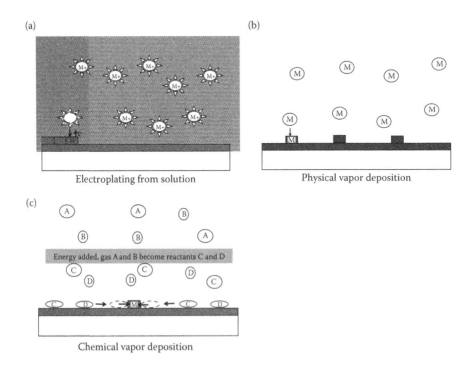

FIGURE 2.1   Principles of (a) electroplating, (b) PVD, and (c) CVD.

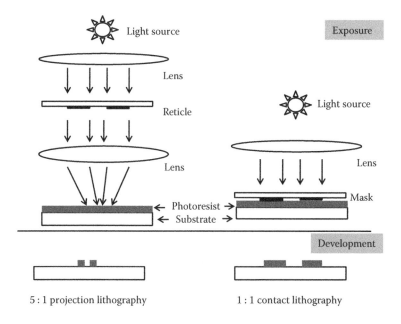

5 : 1 projection lithography   1 : 1 contact lithography

FIGURE 2.2 Comparisons between projection lithography and contact lithography.

etch (RIE) can etch vertically into a thin film. Undercut phenomena and vertical profile are illustrated in Figure 2.3. In wet etch, the etchants travels into the cavity and attacks the thin film in all directions. The RIE process takes place under vacuum, where ions are pulled out of plasma and will travel vertically toward the thin film. The vertical etch can be activated by ion bombardment and the side walls can be protected by passivation (from feed gases and/or by-products), as in the case of the Bosch process for deep reactive ion etch (DRIE).

An example of the top-down complementary metal-oxide semiconductor (CMOS) semiconductor process to form a beam structure on a substrate is shown in Figure 2.4. This simple process flow consists of six steps to make a beam structure; the beam width is defined in the lithography step. NEMS involves structures below 100 nm (sometimes below 20 nm). Using lithography to define line (beam) width

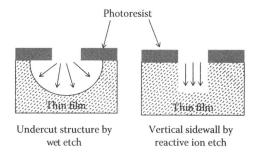

Undercut structure by   Vertical sidewall by
wet etch   reactive ion etch

FIGURE 2.3 Understanding of undercut etch and vertical etch profile.

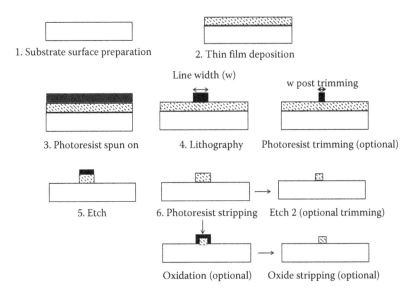

FIGURE 2.4 Forming a beam by top-down CMOS process.

becomes expensive (requires advanced projection tools), and sometimes impossible. Therefore, trimming needs to be done to reduce beam or line size. Trimming is done post-lithography, during dry etch, or later. Integrated circuits (ICs) are fabricated by repeating the processes (shown in Figure 2.4) a number of times. Thin films deposited on patterned substrate, often results in uneven surfaces. With deposition of many layers, the unevenness increases leading to lithography failure at some point when the depth of field cannot handle the roughness. CMP has emerged as a good solution for polishing uneven surfaces. Line width trimming, in case of Si, can also be done by Si oxidation [17,18]. Yang et al. used high-temperature (1150°C) oxidation to trim 160 nm Si pillars to about 20 nm Si nanowires, along with supporting curvature shape (footing) at the bottom of the Si nanowire.

2.3.2 MEMS PROCESSES

MEMS processes have not only relied on semiconductor CMOS methods (described above) but also developed special process variations not found in traditional semiconductor manufacturing. They are micromachining (bulk micromachining and surface micromachining) and LIGA [19]. Bulk micromachining selectively removes material from the bulk to create the desired structure. Wet etching of single crystal silicon is the most important bulk micromachining technology in MEMS/NEMS fabrication. Wet etching of Si can be isotropic by using HNA (HF/nitric/acetic acid) or anisotropic by using KOH and tetramethylammonium hydroxide (TMAH). Figure 2.5 shows an example of anisotropic Si wet etching process used to fabricate 54.7° Si pits with suspended structures. In addition, anisotropic etchants etch heavily doped p-type Si layers very slowly. Therefore, heavily doped p-type Si layer is often used as a etch stop. Also, by exploiting concave and convex corner etching

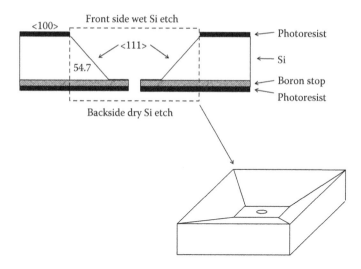

FIGURE 2.5   Illustration of a suspended Si thin film with through wafer hole.

rates, cantilevers can be formed by undercutting convex corners [19]. Surface micromachining process can take advantage of different etch rates on structural and sacrificial layers to create surface micromachined cantilevers. LIGA technology, lift-off, and deep RIE of silicon are also important processes for MEMS fabrication (see Ref. [19] for more details).

2.3.3   Lithography and Other Patterning Techniques

The important goal for MEMS/NEMS is to make the processes and devices smaller and cheaper. The technologies considered for to fulfill this goal have involved, for example, plastic substrate materials, soft and imprint lithography, directly write (with dip-pen), electron beam lithography, FIB nanopatterning, nanosphere lithography (NL), and so on. In the next section, we will discuss these key lithography and patterning methods.

2.3.3.1   Stamp-on-Substrate

The stamp-on-substrate or soft lithography includes microcontact printing [20], molding methods, and hot embossing (imprint) [21–24]. Soft lithography has been demonstrated for patterning less than 100 nm. For example, Chou et al. used nano-imprint to generate 25 nm via array in polymer [24]. Marzolin et al. applied micro-molding to produce silica pyramids array with less than 50 nm tip radius [25]. Xia et al. reviews the potential of soft lithography as a high-resolution, repeatable, low-defect method [26].

2.3.3.2   Write-on-Substrate

This writing process uses one "pen" or array of "pens." "Dip-pen" nanolithography was invented by Chad Mirkin's group in the late 1990s [27]. An atomic force micro-scope (AFM) tip is used to transport molecules through a water meniscus formed at the tip on to a substrate to form a pattern. This technology has evolved in the subsequent years to include a polymer pen in order to improve the writing speed [28]. The soft polymer pen replaced the hard AFM tip so that the writing width can be changed by forced deformation of the soft tips. Another variation in the dip-pen technology is "hard-tip soft-spring lithography" [29]. Here, the hard Si pyramids are installed on top of a polymer layer, so that every pen in a large array can write at the same line width even when the forces on the tips are different. Obermair et al. demonstrated a "mechano-electrochemical pen" technology [30]. By combining electrochemical deposition with AFM, Obermair et al. could write–delete–rewrite Cu nanostructures on an Au film [31].

2.3.3.3 Electron Beam Lithography

In electron beam (e-beam) lithography, e-beam photoresist such as hydrogen silsesquioxane (HSQ) and poly(methyl methacrylate) (PMMA) are exposed and patterned by using streams of electrons. These electron optical systems can offer angstrom-level resolution [31]. Therefore, e-beam lithography has been a popular lithography technology for research and development when ultrafine resolutions are required [31]. There are many factors that affect e-beam lithography resolution such as electron–resist interaction, electron–substrate interaction, and resist–developer interaction. The diameter of the e-beam increases in the resistance; higher voltages can reduce both forward broadening and backscattering [32]. Lee et al., using Monte Carlo simulations, explored the secondary electron–resist interaction and found that secondary electron spot sizes are 5–10 nm at 100–200 V [33]. Even though nm-lines can be written by using e-beam lithography, due to proximity effects, line pitch gets affected. In spite of some challenges, e-beam lithography technique is very popular in NEMS research and development. In order to improve manufacturing throughput, newer systems such as multibeam cluster tools [34] and projection e-beam tools [35] have been developed.

2.3.3.4 Focused Ion Beam Milling

The FIB method has many applications in microscopic analysis, ion implantation, lithography, IC failure analysis, transmission electron microscope (TEM) sample preparation, photo mask repair, and circuit modification. In this context, we will discuss FIB for direct writing (or nanopatterning). FIB direct writing (milling or sputtering) is different from e-beam writing, which is done on a photosensitive layer. FIB can mill a broad range of materials at various depths, sizes, and shapes. The milling resolution strongly depends on ion–solid interactions and depth of the pattern. For example, FIB can be used to etch 3 nm nanopores on 10 nm SiC or SiN thin film [36]. Conventional FIB systems use heavier ions such as Ga^+, which have a larger interaction radius compared to newly invented helium ion tools. Although helium ion beam systems provide better resolution for some conditions, they mill slower compared to Ga ion-based systems [37].

2.3.3.5 Nanosphere Lithography

NL uses capillary force to assemble monolayer of polymer spheres on a substrate to be patterned. Different processes such as convective assembly [38], spraying [39], and spinning [40] have been developed to assemble the polymer spheres. Figure 2.6

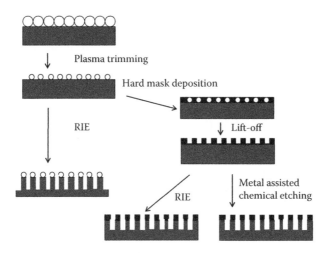

FIGURE 2.6 Typical nanosphere patterning process flow.

illustrates a typical nanosphere assembly method, where the plasma process is used to trim monolayer of polymer spheres into smaller spheres. These smaller spheres can be then used as an etching mask to create the nanowire structure by RIE or as lift-off resist in metal-assisted chemical etching (MACE) process.

2.3.4 NANOWIRE DIRECT ASSEMBLY

2.3.4.1 Nanomanipulation

A direct method to assemble nanowire is by either mechanical manipulation using AFM tip [41], with a probe ("pick and place") [2], or by nonmechanical methods such as optical probe [3,4]. These methods are time consuming (especially for high-volume manufacturing) and are not repeatable.

2.3.4.2 Electrical, Magnetic, and Shear Force Assembly

The method of electrical assembly by dielectrophoresis (DEP) is well known [5,6]. This manipulation process uses a movement of electric dipole in the electric field to assemble nanowires. In a parametric study, Xue et al. concluded that the width of the electrodes and nanotube solution concentration is critical to achieve a small bundle or a single CNT alignment. A single CNT was aligned between 3-μm-wide electrodes at a concentration of 0.0125 mg/mL [6]. Pathangi et al. optimized AC frequency and nanotube concentration to maximize the yield of single nanotube alignment between 12–20 pairs of electrodes [7]. In other works, a combined DEP and manipulation was used to create an automated system to assemble CNTs on chips [8]. In this method, DEP is essentially used to separate metallic and semiconductor nanotubes. Magnetic assembly is similar to DEP and works on ferromagnetic and superparamagnetic magnetic materials [9]. Shear force assembly uses microchannels to provide a uniform shear force to align nanotubes [10,11]. By realignment of the microchannels, assembly of additional layers of nanotubes can be enabled [12].

2.4 NEMS DEVICES

NEMS devices have shown potential to be used across many fields, including sensors, medical devices, and electronics. In this section, we will discuss few specific examples of NEMS devices. As stated earlier, NEMS cantilever-based resonators are an active area of research for high-volume manufacturing. These resonators can be used for high-resolution mass spectrometry [42–44]. The resonance frequency decreases when a particle is adsorbed on the resonator platform. Therefore, the smaller the cantilever, the lower the detection limit (or higher resolution). The mass detector platform can be a suspended beam [42], a suspended nanotube [43], or a nanotube cantilever [44]. Suspended beam resonators can be manufactured in large volume using semiconductor processes to reduce cost. CNT resonators with very small mass (~10^{-21} kg) have reported atomic-level mass detection [43,44]. These NEMS resonators (as mass detectors) can be converted into chemical and biological sensors by just coating a thin functional layer on the resonator platform. This functional layer can specifically adsorb and/or form bonding with the sensing species or molecule. Li et al. applied a thin layer of PMMA on a NEMS resonator operating at 127 MHz to detect 1,1-difluoroethane ($C_2H_4F_2$; DFE) gas pulses at atmospheric pressure and room temperature [15]. This work demonstrated DFE detection at a resolution below 1 attogram (10^{-18} g). The choice of a functional layer is flexible and abundant. With antigen layers, antibodies can be detected and vice versa. DNA strands can be used to detect complementary strand of DNA. Apart from measuring frequency as a measure of mass change, other types of NEMS resonators detect the deflection of the mechanical device using optical beam deflection or piezoresistivity [45,46].

The second area of focus is electronic circuits. An electromechanical resonator can be used in electronics as an oscillator, receiver, filter, and mixer. There are many commercially viable products with M/NEMS in electronic circuits. SiTime (sitime.com) and Discera (discera.com) have devices that operate in the MHz range. The Avago (avagotech.com) Film Bulk Acoustic Resonator (FBAR) operates around 2 GHz. Erbe et al. successfully developed a mechanical mixer in the radio-frequency (RF) regime [47]. Bartsch et al. developed a scalable NEMS channel select receiver for simultaneous signal mixing and filtering [48]. This receiver device combines a high-Q NEMS resonator with transistor-based signal processing. This integrated front-end receiver can operate at 114.4 MHz at near ambient pressure (0.1 atm) and at room temperature (300 K). For signal processing applications, a wide range of frequency and high Q factors is required. Island et al. built a 23–30-nm-long suspended single-wall CNT resonator, which demonstrated fundamental bending frequencies $f_{bend} \approx$ 75–280 GHz, and the calculated Q of about 10^6 [49]. These demonstrations have indicated that mechanical resonator devices can cover the entire RF range from low frequencies (30–300 kHz) to extremely high frequencies (30–300 GHz).

The next example of NEMS in electronic circuits is switching circuits. Currently, the CMOS, which is the building block for ICs, is used in electronic switches. The CMOS switch that includes a complementary and symmetrical pair of metal oxide semiconductor field effect transistors (MOSFET) has certain limitations. To switch on/off the MOSFET, the channel conductance is changed by the gate voltage, which changes the charge carrier distribution in the device. However, MOSFETs, even in their off state, leak current. The $I_d - V_g$ curve plots drain current versus gate voltage. The larger the slope of this curve, the better the switch performance. With the scaling of MOSFETs, this slope is becoming smaller. Other factors such as heat and radiation also affect the charge carrier concentration in MOSFETs, thereby impacting their switching performance. NEMS devices can alleviate some of the problems encountered with CMOS switches. A NEMS switch uses an electrostatic force on the active element to make contact with the "on" terminal, and will switch off when the electrostatic force is removed. Because of their mechanical nature, these switches can have near-zero off leakage current, large $I_d - V_g$ slope, and are not sensitive to heat or radiation [50,51]. Simulations of NEMS switches for thermal management in high-performance computing applications have shown thermal stability and energy savings by taking advantage of a low "OFF" state current [52].

The third area of application for MEMS/NEMS devices is in the biomedical field termed as Bio-MEMS/NEMS. The biomedical applications can be divided into three categories: drug discovery [53], diagnosis [54], and therapeutics [55]. For example, micro/nanoarrays can identify drug targets, and study toxicity of drug candidates with highest efficiency and automation. These arrays can also be used to study genes, their sequences, and expressions. Other examples include polymerase chain reaction (PCR) chips for DNA amplification, micro/nanofluidic devices for sample collection, separation, and transport, micro and nano diagnostic chips. For therapeutics, microneedle devices can control the amount and time for drug delivery; lab-in-a-pill and nanorobots can travel digestive system or circulatory system to find and cure illness. In the next section, we will discuss one specific biomedical NEMS device, which has enormous application for drug delivery and in neuroscience research.

The overarching goal of neuroscience research is to target and discover the relationship between the functional connectivity map of neuronal circuits and their physiological or pathological functions. In the past, extracellular microelectrode arrays (MEAs) have been used to record and stimulate a population of excitable cells *in vivo* [56]. The recorded spikes (signal) by extracellular electrodes, though informative, do not provide the source mechanism for neuron firing because the extracellular recordings do not record synaptic signals (subthreshold). On the other hand, intracellular recordings can help study the functions of "silent" neurons and neuroplasticity [57]. The current intracellular recording technologies include a sharp or patch electrode to measure neurons. For recording a large number of neurons, technologies such as mushroom-shaped microelectrodes [58], vertical nanowire electrodes [59], and nanoFET devices [60] are currently under development. The vertical nanowire electrodes are fabricated by e-beam lithography, oxidation, and etching. NanoFET devices are fabricated by *in situ* nanowire growth and e-beam lithography. These methods have been either noncompatible to some cells or have exhibited high electrode impedance and noise leading to large signal loss. In this regard, we have developed "Fin"-shaped nanoelectrodes (as shown in Figure 2.7) that seek to overcome the limitations between electrode impedance and electrode size. Compared to the 3×3 array of 150 nm diameter nanowire electrodes [59], the "Fin" electrodes (three fin design) reduces impedance by a factor of 10. 150-nm-thick fins are also less damaging to cells compared to mushroom-shaped electrodes. In addition, the incorporation of "Fin" nanoelectrodes as a nano FinFET offers low noise

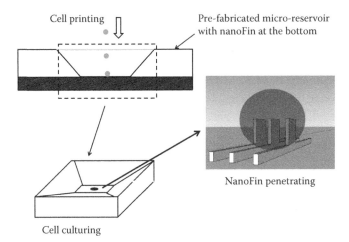

Cell printing

Pre-fabricated micro-reservoir
with nanoFin at the bottom

NanoFin penetrating

Cell culturing

FIGURE 2.7   Schematic illustration of a nanoFin device with cell printed on the electrode.

levels. Furthermore, the use of microprinting technology for cells enables printing in a controlled manner into a confined reservoir [61].

Figure 2.8 shows the nano FinFET arrays fabricated on silicon on insulator (SOI) substrate, with the top silicon patterned to form the metallized contacts and interconnections to the "Fin" nanoelectrodes. The "Fin" nanoelectrodes are created using FIB etching (as shown in Figure 2.9), followed by the fabrication of reservoir enclosing the electrodes. Next, controlled volume and number of cells were printed onto the reservoir on the electrodes. We have demonstrated the ability of microprinting technology to print viable neuronal cells onto predefined areas within nanoelectrode reservoir. The next step is to study the relationship between the electrode geometry

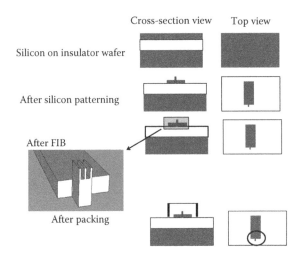

Cross-section view   Top view

Silicon on insulator wafer

After silicon patterning

After FIB

After packing

FIGURE 2.8   Process flow of the nanoFin probe device fabricated on SOI wafer.

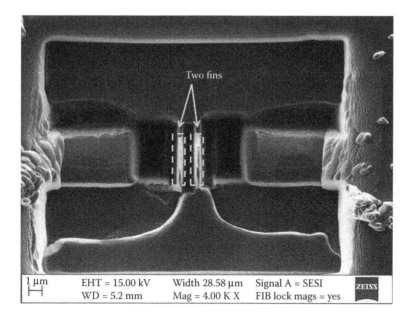

Two fins

| 1 μm | EHT = 15.00 kV | Width 28.58 μm | Signal A = SESI | ZEISS |
| | WD = 5.2 mm | Mag = 4.00 K X | FIB lock mags = yes | |

FIGURE 2.9 Scanning electron microscope (SEM) image showing the top view of "Fin"-shaped nanoelectrodes.

and neuronal cells and intracellular activity (action potential) with and without subthreshold (10–40 mV) electrical stimulus.

2.5 CHALLENGES IN NEMS DESIGN AND FABRICATION

A key challenge in NEMS is the loss of bulk property. When the size of a device approaches less than 20 nm, the familiar bulk properties start to fade away, and the physical constants change with the scaling. The knowledge of nanoscale phenomena is extremely critical for NEMS design and fabrication. Limited experiments have been done to explore physical properties at the nanoscale [62,63]; however, more experimental research is required to understand the behavior of nanoscale materials, processes, and devices. The other consequence of device shrinking is that mechanical deflection/deformation is so small that it is hard to measure. The mechanism to transduce, actuate, and couple nanomechanical motion to macroscale regime becomes more challenging. The detection schemes used in MEMS devices do not work at the nanoscale. Newer methods need to be investigated as scaling continues [64–66]. NEMS devices also face numerous fabrication challenges, particularly in nanowire/nanotube manipulation and assembly and problem of material and process incompatibilities. Most successful NEMS thus far have been manufactured by micromanipulation and e-beam lithography. However, micro/nanomanipulation and e-beam lithography are time consuming and not suited for large-volume manufacturing [67]. Electric, magnetic, or shear force assembly methods have potential in large-volume production, but reliable and repeatable demonstration is lacking.

In order to overcome material and process incompatibilities, hybrid integration technologies have been proposed. Here, the complete devices (e.g., N/MEMS and CMOS ICs) are manufactured on separate substrates with dedicated technologies. Thereafter, the substrates are diced into chips and the chips are integrated using wafer-level heterogeneous integration technologies [68]. Heterogeneous integration technologies combine the advantages of monolithic and chip-level hybrid integration approaches and allow for the manufacturing of complex micro- and nanosystems that are not possible to be manufactured with conventional top-down semiconductor techniques.

2.6 CURRENT TRENDS AND FUTURE OUTLOOK

Because NEMS devices are still at its infancy, NEMS research is very active in both fundamental science and in developing new applications: assembly of nanomaterials, exploring nanoscale physics, development of CMOS-compatible NEMS, and application of NEMS for sensors and electronics are currently underway. In 2000, IBM demonstrated the first very-large-scale integration (VLSI) NEMS device [69]. As stated earlier, Marketsandmarkets.com [1] predicted the total NEMS market would grow 29.69% annually between 2012 and 2022, and reach $108.88 million by 2022. MEMS market, which started in the mid-1980s, reached $2–$5 billion by 2000. Compared to MEMS devices, NEMS is making a slower start. This is partly because MEMS took full advantage of the mature IC technology. Large-volume NEMS fabrication technology (working with less than 100 nm nanostructures) does not exist today. In the future, we believe that fundamental research will be very active. Micro/nanofluidics and nanoenergy harvesting devices will grow faster than other NEMS devices because manipulation of a single nanoparticle or nanowire is not required; CMOS-compatible NEMS will grow at a steady rate as they are the most reliable extension of MEMS technology. Biological NEMS devices will find their commercialization opportunities in lab-on-chip, drug delivery, genomics, proteomics, biodetection, and food safety applications. RF NEMS, including RF switches, will soon become commercially available if the reliability issues are solved. In summary, fundamental research, research on new nanomaterials and processes, will likely drive the NEMS commercialization efforts.

2.7 SUMMARY

In summary, various research efforts have explored fundamental nanoscale phenomena, fabrication technologies, and applications of NEMS devices in the last decade. NEMS devices have been fabricated to improve and understand nanoscale motion, aid drug delivery, and sense chemical and biological species, and for electronic circuits. However, the development of NEMS devices is still largely a bench (lab) activity, with little commercialization. But the importance of NEMS for real-world applications is unambiguous. With the current efforts to address challenges in NEMS development, it is likely that we will soon witness exponential growth in commercialized NEMS devices. Lastly, integration approaches such as wafer-level heterogeneous integration technologies will be required to overcome material, process, and integration incompatibilities.

REFERENCES

1. Markets and Markets, *Nanoelectromechanical Systems (NEMS) Market (2012–2022) by Applications (STM/AFM, Medical, Gas/Flow Sensor, RF), Products (Switches, Cantilevers), Components (Nanotubes, Nanowires, Nanofilms), Materials (Graphene, ZnO, SiC, GaN, SiO2)*, Markets and Markets, USA, October 2012. http://www.marketsandmarkets.com/Market-Reports/global-NEMS-%28Nano-Electro-Mechanical-Systems%29-report-143.html?gclid=COnCoIXd8b0CFcqUfgodrg0A7Q, Accessed April 21, 2014.

2. Kim, P. and Lieber, C.M. Nanotube nanotweezers, *Science*, 286, 2148, 1999.

3. Tan, S. et al. Optical trapping of single-walled carbon nanotubes, *Nano Lett.*, 4, 1415, 2004.

4. Zhang, J. et al. Multidimensional manipulation of carbon nanotube bundles with optical tweezers, *Appl. Phys. Lett.*, 88, 053123, 2006.

5. Liu, Y. et al. Dielectrophoretic assembly of nanowires, *J. Phys. Chem. B*, 110, 14098, 2006.

6. Xue, W. and Li, P. Dielectrophoretic deposition and alignment of carbon nanotubes. In: *Carbon Nanotubes—Synthesis, Characterization, Applications*, Yellampalli, S. Ed., ISBN: 978-953-307-497-9, InTech, 2011.

7. Pathangi, H., Groeseneken, G., and Witvrouw, A. Dielectrophoretic assembly of suspended single-walled carbon nanotubes, *Microelectron. Eng.*, 98, 218, 2012.

8. Lai, K.W.C. et al. Automated nanomanufacturing system to assemble carbon nanotube based devices, *Int. J. Rob. Res.*, 28, 523, 2009.

9. Hangarter, C.M. et al. Hierarchical magnetic assembly of nanowires, *Nanotechnology*, 18, 205305, 2007.

10. Huang, Y. et al. Directed assembly of one-dimensional nanostructures into functional networks, *Science*, 291, 630, 2001.

11. Mathai, P.P. et al. Simultaneous positioning and orientation of single nano-wires using flow control, *RSC Adv.*, 3, 2677, 2013.

12. Wang, M.C.P. and Gates, B.D. Directed assembly of nanowires, *Mater. Today*, 12(5), 34, 2009.

13. Smeets, R.M.M. et al. Translocation of RecA-coated double-stranded DNA through solid-state nanopores, *Nano Lett.*, 9, 3089, 2009.

14. Cohen-Karni, T. and Lieber., C.M. Nanowire nanoelectronics: Building interfaces with tissue and cells at the natural scale of biology, *Pure Appl. Chem.*, 85(5), 883, 2013.

15. Li, M., Tang, H.X., and Roukes, M.L. Ultra-sensitive NEMS-based cantilevers for sensing, scanned probe and very high-frequency applications, *Nat. Nanotechnol.*, 2, 114, 2007.

16. Wolf, S. and Tauber, R.N. *Silicon Processing for the VLSI Era Volume 1: Process Technology*, 2nd ed., Lattice Press, Sunset Beach, 2000.

17. Yang, B. et al. Vertical silicon-nanowire formation and gate-all-around MOSFET, *IEEE Electron Device Lett.*, 29(7), 791, 2008.

18. Ng, E.J. et al. High density vertical silicon NEM switches with CMOS-compatible fabrication, *Electron. Lett.*, 47(13), 759, 2011.

19. Madou, M.J. *Fundamentals of Microfabrication*, CRC Press, Boca Raton, Florida, 1997.

20. Kumar, A. and Whitesides, G.M. Features of gold having micrometer to centimeter dimensions can be formed through a combination of stamping with an elastomeric stamp and an alkanethiol "ink" followed by chemical etching. *Appl. Phys. Lett.*, 63(14), 2002, 1993.

21. Xia, Y. et al. Complex optical surfaces formed by replica molding against elastomeric masters, *Science*, 273, 347, 1996.

22. Zhao, X., Xi, Y., and Whitesides, G.M. Fabrication of three-dimensional micro-structures: Microtransfer molding, *Adv. Mater.*, 8, 837, 1996.

23. Kim, E., Xia, Y., and Whitesides, G.M. Polymer microstructures formed by moulding in capillaries, *Nature*, 376, 581, 1995.
24. Chou, S.Y., Krauss, P.R., and Renstrom, P.J. Imprint of sub-25 nm vias and trenches in polymers, *Appl. Phys. Lett.*, 67, 3114, 1995.
25. Marzolin, C. et al. Fabrication of glass microstructures by micro-molding of sol-gel precursors, *Adv. Mater.*, 10(8), 571, 1998.
26. Xia, Y. and Whitesides, G.M. Soft lithography, *Annu. Rev. Mater. Sci.*, 28, 153, 1998.
27. Piner, R.D. et al. "Dip-Pen" nanolithography, *Science*, 283, 661, 1999.
28. Huo, F. et al. Polymer pen lithography, *Science*, 321, 1658, 2008.
29. Shim, W. et al. Hard-tip soft-spring lithography, *Nature*, 469, 516, 2011.
30. Obermair, C. et al. Reversible mechano-electrochemical writing of metallic nanostructures with the tip of an atomic force microscope, *Beilstein J. Nanotechnol.*, 3, 824, 2012.
31. Manfrinato, V.R. et al. Resolution limits of electron-beam lithography toward the atomic scale. *Nano Lett.*, 13, 1555, 2013.
32. Wu, C.S., Makiuchi, Y., and Chen, C. High-energy electron beam lithography for nanoscale fabrication. In: *Lithography*, Wang, M. Ed., ISBN: 978-953-307-064-3, InTech, 2010.
33. Lee, K.W. et al. Secondary electron generation in electron-beam-irradiated solids: Resolution limits to nanolithography, *J. Korean Phys. Soc.*, 55(4), 1720, 2009.
34. Chang, T.H.P. et al. Multiple electron-beam lithography, *Microelectron. Eng.*, 57–58, 117, 2001.
35. Freed, R. et al. Reflective electron-beam lithography: Progress toward high-throughput production capability, *Proc. SPIE 8323*, Alternative Lithographic Technologies IV, 83230H, 2012.
36. Gierak, J. et al. Sub-5 nm FIB direct patterning of nanodevices, *Microelectron. Eng.*, 84(5–8), 779, 2007.
37. Boden, S.A. et al. Focused helium ion beam milling and deposition, *Microelectron. Eng.*, 88, 2452, 2011.
38. Dimitrov, A.S. and Nagayama, K. Continuous convective assembling of fine particles into two-dimensional arrays on solid surfaces, *Langmuir*, 12, 1303, 1996.
39. Cui, L. et al. Ultra-fast fabrication of colloidal photonic crystals by spray coating, *Macromol. Rapid Commun.*, 30, 598, 2009.
40. Jiang, P. and McFarland, M.J. Large-scale fabrication of wafer-size colloidal crystals, macroporous polymers and nanocomposites by spin-coating, *J. Am. Chem. Soc.*, 126, 13778, 2004.
41. Postma, H.W.C., Sellmeijer, A., and Dekker, C. Manipulation and imaging of individual single-walled carbon nanotubes with an atomic force microscope. *Adv. Mater.*, 12(17), 1299, 2000.
42. Naik, A.K. et al. Towards single-molecule nanomechanical mass spectrometry, *Nat. Nanotechnol.*, 4, 445, 2009.
43. Chiu, H. et al. Atomic-scale mass sensing using carbon nanotube resonators, *Nano Lett.*, 8, 4342, 2008.
44. Jensen, K., Kim, K., and Zettl, A. An atomic-resolution nanomechanical mass sensor, *Nat. Nanotechnol.*, 3, 533, 2008.
45. Fritz, J. et al. Translating biomolecular recognition into nanomechanics, *Science*, 288, 316, 2000.
46. Wu, G. et al. Origin of nanomechanical cantilever motion generated from biomolecular interactions, *Proc. Natl. Acad. Sci. USA*, 98(4), 1560, 2001.
47. Erbe, A. et al. Mechanical mixing in nonlinear nanomechanical resonators, *Appl. Phys. Lett.*, 77, 3102, 2000.
48. Bartsch, S.T., Rusu, A., and Ionescu, A.M. A single active nanoelectromechanical tuning fork front-end radio-frequency receiver, *Nanotechnology*, 23, 225501, 2012.

49. Island, J.O. et al. Few-hundred GHz carbon nanotube nanoelectromechanical systems (NEMS), *Nano Lett.*, 12, 4564, 2012.

50. Loh, O.Y. and Espinosa, H.D. Nanoelectromechanical contact switches, *Nat. Nanotechnol.*, 7, 283, 2012.

51. Jang, J.E. et al. Nanoscale memory cell based on a nanoelectromechanical switched capacitor, *Nat. Nanotechnol.*, 3, 26, 2008.

52. Huang, X. et al. A nanoelectromechanical-switch-based thermal management for 3-D integrated many-core memory-processor system, *IEEE Trans. Nanotechnol.*, 11(3), 588, 2012.

53. Jain, K.K. The role of nanobiotechnology in drug discovery, *Drug Discov. Today*, 10(21), 1435, 2005.

54. Gabig, M. and Wêgrzyn, G. An introduction to DNA chips: Principles, technology, applications and analysis, *Acta Biochim. Pol.*, 48(3), 615, 2001.

55. Patel, G.M. et al. Nanorobot: A versatile tool in nanomedicine, *J. Drug Targeting*, 14(2), 63, 2006.

56. Kipke, D.R. et al. Silicon-substrate intracortical microelectrode arrays for long-term recording of neuronal spike activity in cerebral cortex, *IEEE Trans. Neural Syst. Rehabil. Eng.*, 11(2), 151, 2003.

57. Spira, M.E. and Hai, A. Multi-electrode array technologies for neuroscience and cardiology, *Nat. Nanotechnol.*, 8, 83, 2013.

58. Hai, A. et al. Spine-shaped gold protrusions improve the adherence and electrical coupling of neurons with the surface of micro-electronic devices, *J. R. Soc. Interface*, 6(41), 1153, 2009.

59. Robinson, J.T. et al. Vertical nanowire electrode arrays as a scalable platform for intracellular interfacing to neuronal circuits, *Nat. Nanotechnol.*, 7, 180, 2012.

60. Tian, B. et al. Three-dimensional flexible nanoscale field-effect transistors as localized bioprobes, *Science*, 329, 830, 2010.

61. Woosley, S. et al. Controlled toxicity studies using micropatterned cells and nanomaterials, *Proc. BMES 2013*, Seattle, WA, September 25–28, 2013.

62. Bhushan, B. and Utter, J. Nanoscale adhesion, friction and wear of proteins on polystyrene, *Colloids Surf. B*, 102, 484, 2013.

63. Cuffe, J. et al. Phonons in slow motion: Dispersion relations in ultrathin Si membranes. *Nano Lett.*, 12, 3569, 2012.

64. Chew, X., Zhou, G., and Chau, F.S. Nanomechanically suspended low-loss silicon nanowire waveguide as in-plane displacement sensor, *J. Nanophotonics*, 6, 063505, 2012.

65. Basarir, O., Bramhavar, S., and Ekinci, K.L. Motion transduction in nanoelectromechanical systems (NEMS) arrays using near-field optomechanical coupling, *Nano Lett.*, 12, 534, 2012.

66. Passi, V. et al. High-throughput on-chip large deformation of silicon nanoribbons and nanowires. *J. Microelectromech. Syst.*, 21(4), 822, 2012.

67. Cullinan, M.A. et al. Scaling electromechanical sensors down to the nanoscale, *Sens. Actuators*, A, 187, 162, 2012.

68. Lapisa, M. et al. Wafer-level heterogeneous integration for MOEMS, MEMS, and NEMS. *IEEE J. Sel. Top. Quantum Electron.*, 17(3), 629–644, 2011.

69. Despont, M. et al. VLSI-NEMS chip for parallel AFM data storage, *Sens. Actuators A*, 80(2), 100, 2000.

3 A Study of Ga-Assisted Growth of GaAs/GaAsSb Axial Nanowires by Molecular Beam Epitaxy

Shanthi Iyer, Lew Reynolds, Thomas Rawdanowicz, Sai Krishna Ojha, Pavan Kumar Kasanaboina, and Adam Bowen

CONTENTS

3.1 INTRODUCTION

The microelectronics revolution has given rise to the concept that smaller device dimensions provide enhanced performance with an increased number of components in a circuit, higher operating speeds, and lower power consumption at a reduced cost [1]. This research has led to the vast range of semiconductor electronic and photonic devices and phenomena with which we are familiar today, such as high electron mobility and complementary metal oxide field effect transistors (FETs), lasers, light-emitting diodes (LEDs), and quantum Hall effects. Their beneficial impact on society in general and the military in particular thus provides motivation for further downscaling to nanometer dimensions. It has been suggested that the new phenomena associated with novel nanoscale materials and devices offer the opportunity for engineering unique material properties and bottom-up assembly and to serve as the building blocks for the next generation of integrated nanosystems. In nanometer-scaled structures, radial and longitudinal quantum confinement in conjunction may also provide the ability to realize the control of electronic, optical, and magnetic properties of the materials in functional devices. Further, the relaxation of the lattice mismatch constraints, a major impediment encountered in thin-film heterostructures, provides the flexibility to integrate nanoscale heterostructures of a wide range of materials with engineered features that could lead to a new class of multifunctional devices, having high impact on the optoelectronic, nanoelectronic, and energy applications.

The current state of the art for photonic devices, namely lasers and detectors, is based on either III–V or II–VI compound semiconductors. The integration with Si-based devices poses problems, due to the large lattice mismatch, differences in thermal expansion coefficients, and polar/nonpolar issues between the two material systems. Although there are methods that permit the integration of the two different systems, the techniques are quite involved and add expense. Relaxed mismatch requirements, in conjunction with high-quality quantum heterostructures that can be synthesized in the nanowires (NWs), lend themselves to the integration of compound semiconductor-based optical devices with Si-based microelectronics, providing a multitude of functionalities, potentially leading to efficient, inexpensive, tunable infrared (IR) lasers for IR countermeasures, integrated sensor/detection systems, and other areas of photonics. That is, nanostructures enable the management of stresses and strains that allow fabrication of integrated materials systems that otherwise might not be possible.

3.2 DISTINGUISHING FEATURE OF NWs OVER THE BULK

NWs are particularly attractive for altering the physical properties of semiconductor materials in view of the high surface-to-volume ratio and the possible quantum confinement effects. Phonon transport is significantly degraded in Si NWs with diameters less than the phonon mean-free path. The measured thermal conductivity is two orders of magnitude less than that of the bulk, and more importantly, it departs from the Debye T^3 law for boundary scattering as NW dimensions become <40 nm [2]. While reduced lattice thermal conductivity may impede progress toward nanoelectronic applications, it could be an advantage for thermoelectric materials [3–5]. Higher strength and stiffness of NWs compared to bulk materials, attributed to a lower density of line defects with decreasing lateral dimensions, may have unique applications as sensors and actuators based on nanoscale cantilevers. The size dependence of the melting temperature in crystals with grains <50 nm is a well-known phenomenon [6], one, which may imply that annealing temperatures required for NWs are a small fraction of that required for bulk materials. This may be particularly relevant to dopant activation in material systems such as p-type GaN.

Quantum confinement in semiconductor structures manifests itself in a size dependence of the band gap. While a simple particle-in-the-box model predicts a d^{-2} dependence of the shift in band gap above the bulk value for semiconductor materials, InP NWs have exhibited a $d^{-1.45}$ dependence [7]. Thus, a higher effective band gap can be achieved in quantum-confined NWs with larger surface-to-volume ratios. A polarization dependence of light absorption/emission has also been observed in NWs in which the PL intensity is maximized in a direction parallel to the long axis [8]. Crystallographic direction-related luminescent characteristics have been reported for GaN NWs [9]. Furthermore, NW lasers have been achieved in both ZnO [10] and GaN [11] systems.

Charge transport in one-dimensional nanostructures is a developing research area and it has been suggested that Si NWs become insulated at sufficiently small diameters. On the other hand, GaN NWs with diameters up to ~18 nm still display semiconductor behavior. N-doped InP NWs have exhibited linear I–V behavior at room temperature and zero-bias suppression of conductance at low temperatures, which is indicative of coulomb-blockade transport [12]. Rectifying behavior has been observed in single GaN NWs with an internal p–n junction [13]. Cross-NW p–n junctions have led to the fabrication of FETs in numerous material systems, as well as LEDs, logic gates, resonant tunneling diodes, and a memory effect [1,14]. Huang et al. [15] reported an NW photonic LED array in which the emission wavelength could be tuned from 350 to 700 nm by using NWs fabricated from semiconductors of the appropriate material composition. In addition, they reported an NW-integrated optoelectronic circuit, in which an Si NW FET was employed for the drive current of a GaN-based LED. Qian et al. [16] have fabricated a tunable InGaN-based LED from a core-shell NW. Recently, a GaAs/AlGaAs core-multi-shell NW-based LED integrated on an Si substrate for operation in the near-IR region has been reported [17].

On the basis of data available in the literature for a host of material systems, from Si to III-nitrides, for example, it is clear that nanostructures offer considerable opportunities for altering the properties of bulk materials and for fabricating novel optoelectronic devices, multifunctional devices, and smart sensors.

3.3   GROWTH MECHANISM OF NWs

Among the different growth mechanisms, the vapor–liquid–solid (VLS) mechanism, originally reported by Wagner and Ellis for the growth of Si whiskers [18], is the most commonly used growth mechanism. Central to this process is the liquid metal catalyst on which the desired component of the source semiconductor in the vapor phase is soluble. The solubility of the species of interest in the liquid creates a concentration gradient in the liquid. The transport of these solutes to the metal–substrate interface causes supersaturation, leading to nucleation and subsequent NW growth. The droplet size defines the diameter of the NW growth and hence, any parameter that affects the equilibrium composition of the droplet would naturally influence the droplet size. The two important parameters that have been found to affect the droplet size are the growth temperature as well as the vapor pressure of the constituent source species [19]. When the reactant species are incorporated as the catalyst, a one-dimensional NW grows. By changing the reactant species, one can grow axial or radial heterostructures by altering the growth conditions to achieve preferential incorporation on the catalyst or uniformly along the surface of the growing NW [20].

Among the several metals that have been used as the catalyst, Au has been the most common. However, the deep trap formation along with the high diffusivity of Au in most of the semiconductors leads to contamination of the NW and in addition, the trend toward the elimination of Au in most of the device processes in the electronics industry makes it unattractive. Hence, recently, there has been an increasing interest on the self-catalyzed growth of NWs using the low melting point of the source element as the catalyst such as Ga melt for GaAs NWs [19].

The growth of thin NWs is more desirable as it is rich in novel physics and these unusual properties can be suitably exploited for various applications. However, the downsizing of the NW diameter poses challenges, as the droplet size decreases due to (a) the increasing difficulty in the realization of supersaturation at the droplet NW interface, (b) presence of strong van der Waals interatomic forces, and (c) Ostwald ripening favoring larger-diameter droplets [21]. In addition, in the case of GaAs and InAs, there is an added complexity due to the very small internal energy difference between their zinc-blende (ZB) and wurtzite (WZ) structures [22]. Hence, the transformation from one structure to the other in these NWs is found to be a strong function of the NW diameter. Thin GaAs NWs had been grown using Au catalyst, until recently [23] where the growth of thin GaAs NWs with diameters <15 nm have been demonstrated using self-catalyzed Ga under high As vapor pressure.

The following section is a brief review of the ongoing work on Sb-based NWs with emphasis on the GaAsSb NWs, which covers an important wavelength region of communication near 1.3 μm, as this region corresponds to low losses in the optical fiber.

3.4   BACKGROUND OF Sb-BASED BINARY NWs AND GaAsSb NWs

The low band gaps associated with InSb (0.17 eV), GaSb (0.7 eV), and AlSb (1.6 eV) make Sb-based NWs of high interest for IR optical applications and high-speed electronic devices. Vaddiraju et al. [22] reported both GaSb and InSb NWs using self-catalyzed antimonidization/reactive vapor transport methods, resulting in room temperature PL emission of 1.72 μm in GaSb NWs. Chen et al. [24] reported an InSb NW-based IR photodetector grown by Au catalyst-assisted molecular beam epitaxy (MBE).

Recently, GaSb NWs were also synthesized on c-plane sapphire substrates by gold-mediated VLS growth using a metal organic vapor epitaxy (MOVPE) process with a higher growth rate and varying the V/III ratio [25]. They have reported [25] 3.5 K PL emission from 1.5 to 1.65 μm for V/III ratio ranging from 0.5 to 2. These NWs exhibited a ZB structure, as in the bulk, in contrast to the GaAs NWs that occur more commonly in a hexagonal WZ structure, due to the very small internal energy difference between its ZB and WZ structure [26]. The GaAs crystal structure has been shown to change from WZ to ZB by insertion of either GaSb [27] or GaAsSb layers [28,29], when the growth is appropriately tailored. The lower interface of GaAs/GaAsSb is found to be independent of the GaAs structure, while the upper interface of GaAsSb/GaAs is found to consist of either stacking faults or twins, depending on the GaAs structure being WZ [28,29] or ZB [30], respectively.

The band alignment, correspondingly, changed from type II, as in the bulk, with emission at 0.98 μm [28,31] to weak type I [28] with emission at 0.8 μm due to the conduction band edge of the WZ structure being slightly higher in the latter type. The Sb composition in these films was 25%. The low-temperature PL peaks seem to be distributed between 1.275 and 1.45 eV and the peak that is dominant seems to be strongly influenced by the type (ZB or WZ) of GaAs phase present above and below the GaAsSb segment [30]. Dheeraj et al. [30] have also reported on GaAsSb segments, of a few nanometers in length, sandwiched between GaAs segments axially. It is observed that the average growth rate was found to be dependent on the growth sequence of the segment on the axial wire, that is, with that of the third segment being higher than the second segment while the growth rate of the fourth segment is essentially the same or lower than that of the third segment. Similar trends for GaAs segments were observed although the structural phase for the GaAs segment was WZ as opposed to ZB in the case of GaAsSb segment. Low-temperature micro-PL exhibited a weak peak at 1.27 eV, which was attributed by the authors to the type II band alignment between the WZ phase of GaAs and ZB phase of GaAsSb.

The GaAsSb NW by itself [32] grown on a GaAs (111)B substrate, exhibits rotational twins around its (111)B growth axis, with an equal amount of twinned and untwinned orientations. More recently, Plissard et al. [29] demonstrated the growth of a GaAs/GaAsSb core with AlGaAs shell without Au as the catalyst, but instead, using the group III element Ga to induce the NW growth. The Sb composition used in the GaAsSb core was 30%, while in the shell, it was 22%. Unlike other III–V alloys such as AlGaAs, GaAsSb requires more accurate control in the composition due to the presence of the two competing group V species with differing sticking coefficients. For instance, Ga growth rate was reported [33] to affect the Sb incorporation in the NW due to both the differences in the sticking coefficients and the bonding, although its effect on the PL results was not reported. Thus, the composition of GaAsSb alloy not only seems to depend on the flux ratios but also on the individual fluxes.

3.5 STRUCTURAL AND OPTICAL CHARACTERISTICS OF MBE GROWN GaAsSb-SEGMENTED NWs

Although different techniques have been used for the growth of GaAs NWs [34], GaAsSb NWs have been grown by MBE [29], and only chemical vapor transport [35]. MBE is an ideal technique as the low deposition rate as well as rapid opening and closing of the source shutter lead to better control of the growth on an atomic level with abrupt interfaces. In addition, the presence of *in situ* monitors and the fact that the growth process is governed by the surface kinetics allowing the growth to occur under nonequilibrium conditions, makes it an attractive research tool to demonstrate the conceptual ideas prior to commercialization using much less expensive mass production tools.

In this chapter, we present our work on Ga-assisted growth of GaAsSb segments grown on the GaAs NW stems by MBE. GaAsSb/GaAs segments up to three have been realized, although only the work on these growths to two segmented

GaAsSb/GaAs is presented here. Structural and optical properties as a function of the number of segments in the NWs are presented using a variety of characterization techniques.

3.5.1 EXPERIMENTAL PROCEDURE

NW growths were carried out in the EPI 930 solid source MBE system with valved As and Sb cracker sources. All the NWs were grown on EPI ready (111) Si substrates. Ga-assisted NW growths were initiated by impinging Ga flux on the substrate for 6–8 s prior to the opening of the As flux. The growth temperature for Ga-assisted growth was investigated in the range of 580–620°C. The As/Ga flux ratio was kept constant for all the samples at an As beam equivalent pressure (BEP) of 2.5×10^{-6} Torr. The NWs were characterized using *in situ* reflection high-energy electron diffraction (RHEED) and scanning electron microscopy (SEM) images from a Zeiss EVO 10 for determining NW physical dimensions and surface features. PL measurements were conducted using a He–Ne laser as the excitation source with a 0.32 m double grating monochromator for wavelength dispersion. A Si detector was used with a conventional lock-in amplifier system. A closed-cycle three-stage APD cryogenic system was used to study the variation of PL characteristics in the 10–300 K temperature range. Raman spectroscopy was carried out using HeNe laser with 633 nm excitation wavelength and a Horiba Jobin Yvon LabRam ARAMIS with a spectral resolution of 0.6 cm^{-1} for spectral dispersion. The Raman signal was detected using multichannel air-cooled charge-coupled device and the peak positions were obtained from the Lorentzian fit to the data. NW microstructure investigation and analysis were performed using conventional and high-resolution transmission electron microscopy (CTEM and HRTEM, respectively), as well as scanning transmission electron microscopy (STEM) on a JEOL2000FX and a JEOL2010F. Compositional analysis was performed using x-ray energy dispersive spectrometry (XEDS) with a super X windowless silicon drift detector (SDD) and Bruker AXS multichannel analyzer on an FEI Titan G2 60–300. The TEM sample preparation consisted of creating an NW suspension by sonication in methanol for 1 min and pipetting droplets onto a lacy Formar/carbon copper TEM grid.

The heterostructured NWs consist of a GaAs stem followed by a GaAsSb segment that was then terminated with a GaAs cap. Although the As to Sb flux ratio was 10, the Sb composition fraction was much less than 10% as determined from electron diffraction spectroscopy (EDS) characterization. The growth schematic of GaAs/GaAsSb/GaAs single-segment NWs referred to as N1-620 and N1 for the NWs grown at 620°C and 600°C, respectively, is shown in Figure 3.1a. N2 refers to two segments of GaAsSb sandwiched between GaAs sections as shown in the schematic in Figure 3.1b, grown at 600°C.

3.5.2 SCANNING ELECTRON MICROSCOPY

Figure 3.2a displays the SEM image of NW1-620 NWs exhibiting a hexagonal shape with well-delineated surface planes and terminating with a flat top. Typically, 80% of the wires were found to be vertically oriented to the substrate (Figure 3.2b). Typical

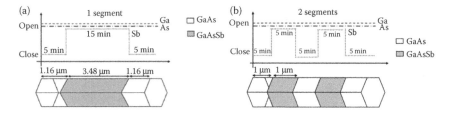

FIGURE 3.1 Growth schematics of (a) one segment and (b) two-segment NWs.

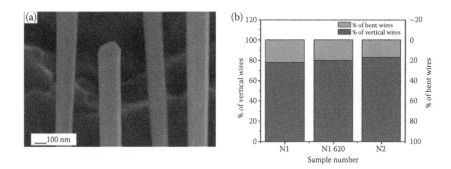

FIGURE 3.2 (a) Hexagonal-shaped GaAs NWs grown on Si substrate and (b) comparison of the percentage of vertical and bent wires in different samples taken over an area of 200 μm².

NW diameters were in the range of 150–180 nm and ~4–5.5 μm in length, and the NW density ranged from 4×10^6 to 3×10^7/cm². It is to be noted that to achieve the flat top, the As flux was left "on" after the termination of the NW growth until the substrate temperature reached 500°C.

Simultaneous closing of the As and Ga shutters results in a Ga droplet on the top of the NW as shown in Figure 3.3a. Leaving the As "on" after termination of the growth resulted in a tapered top as shown in Figure 3.3b. Thus, the tip of the NW can be suitably controlled by the As flux caused by varying Ga consumption consistent with the VLS growth mechanism.

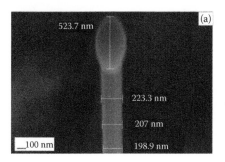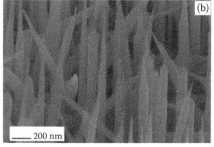

FIGURE 3.3 (a) N1 wire with Ga droplet on the top and (b) NWs with tapered top.

3.5.3   TRANSMISSION ELECTRON MICROSCOPY

Transmission electron microscope (TEM) images and associated selected area electron diffraction (SAED) patterns of NW N1-620 with a single GaAsSb segment as depicted in Figure 3.1 are shown in Figure 3.4. This NW with Ga droplet formation at the NW tip exhibited pure ZB phase verified by SAED (see Figure 3.4b) with randomly spaced stacking faults present as multiple twins in the upper-third to one-quarter of the NW in the form of lamellar parallel to the ($1\bar{1}1$) twin composition plane. This lamellar twinned region corresponded to the cessation of

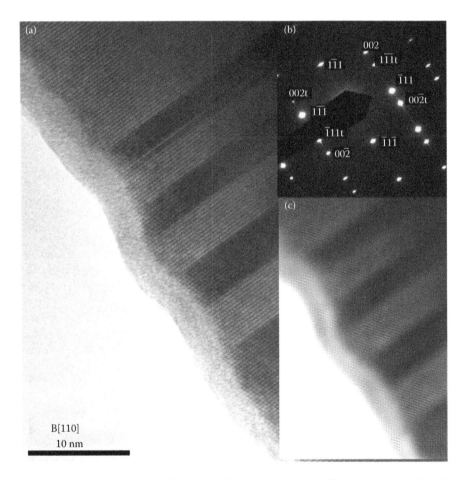

FIGURE 3.4   TEM image (a) of NW N1-620 consisting of pure ZB phase along with multiple twin boundaries. The HRTEM image (b) of twin boundaries with SAED pattern (inset) at the twin (111) plane. Fast Fourier filter of HRTEM image (c) applied to the (002) reflections of both ZB twin phases illustrates the mirroring of the (002) and (002t) planes across the twin boundary.

Sb incorporation. SAED pattern indexing confirms the ZB NW crystal growth to be [111] oriented and the twin composition planes (111) oriented.

For these GaAs/GaAsSb NWs, the twin boundary separates the crystal domains by rotation about the composition plane symmetry axis. Since the GaAs ZB crystal structure is in the $F\bar{4}3\,m$ space group, the [111] symmetry axis belonging to the $\bar{4}3\,m$ point group is parallel to the NW crystal growth zone axis. Subsequently, the ZB phases mirror one another across the twin boundary with the ZB zone axes $[1\,\bar{1}\,1]$ and $[\bar{1}\,1\,\bar{1}]$ changing directions across the twin composition plane. For ZB GaAs, the symmetry operation for a rotation about the [111] axis is a triad. Thus, the twinning operation may be a 60°, a 180°, or a 240° rotation about the twin axis. However, it is conventional to describe the twinning operation as a rotation of 180°.

For the N1-620 NWs exhibiting Ga droplets at the tip, there was no evidence of a WZ phase at any point along the NW. However, for the N1 and N2 NWs in which the tips were devoid of Ga droplets, a region of WZ phase would occur just below the tip, as shown in Figure 3.5. However, at the tip, these NWs without Ga droplets consistently terminated in the ZB phase.

The XEDS line scan shown in Figure 3.6 confirms the incorporation of Sb alloying in the NW. The Sb composition and location along the N1-620 NW was established using STEM and XEDS elemental mapping of the entire NW. This enabled quantification of the atomic weight percent of Ga, As, and Sb in selected regions with

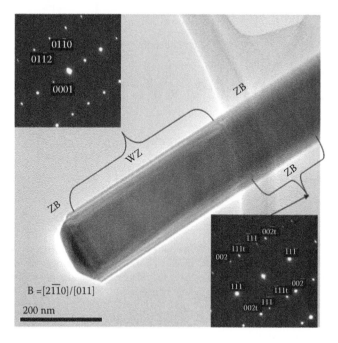

FIGURE 3.5 N2 NW without a Ga droplet at the tip and with a 400 nm region with a WZ structure just below the tip. The very tip end itself and the remaining NW below the WZ region are pure ZB with multiple twinning.

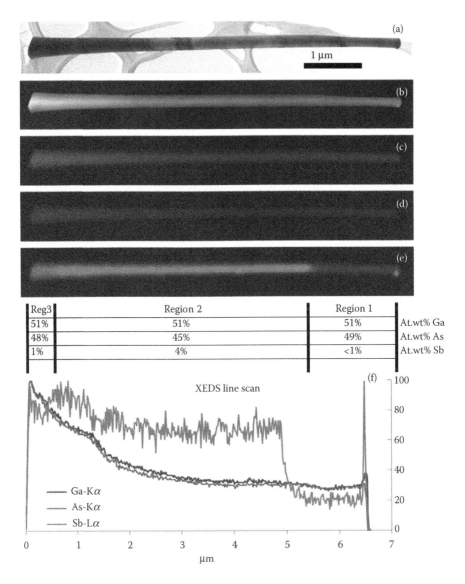

FIGURE 3.6 **(See color insert.)** A NW with a single GaAsSb band (Region 2 (e)) imaged with TEM (a), HAADF–STEM (b), and XEDS–STEM mapping (c) through (e) of Ga, As, and Sb, respectively. An XEDS line scan (f) shows the relative elemental x-ray count distribution along the NW axis. The collective x-ray count increases with NW diameter in the direction of the NW base.

and without intentional Sb incorporation. Although Sb was not intended to be present in Region 3 of Figure 3.6e, XEDS quantification reveals the presence of Sb due to coaxial deposition during the growth of GaAsSb segment in Region 2. One should note that the growth rate of both GaAs and GaAsSb in the NWs is ~20–25% greater than anticipated compared to the schematic in Figure 3.1a.

3.5.4 PHOTOLUMINESCENCE

Photoluminescence is a powerful tool to investigate the optical properties of the NWs, which are greatly influenced by the surface and morphological effects due to the inherent large surface-to-volume ratio. For instance, the WZ–ZB interface in GaAs is reported to yield PL peak energies lower than the corresponding polytypes due to the two phases at the interface forming a type II band alignment [36]. Similarly, the presence of defect levels often leads to luminescence at energies below that of the band gap at low temperatures. Thus, a clear interpretation of luminescence spectra of these NWs requires more detailed investigations. The temperature dependence of photoluminescence determines the nature of defects and nonradiative recombination centers and hence is indicative of the lifetime of the carriers. In this work, we have carried out a detailed study of the temperature dependence of the PL spectra for the three NWs. Figure 3.7a displays the low-temperature PL of all the three NWs. The N2 sample exhibits the highest PL peak intensity with the spectra having a slightly asymmetric line shape with a relatively sharp high-energy cutoff.

In Figure 3.7b, the temperature dependence of the PL peak energy is displayed for all the samples. In the low-temperature regime, single-segment wires exhibit a slight red shift with decreasing temperature compared to the double-segmented NW.

The temperature dependence of the PL peak position was fitted using the Varshni equation

$$E_g(T) = E_g(0) - \frac{\alpha T^2}{\beta + T} \tag{3.1}$$

where T is the absolute temperature, $E_g(0)$ is the band gap at 0 K, and α and β are the fitting parameters. The values of these parameters for all the three samples are listed in Table 3.1. The values of α and β of the NWs are comparable to those of GaAs NWs reported in the literature. An excellent fit is obtained with the Varshni formula

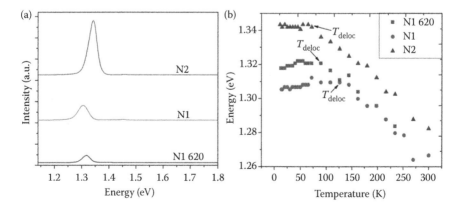

FIGURE 3.7   (a) 11 K PL spectra of the NWs and (b) temperature dependence of PL peak energy.

TABLE 3.1

Summary of Samples and Pertinent PL Parameters

Sample	Segments	Growth Temperature (°C)	α (eV/k)	β (k)	E_g (eV)	Energy at 11 K	E_a (meV)	E_b (meV)	E_{local}^{max} (meV)	T_{deloc} (°C)
N1-620	1	620	3.87×10^{-4}	251	1.328	1.31	51.4	4.5	9.8	90
N1	1	600	3.75×10^{-4}	247	1.326	1.31	47.6	6.8	20	172
N2	2	600	3.7×10^{-4}	230	1.345	1.34	84.7	10.8	0	72

for sample N2 over the entire temperature range, while for the single-segmented samples, there is considerable difference between the experimental values and those obtained using Varshni's formula for temperatures below about 110 K. That is, in these latter samples, the temperature-independent portion that is normally observed at the lowest temperatures extends to considerably higher temperatures.

In Table 3.1, we also list the values of E_{loc}^{max} and T_{deloc}, which are defined as the maximum localization energy measured as the largest energetic difference between the experimental PL peak energy and the value of the energy predicted by the Varshni relation, and the temperature at which delocalization of the carriers is complete, respectively.

These quantities are shown in Figure 3.8. The values achieved are comparable in the two single-segmented NWs, but they are considerably lower for the double-segmented NW. To understand the nature of the recombination mechanisms, the temperature dependence of the total integrated PL intensity (I_{PL}) was also measured in these samples, as shown in Figure 3.8. The integrated intensity rapidly decreases with increasing temperature, in particular, for temperatures above 80 K. A best fit to the temperature dependence of the I_{PL} was obtained using the following phenomenological expression [37]:

$$I_{PL}(T) = \frac{I_0}{1 + A\exp(-(E_a/kT))} \tag{3.2}$$

in which the presence of only one nonradiative recombination channel has been assumed in a given temperature range of investigation. I_0 is the PL intensity at $T = 0$ K and E_a represents the thermal activation energy of the first nonradiative channel, respectively.

The values of I_0, E_a, and E_b that resulted in the best fit to the experimental data in all three samples are also listed in Table 3.1. The values of E_a and E_b in the N1 and N1-620 are comparable, but are ~45% smaller than those for the double-segmented

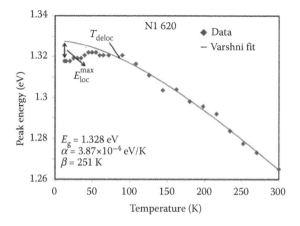

FIGURE 3.8 Varshni fit for the data of sample N1-620.

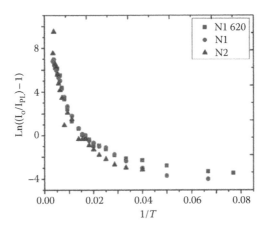

FIGURE 3.9 Temperature dependence of the integrated intensity.

sample N2. The difference between the experimental values and Varshni-predicted PL peak energy, as well as the invariance of the PL peak energy with temperature from 11 up to 80–100 K and the corresponding very-low-temperature variation of the full width half maxima (FWHM) (see Figure 3.9) in this temperature range, can be considered as a strong evidence for exciton localization. The variation in the PL peak energy as a function of temperature in these NWs can be quantitatively explained as follows: At low temperatures, the excitons are localized either at the defects induced in the NWs by surface irregularities or impurities or at the band-tail states in the density of states (DOS). The latter is more likely as the FWHM significantly varies with temperature.

As the temperature is increased above the exciton localization energy identified as T_{deloc}, the excitons become delocalized due to dissociation into electron–hole pairs. Above this, the emission energy decreases as a function of temperature due to the band gap shrinkage following the Varshni-like relation. The differences in the PL peak energies observed at low temperature (10 K) and at room temperature in all these samples seem to be similar, ~55–60 meV, and, thus, independent of the GaAsSb segment length. These values are somewhat smaller than 76 meV reported by Chiu et al. [38] in GaAsSb quantum wells (QWs) and 81 meV from the bulk values. Thus, our data suggest that downsizing radially reduces the temperature-induced band gap variation.

The delocalization temperature is the lowest for the double-segmented NW with the smallest localization energy and the effect of axial confinement results in opening of the band gap, blue shifting the PL peak energy in the entire temperature range.

The value of the FWHM of an excitonic transition in these NWs is representative of the quality of the alloy segments, as well as the extent of the interface roughness [32]. The FWHM of the N2 is the lowest in the entire temperature region, which is an indication of better optical quality of these NWs. The observed variation of the FWHM with temperature in these NW structures can be qualitatively explained as follows: At low temperatures, FWHM reflects the energy distribution of the exciton states in the localizing potentials. As the temperature is raised, the excitons begin to become delocalized. For temperatures above T_{deloc}, almost all the excitons have become delocalized and the exciton–optical phonon interaction becomes dominant with the value of the FWHM increasing with temperature. As shown in Figure 3.9, the variation in FWHM with temperature is less and T_{deloc} is smaller in sample N2 due to the smaller range of values of the localization energies. In sample N1, the FWHM exhibits a small inverted S-curve, that is, a mild dip in the FWHM curve as the temperature approaches T_{deloc}, which is a signature of strong exciton localization normally observed in dilute nitride system [39–41]. The T_{deloc} is considerably larger for this sample. The large value of T_{deloc} is consistent with the large $E_{\text{loc}}^{\text{max}}$ also observed in this sample.

We have analyzed the variation of I_{PL} as a function of temperature using Equation 3.2 and the values of the various fitting parameters for all three samples are listed in Table 3.1. We find that the values of E_a and E_b are comparable for the two single-segment wires, but are considerably larger for the N2 sample. For the sample N1-620, the E_b is 4.5 meV, which is in excellent agreement with exciton binding energy in GaAs. The low activation energy (E_b) and the high activation energy E_a correspond to nonradiative channels that are responsible for determining the quenching of PL intensity at temperatures below 100 K and higher than 100 K, respectively. The low values of 4–10 meV for E_b indicate that the weak excitons are bound to shallow defects. Higher E_a at 77 meV has been observed in GaAs NWs grown by Au-assisted catalyst and has generally been assigned to a deep center associated with Au-induced defects [42], which cannot be the case in our NWs since our NWs are Ga catalyzed. Such a deep center has been observed in CdS NWs [43]. The observation of such a deep exciton suggests the presence of the defect complexes in these NWs.

The room temperature PL is realized for the samples grown at 600°C and also, the low-temperature PL peak intensity was considerably higher; particularly, the N2 exhibited the highest PL peak intensity almost threefold higher than the next largest PL intensity observed in N1. Sample N1-620 had the lowest PL intensity. The lowest FWHM and smallest variation in the FWHM observed for the double-segmented wire is consistent with the better quality of this NW. Conversely, N1-620 showed significant variation in FWHM with temperature and is characterized by low PL peak intensity.

The foregoing discussion of the behaviors of the PL peak energy, excitonic linewidth, and integrated PL intensity as a function of temperature in our samples suggests that among the three NWs investigated, the double-segmented GaAsSb NW is the better quality NW with reduced defect density of localized states and of nonradiative recombination centers. The temperature dependence of the PL can be well described by a Varshni-like relation for temperatures above 100–150 K. The low-temperature behavior appears to be dominated by excitons bound to shallow defects and two nonradiative channels were found, one weakly bound exciton related and the other related to a deep center, which appears to be influenced by the NW structure.

3.5.5 RAMAN SPECTROSCOPY

Raman spectra were taken at different locations and representative Raman spectra are shown in Figure 3.10 and summarized in Table 3.2 for the NWs N1 and N2 and

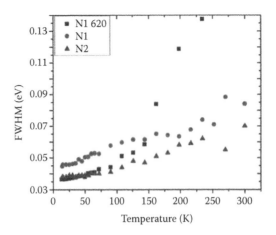

FIGURE 3.10   Temperature dependence of FWHM.

a reference GaAsSb thin film grown on a (001) GaAs substrate. For the reference sample, one should note that the GaAs-like longitudinal optical (LO) and transverse optical (TO) modes occur at 279 and 255 cm^{-1}, respectively, and the intensity of the LO mode is greater than that of the TO mode. This is not unexpected since the TO mode is forbidden for the (100) surface. Its presence is most likely associated with relaxation of the selection rules due to the presence of defects. Very weak GaSb-like TO and LO modes are also observed at 227 and 240 cm^{-1} in this reference sample. In the NW samples, the spectra exhibit GaAs-like TO and LO phonon modes at 266 and at 289 cm^{-1}, respectively, which are in good agreement with the corresponding bulk GaAs values. Compared to the reference epitaxial film, however, the GaAs-like TO and LO modes in the NWs are shifted upward by 10–11 cm^{-1}, which are opposite to the downshift in energy reported [44] in GaAs NWs. The FWHM of the GaAs-like TO and LO modes in NWs N1 and N2 are 8/8 cm^{-1} and 8/11.1 cm^{-1}, respectively. These values and the peak positions indicate good optical quality of the NWs. In addition, the GaSb-like TO and LO phonon modes were also observed at 222 and 237 cm^{-1}, respectively, in reasonable agreement with the corresponding bulk values for GaSb. Note that the GaSb-like modes are downshifted in energy 3–5 cm^{-1} compared to the reference sample. The GaSb peak intensity scales to the anticipated

TABLE 3.2

Raman Mode Data for N1, N2, and the Reference GaAsSb Epilayer Grown on GaAs

Sample	GaAs TO (cm^{-1})	GaAs LO (cm^{-1})	GaSb TO (cm^{-1})	GaSb LO (cm^{-1})	GaAs TO/ LO	GaSb TO/ LO
N1	266	289	222	235	2.2	0.8
N2	266	289	223	237	1.6	0.73
Reference	255	279	227	240	0.73	0.94

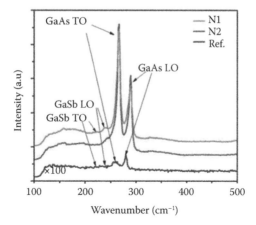

FIGURE 3.11 Raman spectra depicting the GaAs- and GaSb-like modes of NWs N1 and N2 compared to an epitaxial $GaAs_{0.95}Sb_{0.05}$ epitaxial film reference.

thickness of the GaAsSb layers in the NWs. The high-intensity ratio of TO to LO phonon modes of GaAs has been commonly observed in GaAs NWs, being the highest for N1 and is attributed [45] to the LO mode being forbidden from certain surface facets as opposed to no such restrictions imposed on the TO phonon modes. A relatively low TO/LO GaAs phonon peak ratio as well as a lower FWHM of these two in NW N2 is also indicative of better quality layers consistent with the PL data discussed earlier (Figure 3.11).

3.6 CONCLUSIONS

Segments of GaAsSb NWs on GaAs stem have been successfully grown by MBE, using a Ga-catalyzed VLS technique. The two-segmented GaAsSb NWs were found to be of excellent optical and structural quality as attested by low FWHM and their temperature variation in the corresponding PL spectra, as well as low TO/LO ratio observed for both GaAs and GaSb peaks. TEM of these NWs revealed both GaAs and GaAsSb to be of ZB phase for the wire terminating without any Ga droplet. The stacking faults and twins were found to be present in all the NWs. These are preliminary data, and further growth optimization will be initiated. Hence, a fairly good optical and structural quality of the GaAsSb NWs grown so far show great promise for applications in the 1.3 μm region and longer wavelength for optical communications.

ACKNOWLEDGMENTS

This work is supported by the Army Research Office (Grant No. W911NF-11-1-0223, technical monitor—William Clark). The authors acknowledge Professor Yuntian Zhu for the use of his Raman system, Dr. Judith Reynolds for carrying out the Raman measurements, and Dr. Ryan White for performing the XEDS measurements on the FEI Titan G2 at NC State University's Analytical Instrumentation Facility.

REFERENCES

1. J. X. Serge Luryi and A. Zaslavsky, *Future Trends in Microelectronics: The Nano Millennium,* (Wiley-IEEE Press, New York, 2002).
2. D. Li, Y. Wu, P. Kim, L. Shi, P. Yang, and A. Majumdar, *Appl. Phys. Lett.* 83(14), 2934–2936, 2003.
3. M. Law, J. Goldberger, and P. Yang, *Annu. Rev. Mater. Res.* 34, 83–122, 2004.
4. L. J. Lauhon, M. S. Gudiksen, and C. M. Lieber, *Phil. Trans. Roy. Soc. Lond. A* 362, 1247, 2004.
5. Y. Xia, P. Yang, Y. Sun, Y. Wu, B. Mayers, B. Gates, Y. Yin, F. Kim, and H. Yan, *Adv. Mater.* 15(5), 353–389, 2003.
6. K. F. Peters, J. B. Cohen, and Y.-W. Chung, *Phys. Rev. B* 57(21), 13430, 1998 and references therein.
7. H. Yu, J. Li, R. A. Loomis, L.-W. Wang, and W. E. Buhro, *Nat. Mater.* 2(8), 517–520, 2003.
8. J. Wang, M. S. Gudiksen, X. Duan, Y. Cui, and C. M. Lieber, *Science* 293(5534), 1455–1457, 2001.
9. T. Kuykendall, P. J. Pauzauskie, Y. Zhang, J. Goldberger, D. Sirbuly, J. Denlinger, and P. Yang, *Nat. Mater.* 3(8), 524–528, 2004.
10. M. H. Huang, S. Mao, H. Feick, H. Yan, Y. Wu, H. Kind, E. Weber, R. Russo, and P. Yang, *Science* 292(5523), 1897–1899, 2001.
11. J. C. Johnson, H.-J. Choi, K. P. Knutsen, R. D. Schaller, P. Yang, and R. J. Saykally, *Nat. Mater.* 1(2), 106–110, 2002.
12. S. De Franceschi, J. A. van Dam, E. P. A. M. Bakkers, L. F. Feiner, L. Gurevich, and L. P. Kouwenhoven, *Appl. Phys. Lett.* 83(2), 344–346, 2003.
13. G. Cheng, A. Kolmakov, Y. Zhang, M. Moskovits, R. Munden, M. A. Reed, G. Wang, D. Moses, and J. Zhang, *Appl. Phys. Lett.* 83(8), 1578–1580, 2003.
14. C. Thelander, H. A. Nilsson, L. E. Jensen, and L. Samuelson, *Nano Lett.* 5(4), 635–638, 2005.
15. Y. Huang, X. Duan, and C. M. Lieber, *Small* 1(1), 142–147, 2005.
16. F. Qian, S. Gradecak, Y. Li, C.-Y. Wen, and C. M. Lieber, *Nano Lett.* 5(11), 2287–2291, 2005.
17. K. Tomioka, J. Motohisa, S. Hara, K. Hiruma, and T. Fukui, *Nano Lett.* 10(5), 1639–1644, 2010.
18. R. S. Wagner and W. C. Ellis, *Appl. Phys. Lett.* 4, 89, 1964.
19. X. Sun, S. Wang, J. S. Hsu, R. Sidhu, X. G. Zheng, X. Li, J. C. Campbell, and A. L. Holmes, *IEEE J. Sel. Top. Quantum Electron.* 8(4), 817–822, 2002.
20. L. J. Lauhon, M. S. Gudiksen, D. Wang, and C. M. Lieber, *Phil. Trans. R. Soc. Lond. A* 420, 57–61, 2002.
21. P. W. Voorhees, *J. Stat. Phys.* 38(1–2), 231–252, 1985.
22. S. Vaddiraju, M. K. Sunkara, A. H. Chin, C. Z. Ning, G. R. Dholakia, and M. Meyyappan, *J. Phys. Chem. C* 111(20), 7339–7347, 2007.
23. G. E. Cirlin, V. G. Dubrovskii, Y. B. Samsonenko, A. D. Bouravleuv, K. Durose, Y. Y. Proskuryakov, B. Mendes et al., *Phys. Rev. B* 82(3), 035302, 2010.
24. H. Chen, X. Sun, K. W. C. Lai, M. Meyyappan, and N. Xi, Traverse City, MI, United States, 2009 (unpublished).
25. R. Burke, X. Weng, M.-W. Kuo, Y.-W. Song, A. Itsuno, T. Mayer, S. Durbin, R. Reeves, and J. Redwing, *J. Electron. Mater.* 39(4), 355–364, 2010.
26. C. Y. Yeh, Z. W. Lu, S. Froyen, and A. Zunger, *Phys. Rev. B* 46(16), 10086, 1992.
27. M. Jeppsson, K. A. Dick, J. B. Wagner, P. Caroff, K. Deppert, L. Samuelson, and L.E. Wernersson, *J. Cryst. Growth* 310(18), 4115–4121, 2008.
28. A. F. Moses, T. B. Hoang, D. L. Dheeraj, H. L. Zhou, A. T. J. van Helvoort, B. O. Fimland, and H. Weman, *IOP Conf. Ser.: Mater. Sci. Eng.* 6, 012001, 2009.

29. S. Plissard, K. A. Dick, X. Wallart, and P. Caroff, *Appl. Phys. Lett.* 96, 121901, 2010.
30. D. L. Dheeraj, H. L. Zhou, A. F. Moses, T. B. Hoang, A. T. J. van Helvoort, B. O. Fimland, and H. Weman, ISBN 978-953-7619-79-4, pp. 414, 2010.
31. D. L. Dheeraj, G. Patriarche, H. Zhou, T. B. Hoang, A. F. Moses, S. Grnsberg, A. T. J. Van Helvoort, B.-O. Fimland, and H. Weman, *Nano Lett.* 8(12), 4459–4463, 2008.
32. D. L. Dheeraj, G. Patriarche, L. Largeau, H. L. Zhou, A. T. J. Van Helvoort, F. Glas, J. C. Harmand, B. O. Fimland, and H. Weman, *Nanotechnology* 19(27), 27505, 2008.
33. J. Todorovic, H. Kauko, L. Ahtapodov, A. F. Moses, P. Olk, D. L. Dheeraj, B. O. Fimland, H. Weman, and A. T. J. van Helvoort, *Semicond. Sci. Technol.* 28, 115004, 2013.
34. K. A. Dick, *Prog. Cryst. Growth Ch.* 54, 138–173, 2008 and references therein.
35. K. J. Kong, C. S. Jung, G. B. Jung, Y. J. Cho, H. S. Kim, J. Park, N. E. Yu, and C. Kang, *Nanotechnology* 21, 435703(6pp), 2010.
36. C.-Y. Yeh, Z. W. Lu, S. Froyen, and A. Zunger, *Phys. Rev. B* 46(16), 10086, 1992.
37. T. Torchynska, J. Aguilar-Hernandez, M. M. Rodriguez, C. Mejia-Garcia, G. Contreras-Puente, F. G. B. Espinoza, B. M. Bulakh et al., *J. Phys. Chem. Solids* 63, 561–568, 2002.
38. Y. S. Chiu, M. H. Ya, W. S. Su, and Y. F. Chen, *J. Appl. Phys.* 92, 5810, 2002.
39. J. Li, S. Iyer, S. Bharatan, L. Wu, K. Nunna, W. Collis, K. K. Bajaj, and K. Matney, *J. Appl. Phys.* 98(1), 13701–13703, 2005.
40. K. Nunna, S. Iyer, L. Wu, J. Li, S. Bharatan, X. Wei, R. T. Senger, and K. K. Bajaj, *J. Appl. Phys.* 102(5), 053106, 2007.
41. S. Bharatan, S. Iyer, K. Nunna, W. J. Collis, K. Matney, J. Reppert, A. M. Rao, and P. R. C. Kent, *J. Appl. Phys.* 102(2), 23501–23503, 2007.
42. S. Breuer, C. Pfuller, T. Flissikowski, O. Brandt, H. T. Grahn, L. Geelhaar, and H. Riechert, *Nano Lett.* 11, 1276–1279, 2011.
43. T. B. Hoang, L. V. Titova, H. E. Jackson, and L. M. Smith, *Appl. Phys. Lett.* 89, 123123, 2006.
44. N. Begum, A. S. Bhatti, M. Piccin, G. Bais, F. Jabeen, S. Rubini, F. Martelli, and A. Franciosi, *Adv. Mater. Res.* 31, 23, 2008.
45. N. Begum, M. Piccin, F. Jabeen, G. Bais, S. Rubini, F. Martelli, and A. S. Bhatti, *J. Appl. Phys.* 104, 104311, 2008.

4 Application of Micro-/Nanotechnology in the Design and Control of Neural Interfaces

Joseph M. Starobin, Syed Gilani, and Shyam Aravamudhan

CONTENTS

4.1 INTRODUCTION

Peripheral nerves (PNs) are the communication links of the human body. Neural signals, called action potentials (APs) travel along these PNs to and from the central nervous system (CNS) to ensure reliable innervations of the skin, to control skeletal muscle contractions, and for rapid signaling of sensory events. Injuries to the PNS, known as peripheral nerve injury (PNI) are common and are a major source of disability, impairing the ability to move muscles and/or feel normal sensations, and may result in painful neuropathies. This injury or amputation also disrupts the bidirectional communication link to the CNS.

After PNI, the clinical procedure is to appose the two nerve ends and suture them together without creating tension, if possible. Even with an end-to-end surgical nerve repair (by suture), functional recovery has been largely disappointing due to the lack of selectivity [1,2]. If the gap is large, such that tensionless apposition is not possible, a nerve autograft, typically the patient's own sural nerve, is used as a bridge. Autografts represent the best clinical option available today, with several advantages including biocompatibility, nontoxicity, and availability of a support structure to promote axonal adhesion, extension, and extracellular matrix (ECM) formation. The technique also has many drawbacks, including a lack of availability, donor site morbidity, and loss of function [3]. A serious drawback to the use of autografts is that the availability of disposable nerve segments is limited, and multiple lengths of nerve graft might be needed to bridge the gap between the injured nerve stumps [4]. In addition, autografts contain inhibitory chondroitin sulfate proteoglycans (CSPGs), which may reduce their performance [5]. Another serious drawback is that there may be a dependence on the type of autograft used for the modality of the nerve regenerated. Using sural nerve grafts (which are the most commonly used autograft in humans) may exclude regenerating motor nerves [6]. Therefore, a clear and urgent clinical need to find alternative approaches that match or exceed the performance of autografts exists.

Driven by this need, several strategies are being pursued, primarily involving the use of hollow polymeric guidance channels (NeuroGen and Integra Life Sciences) alone, or using guidance channels as carriers for hydrogel-based scaffolds for nerve repair [7]. However, these have attained only limited success, and present an opportunity for a combination of prostheses, regenerative medicine, and tissue engineering to fill this void [8]. These guidance channels present more challenges including misrouting of regenerating axons at the lesion site, slow regeneration (<1 mm/day), and progressive reduction in the capacity of supporting Schwann cells to aid axon regrowth [9]. Schwann cells are critical in determining the regenerative capacity of PNs. These interface technologies also fall short in terms of the number of functional electrodes and their recording reliability [10]. In particular, a single hollow conduit cannot accurately identify and guide thousands of regenerating nerves. Therefore, the important challenges such as limited or "unidirectional" prosthetic control, lack of proprioception, tactile feedback, and other sensations still remain to be addressed. Therefore, an ideal PN interface should reliably and efficiently interface with a large number of nerves at high resolution and selectivity, eventually needing to provide a means to restore full nerve function. This is a formidable task given the large number of required nerve connections and extensive high-density signal processing. For example, sieve electrodes have only 10 electrode connections [11] and Utah arrays have about 100 electrode connections or so [12]. Newer peripheral interfaces such as penetrating electrodes (single needles [13], needle arrays [14], longitudinal intrafascicular electrodes (LIFEs) [15], or multielectrode array [16]) offer greater sensitivity and selectivity, but are too invasive and often damage nerves [10]. In most cases, there exists a trade-off between the ability to selectively interface with single or few axons and the degree of injury to the nerve caused by interfacing [10].

In this chapter, we will pave a way to address these critical challenges by encouraging nerves to regenerate into specific, high-density 3D (three-dimensional) microchannel–electrode structures so as to improve spatial selectivity of the sense and control functions. The objective of this work is to investigate the regenerative microchannel-based electrode interface (ReME), as shown in Figure 4.1 to establish high-density, bidirectional communication link (rolled-up version) using 3D bundle of parallel 2D (two-dimensional) microchannels with embedded longitudinal (sense) and radial (control) electrodes.

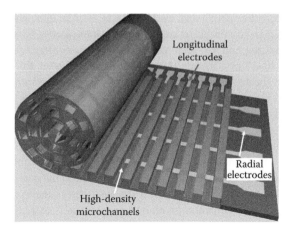

FIGURE 4.1 (**See color insert.**) ReME interface is shown as both rolled 3D structure and partially unrolled 2D microchannels with electrodes.

4.2 CURRENT STATUS OF PERIPHERAL NEURAL INTERFACES

In the United States, over 1.7 million people suffer from peripheral injury or limb loss. PN gaps are often created from severe traumatic injuries and invasive surgical procedures such as tumor resection [17]. Typically, traumatic nerve injuries resulting from collisions, motor vehicle accidents, gunshot wounds, fractures, lacerations, and other forms of penetrating trauma, affect more than 250,000 U.S. patients per year. Owing to the difficulties in treating such injuries, more than 200,000 patients are left without any benefit of medical intervention. Even among the patients who receive treatment for traumatic PNIs, more than 50% show no measurable signs of recovery or suffer from drastically reduced muscle strength. Owing to poor recovery, a majority of patients are required to return for follow-up care to address the continued functional deficits. Surgical injuries can also contribute to such undesirable functional deficits. For example, prostatectomy procedures most often require sacrificing cavernosal nerves while removing the tumor. This complication adversely affects erectile function and bladder control in more than 260,000 U.S. patients per year. The clinical "gold standard" for bridging PN gaps is autografts (typically the sensory sural nerve). However, the use of autografts is limited by several practical problems, including (1) a requirement for multiple nerve segments [18], (2) mismatch between the injured nerve and the nerve graft due to size/length/modality [19], (3) functional loss at the donor sites [20], and (4) donor site numbness, hyperesthesia, or formation of painful neuroma [21]. Moreover, only 40 ~ 50% of patients regain useful function after receiving autografts [22]. Therefore, it becomes critical to develop alternative bioengineered approaches that match or exceed the performance of autografts. The aim of the peripheral prosthetics is to interface with the intact portion of the original limb's nervous system, which may remain functional for years after injury.

Biomaterial-based tubular nerve conduits (or guidance channels) have been clinically used for repairing PNIs [23]. These nerve conduits bridge the injured nerve stumps and help to form a fibrin cable that provides a substrate for the ingrowth of

Schwann cells and other cells, such as fibroblasts or macrophages. However, this approach is limited in its ability to enable regeneration across nerve gaps that are more clinically relevant, and has been unsuccessful in promoting regeneration across gaps longer than 10–15 mm in rats [1]. Several other types of PN interfaces have been developed in the recent years including nerve cuffs, penetrating electrodes of various types, and regeneration sieves. Nerve cuffs are regarded as the most mature technology as they have reached clinical use in many applications including phrenic nerve stimulation, peroneal nerve stimulation, sacral root stimulation, vagus nerve stimulation, and functional electrical stimulation (FES) systems [24]. Nerve cuffs are well suited for the above-mentioned applications because they do not demand highly detailed spatial selectivity of stimulation or recording within the nerve. In nerve cuffs, spatial selectivity tends to be limited because the electrical contacts are at the periphery of the nerve and communicating with deeper fascicles is extremely challenging. To address the issue of obtaining higher selectivity, penetrating and sieve electrodes can be dispersed within the nerve. Thus, the electrodes can selectively interact with a larger population of axons. However, in these cases, the extracellular signal available for recording and stimulation as AP passes an electrode is extremely small.

Posttraumatic adjustment of pulse propagation in nerves with reduced excitability is a challenging biomedical and technological problem. In many cases, such adjustments can be achieved with surgical interventions. However, surgery does not necessarily restore nerve conductivity to its pretrauma levels; therefore, still impaired conduction may cause partial or complete muscular paralysis. In this case, propagation of excitation waves can be enforced only by applying external FES using implantable [25] or surface stimulation electrodes [26,27]. This method has been confirmed as an effective tool for restoration of the movement of paralyzed muscles in individuals with a variety of neurological impairments [28]. Computerized systems for the control of FES can deliver sequences of electrical stimuli with different frequency, amplitude, and duration. Commonly, these systems include a variety of control units and electrical leads with multiple arrays or patch-type stimulation electrodes. These electrodes deliver programmable stimuli, which are designed to maximize the effect of stimulation based on the configuration of a particular stimulation field [29]. Successful FES process of functional restoration of muscle contraction depends on the ability of the nerve tissue to adequately conduct AP. It also depends on excitation–contraction coupling in neuromuscular junctions, which transmit nervous impulses to muscle fibers. If any of these steps is impaired, the muscle does not contract normally. Usually, after severe neuromuscular injuries, nerve conductivity is significantly reduced, which, in turn, prevents the passage of excitation waves through neuromuscular transmitters. Under these circumstances, propagation of excitation pulses is marginally stable and implementation of FES necessitates a significant increase of frequencies and amplitudes of FES. This can facilitate conduction blocks (instead of stabilization of propagation) and may completely disrupt the process of training paralyzed muscles.

In summary, PN regeneration and prostheses, especially over long nerve gaps (>15 mm in rats), are difficult to achieve using the current gold standard autografts. Alternative strategies that can match or exceed the autograft performance are being explored in challenging clinically relevant models. Currently, tubular hydrogel nerve scaffolds are promising, although designing scaffolds that match the performance of the autografts still remains elusive. The other unaddressed challenge is the need for highly sensitive neuroprosthetic interfaces that can record or stimulate the extracellular signal as APs pass through them (independently of Nodes of Ranvier). Therefore, it becomes imperative that both these challenges for high efficient and reliable sensory and motor nerve regeneration are addressed. In this work, we will explore a few of nano-/microregenerative electrode designs, including regenerative interface that seeks to tackle these critical requirements including improved posttrauma FES techniques.

4.2.1 REGENERATIVE ELECTRODE DESIGNS

The alternative to penetrating electrodes of various types is based on the inherent capacity of PNs to regrow after injury such as through a tubular conduit [30]. In the first method, the conduit is embedded with either a sieve or is needle shaped. Silicon sieve electrode arrays have been used to obtain both neural recordings and electrical stimulation [31]. Recently, polyimide sieve electrodes with seven integrated ring electrodes have been developed [32]. As stated earlier, in these cases, the activity was only recorded from few electrodes in the sieve, due to compression injury of the regenerated nerve [33]. Garde et al. [16] recently developed regenerative multielectrode interface (REMI) that was successful in obtaining single and multicellular recordings. The REMI allowed for unrestricted nerve growth without compression injury, along with recordings with high signal-to-noise ratio. An alternative design is the microchannel roll electrode (MCRE), which records from axons as they regrow through microchannels with embedded electrodes. The aim of this approach is to maximize the contact between regenerating axons and the embedded electrodes [34–39]. Our design, which we will discuss in the next section, is similar to the MCRE design. However, in addition to recording (sense), our design will also provide for control electrodes [37]. These control electrodes will enable posttraumatic adjustment of pulse propagation in regrown nerves. In both designs, the 2D microchannels are rolled into 3D bundle of microchannels, which will fit the transected PNs. In summary, the selectivity of regenerative electrode designs is far superior compared to other interface designs, as they are in contact with the individual axons and are able to record single APs [38,39].

4.3 REGENERATIVE MICROCHANNEL-BASED ELECTRODE INTERFACE

The PN interface capable of long-term, reliable and high-resolution stimulation and recording is yet to be developed. In this work, we investigate (using numerical simulations) an *in vitro* model of a high-density PN electrode interface, known as "regenerative microchannel-based electrode interface" (ReME). As shown in Figure 4.1, the ReME interface contains a 3D architecture of parallel microchannels with embedded (control and sense) electrodes for nerve growth and for posttraumatic adjustment of pulse propagation in regrown nerves. The objectives of this work are to (1) design and fabricate a conformable 2D/3D ReME interface with specific substrate properties, dimension, and electrode material, and (2) model and optimize the entrainment of marginally stable conduction in the 3D ReME architecture.

4.3.1 Design and Fabrication

The central hypothesis of the design is that axon regeneration into confined high-density 3D microchannels can facilitate a higher degree of electrical and spatial isolation [24,36], and provide a means for selective and efficient stimulation/recording, which is required for the entrainment of marginally stable excitation. As described in the previous works [34,37], the rationale behind this hypothesis is that confinement of axons in microchannels provides a greater degree of volume restriction (or increased extracellular resistance) and thus, a decrease in the stimulus or recording current. The reduction of current is highly desirable because redox reactions, leading to electrode corrosion, pH changes, release of toxic metal ions, or generation of free radicals are eliminated in the process [24]. On the contrary, other interface approaches were forced to endure the low impedance of extracellular volume and high ionic diffusion/dispersion [34], which resulted in very small extracellular potential (AP <10 μV). It is possible to circumvent this problem by recording or stimulating at the Nodes of Ranvier, which occurs at least every 2 mm [40], where the extracellular potential is at its highest value. However, this approach imposes a challenging spatial dependence problem, which interfaces such as penetrating and sieve electrodes do not address, leading to poor signal-to-noise ratios.

Figure 4.2 illustrates the top (a) and cross-sectional (b) view of ReME interface design. The longitudinal electrodes in the microchannel are responsible for the SENSE function, while radial electrodes provide the CONTROL function. Table 4.1 describes the design parameters and microchannel configurations studied in this work. The dimensions (W and L) are used to understand the structural and electrical properties of the interface, while spacing (S) variations are used to understand the effect of axonal growth density on their properties. This design is chosen to demonstrate that (1) minimally invasive, 3D high-density, microchannel-based electrodes (SENSE) independent of Nodes of Ranvier can be developed [38], and (2) marginally stable or blocked AP propagation in regenerated nerves can be stabilized (CONTROL) using a set of additional spatially distributed subthreshold stimuli [37,41,42]. As stated earlier, to our knowledge this is the first demonstration where sense and control functions are incorporated in a single device, namely (1) high-density, bidirectional communication link between an amputated nerve and a prosthetic limb and (2) posttraumatic adjustment of pulse propagation in nerves. The key outcomes of this work are that the microchannel–electrodes can stimulate and record with the highest selectivity and can provide effective "contact guidance" (directing regenerating axons toward the appropriate target).

The fabrication details of the ReME interface are described as follows. First, a layer of parylene (10 μm) is deposited on a thin photoresist-coated Si/glass substrate. Parylene is a soft polymer, which is biostable, highest United States Pharmacopoeia (USP) class IV biocompatible, and FDA (Food and Drug Administration) approved for clinical applications. Next, the longitudinal electrodes (Ti/Au) are patterned.

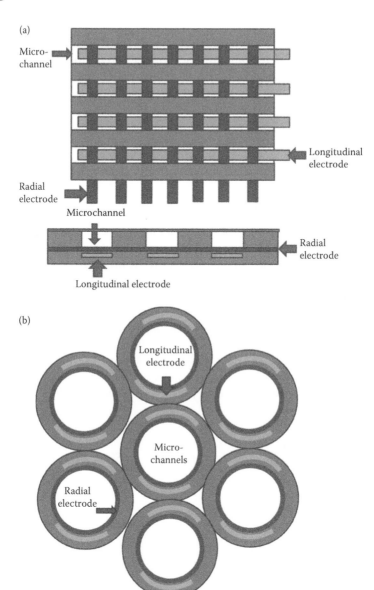

FIGURE 4.2 **(See color insert.)** (a) Top view of the 2D ReME with parylene as the structural material and Au electrodes, and (b) cross-sectional view of 3D ReME microchannels with embedded stimulation and recording electrodes.

Then, another layer of polymer is coated followed by patterning of radial electrodes. Then, 2D ReME is completed by depositing the walls, followed by dry etching using reactive ion etching (RIE) to open trenches of 2–10 µm in depth. Finally, the 2D parylene microchannels are rolled to form the 3D ReME interface, as shown in Figure 4.1 with 10 µm bonding layer of polydimethylsiloxane (PDMS).

TABLE 4.1

Design Parameters and Microchannel Configurations Used in This Study

Type/Channel	Width (μm)	Length (μm)	Spacing (μm)
I	2	5	5
II	5	5	5
III	10	15	10
IV	10	15	20

4.3.2 MODELING OF CONTROL FUNCTIONS

In this section, the numerical model used to optimize the stimulation protocols for nerve growth and for entrainment of marginally stable conduction for posttraumatic adjustment of pulse propagation in regrown nerves is discussed. The dynamics of conduction in nerves can be quite accurately simulated using Fitzhugh–Nagumo-type reaction–diffusion equations with just two fundamental excitation and post-excitation recovery variables [43,44]. Unlike the previous work, where we studied the dynamics initiated by a single excitation source [37], here, we consider an adjusted model (Equation 4.1) that is set up to reflect the interference of several sources to deliver multiple stimuli of different amplitudes. In particular, the main source delivers localized overthreshold stimuli at the beginning of the nerve. A set of additional sources originates secondary low-energy subthreshold stimuli that are extended throughout the nerve.

$$\frac{\partial u}{\partial t} = D\frac{\partial^2 u}{\partial x^2} - i(u,v) + P(x,t) + \sum_{i=1}^{n} S(x - x_i, t)$$

$$i(u,v) = \begin{cases} \lambda u, u < v \\ u - 1, u \geq v \end{cases}$$

$$\frac{\partial v}{\partial t} = \varepsilon(u + v_r - v).$$

(4.1)

Here, u and v are dimensionless excitation and recovery functions, respectively. D is an unvarying dimensionless diffusion coefficient that is set to 1. ε is a small parameter and v_r is the excitation threshold. λ and ζ control the rates of changes of excitation and recovery variables, respectively. The system of Equation 4.1 was numerically solved in a one-dimensional (1D) nerve cable of finite length using a second-order explicit difference scheme with zero-flux boundary conditions [45]. Spatial, Δx and temporal, Δt steps used in the numerical integration were equal to 0.23 and 0.0072, respectively. The cable length L was equal to $150\Delta x$. Primary forcing, $P(x,t)$, was a train of rectangular pulses with duration $100\Delta t$, an overthreshold amplitude A_0, and period T_0. Primary stimuli were applied near the left end of the cable between $x = 2\Delta x$ and $x = 15\Delta x$. Secondary forcing $\{S(x - x_i, t)\}$, $i = 1, 2, \ldots, n\}$, also a rectangular pulse train with duration $100\Delta t$ was delivered at n equidistant locations between $x = 40\Delta x$ and $x = 140\Delta x$. The secondary stimuli were simultaneously activated after the wave front resulted from the first primary stimulation arrived at the end of the cable. The amplitude and period of secondary stimuli were equal to A and T, respectively. The values of other parameters in all computations were $\varepsilon = 0.1$, $\lambda = 0.4$, $\zeta = 1.2$, $A_0 = 1.4$, $\alpha = 0.31$, $\beta = 0.0025$, and $n = 6$.

As described in Ref. [44], we used a simplified primary rate-dependent threshold given by the linear equation $v_r = \alpha - \beta T_0$, $\alpha > 0$, $\beta > 0$. The duration of a pulse, T_h, was measured as the time interval between consecutive intersections of u and v near their rest and excited states, respectively. The steady-state value of T_h was computed after 80 primary stimulation periods. In the absence of secondary stimulation $(A = 0)$, at long periods T_0, the variable v has enough time to reach its ground state v_r before the next stimulus is applied. However, as T_0 approaches a critical limit, T_{end}, the system does not respond to every stimulus and exhibits unstable $M{:}N$ $(M > N)$ excitation blocks that occur due to the incomplete recovery of the control variable v.

An analysis of pulse duration T_h as a function of secondary frequency $F = T^{-1}$ and forcing amplitude $A \geq 0$ revealed a variety of entrainment regimes (Figure 4.3). We found that the system did not respond to secondary stimuli at amplitudes, which were smaller than critical values depicted by the curve with circular markers. For the amplitudes above this curve and frequencies smaller than F_0, we observed intermediate $M{:}M$ responses with $M > 1$. For even greater amplitudes, above the upper curve with square markers, the system locked to secondary stimuli with consistent 1:1 responses. It should be noticed that such locking occurred over a wide range of secondary frequencies, and that secondary amplitudes were 5 times smaller than the

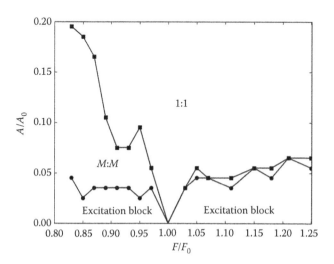

FIGURE 4.3 Locking margins for different frequencies and amplitudes of secondary stimuli for $T_0 = 30$, $A_0 = 1.4$, and $x = L/2$. (Reprinted from Starobin, J.M. and Varadharajan, V., *Nonlinear Biomed Phys*, 5, 8, 2011.)

amplitudes of primary stimulations. We also observed that for frequencies greater than F_0, the entrainment of blocked excitation occurred without intermediate $M{:}M$ responses (Figure 4.3).

Spatiotemporal contours of u shown in Figure 4.4a demonstrate expected unstable responses to primary stimulation at $T_0 < T_{end}$. Temporal dynamics of u and v, as well as u-contours, show 3:2 excitation blocks (Figure 4.4b). However, in the presence of secondary stimulations, such unstable responses can be entrained and stabilized by secondary driving even at $T_0 < T_{end}$. Indeed, Figure 4.5a demonstrates that 3:2 blocks

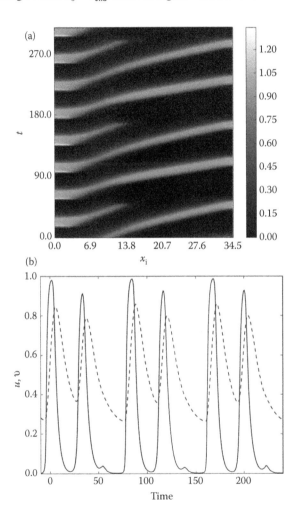

FIGURE 4.4 Panel (a) shows gray-scale striations depicting dynamics of u as a function of time, t, and spatial location, x_i. A corresponding scale bar is shown on the right. Panel (b) demonstrates temporal dynamics of u (solid line) and v (dashed line) for three consecutive cycles at $T_0 = 28$, $A = 0$, and $x = L/2$. Interval between intersections of solid and dashed lines in the first pulse is T_h. (Reprinted from Starobin, J.M. and Varadharajan, V., *Nonlinear Biomed Phys*, 5, 8, 2011.)

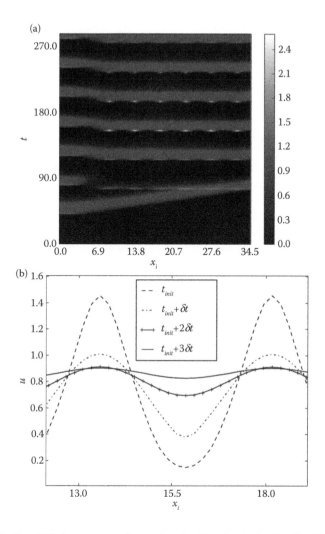

FIGURE 4.5 Panel (a) shows gray-scale synchronized bands of u for $T_0 = T = 28$ and $A = 0.1$ A_0. A corresponding scale bar is shown on the right. Panel (b) depicts temporal evolution of a spatial profile of u at the bottom of the second horizontal band for three equidistant moments of time, starting from $t_{init} = 136$, $\delta t = 0.36$. (Reprinted from Starobin, J.M. and Varadharajan, V., *Nonlinear Biomed Phys*, 5, 8, 2011.)

can be transformed into stable 1:1 responses that homogeneously evolve in the entire cable except for short segments located near the site of primary stimulation. The formation of these fully synchronized responses is preceded by very short (~0.02 T_0) transient periods during which standing-wave-type oscillations of u rapidly saturate at constant excitation levels (Figure 4.5b).

The analysis described in the previous section is quite robust and sufficiently accurate for relatively large axons with constant diffusion coefficient. Indeed, for the axons of a large diameter, the amount of charge, Q_m, moved through the axon during each AP is very small compared to intracellular ionic charge, Q_i. Using a membrane capacitance, C, and an AP amplitude, U, one can estimate the relative charge depletion per 1 cm of an axon length given by Equation 4.2

$$\rho = \frac{Q_m}{Q_i} = \frac{2 \cdot 10^3 \cdot CU}{\alpha_M \cdot F_k \cdot r} \tag{4.2}$$

where α_M, F_K, and r are the extracellular sodium molar concentration, Faraday constant, and axon radius, respectively. If the nominal values for each of these parameters are equal to 20 μF/cm^2, 150 MV, and 5 mM/L ($F_k = 9.65 \times 10^4$ C/mol), then for a radius of 0.1 cm, the value of ρ will be negligibly small, $\rho = 0.124 \times 10^{-3}$. Under these conditions, there is basically no change between intracellular and extracellular sodium concentrations and diffusion coefficient may be indeed considered as a constant value. However, for very small PNs of radii smaller than 1 μm, the value of ρ may exceed 10% and consequently, the dependence of the diffusion coefficient on AP should be taken into account [46,47]. Evidently, such changing diffusion may significantly alter the stability of propagation of excitation waves.

To quantify the degree of these changes, we refined the propagation of excitation pulses in 1D excitable nerve cable and studied the influence of different diffusion functions $D(u)$ on the stability of propagation. Unlike the previous simulations for constant diffusion, we used the Fitzhugh–Nagumo model with a nonlinear diffusion coefficient and explicitly included source amplitude, A:

$$\frac{\partial u}{\partial t} = D(u)\frac{\partial^2 u}{\partial x^2} - A(u - m_1)(u - m_2)(u - m_3) + P(x,t);$$
$$\frac{\partial v}{\partial t} = \varepsilon(u - v), \tag{4.3}$$

where constants m_1, m_2, m_3 determine source equilibriums and other constants and functions are similar to those introduced in Equation 4.1. Diffusion function was determined by the equation

$$D(u) = D_0 + 0.4u^2, \tag{4.4}$$

where D_0 was set to 0, 3.8, or 6.8 and constants m_1, m_2, m_3 were equal to 0, 1.7, and 3.5, respectively. The system in Equations 4.3 and 4.4 was numerically solved using the same as for Equation 4.1 explicit difference scheme. To provide accurate calculations of the steady-state wave front speed, we used reduced spatial and temporal steps that were equal to 0.02 and 0.001, respectively.

We found that a nonlinear component of the diffusion function played a significant role in the stabilization of propagation even without additional stimuli. Indeed, for $D_0 = 0$ exclusively nonlinear diffusion term stabilized wave propagation at noticeably lower amplitudes of the reaction source. On the contrary, for $D_0 = 3.8 - 6.8$ and less significant contribution of nonlinear diffusion, waves of the same speed were stable at fourfold greater amplitudes compared to the corresponding values at $D_0 = 0$ (Figure 4.6). We also observed that for constant diffusion coefficient, the amplitudes required for stable propagation increased on the order of magnitude, but were still insufficient to maintain the required stability at speeds higher than 0.24 (inset in Figure 4.6).

To determine $D(u)$ that characterizes diffusive properties of a conductive channel, one needs to measure excitation wave front speeds, C, and electric potential, u, at the sites of each sense electrode (Figure 4.1). The equation for a diffusive wave front speed is based on Fick's law

$$C = \chi \partial u_e / \partial x, \tag{4.5}$$

where χ and u_e represent extracellular ionic mobility and potential. The dependence between extracellular ionic mobility and diffusion coefficient D is given by Einstein's equation [46]

$$D = \chi RT / ZF_k \tag{4.6}$$

where R is a gas constant, Z is an ionic valence, T is temperature of the conduction unit, and F_k is Faraday's constant. Since intracellular charge mobility is much smaller than the value of χ, one can use D from Equation 4.6 as a good estimate for the diffusion term shown in Equations 4.1 and 4.3. Thus, measurements of C and u_e along with Equations 4.5 and 4.6 will be sufficient for computing the diffusion of

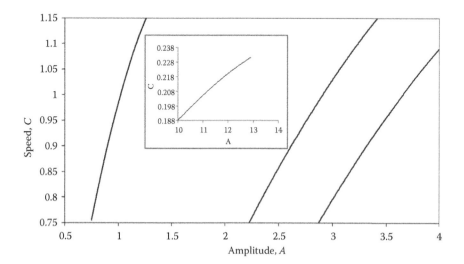

FIGURE 4.6 Dependence of a steady-state wave speed on the amplitude of the reaction source at D_0: 0—left, 3.8—center, and 6.8—right curves. The inset depicts the curve computed for $D(u) \equiv 1$.

excitation in each of the conduction units. One has to acquire sequences of action potential durations (APDs) and refractory intervals (RIs) to customize the model using Equations 4.4 through 4.6. Values of APD and RI can be experimentally acquired using electrodes embedded in the ReME interface. Computational APD can be determined as the interval T_h (Figure 4.4) and RI can be computed as the difference between the period of stimulation and APD. Dimensionless APD and RI values from the model should be converted into dimensional APD^m and RI^m intervals by

$$\text{APD}^m = \frac{c_m}{\sigma_{Na}}\text{APD}; \quad \text{RI}^m = \frac{c_m}{\sigma_{Na}}\text{RI}, \tag{4.7}$$

where c_m and σ_{Na} are characteristic values of membrane capacitance and sodium membrane conductance. The ratio of c_m and σ_{Na} for normal axons is approximately equal to 1 ms [46,48].

Experimental values of APD^e are measured as the time interval between the maximum positive and negative slopes of u_e. Our model has four parameters ε, ζ, λ, and v_r that need to be determined for each stimulation rate. The fitting procedure minimizes the difference Δ_{ss} between the steady-state values of the APD^m intervals computed from the model and the APD^e intervals measured during stimulations. The initial estimates for all parameters are obtained by searching over a sparse four-dimensional parameter space grid. Parameter estimates are continuously refined using Powell's optimization technique [49]. For every iteration, a new value of APD^m is obtained from the model using new parameter estimates. The iterations stop when Δ_{ss} decreases below 1%. In summary, the methodology/simulations described above are implemented for sources distribution optimization in the ReME interface. The goal of this optimization is to allow for maximum stabilization of nerve conduction using a minimal number of low-amplitude subthreshold electrical stimuli to train and adjust pulse propagation in growing/regrown nerves.

4.4   OUTLOOK AND SUMMARY

This chapter presented preliminary results for the design and control of ReME peripheral interface with high recording and stimulation sensitivity. The long-term goal of this work is to experimentally optimize and evaluate the sense and control functions of the 3D ReME interface, including the ability to record APs using longitudinal electrodes and control pulse propagation in regenerated nerves using radial electrodes. The successful accomplishment of the above-stated objective will have far-reaching implications in the development of artificial bionic or neuroprosthetic systems, along with full neural control and sensory feedback. In particular, our study will provide enhancements to sense, control, spatial selectivity (being able to interface with a large number of nerves), and long-term stability to both acute and chronic injured PNs. In clinical terms, these regenerative neural interfaces might offer high therapeutic value and quality of life for patients with muscle or nerve injuries, including pain management and bladder control restoration.

REFERENCES

1. Hudson, T.W., Evans, G.R., and Schmidt, C.E., Engineering strategies for peripheral nerve repair, *Clin Plast Surg*, 26(4), 617, 1999.
2. Pfister, B.J. et al., Biomedical engineering strategies for peripheral nerve repair: Surgical applications, state of the art, and future challenges, *Crit Rev Biomed Eng*, 39(2), 81, 2011.
3. Lee, S.K. and Wolfe, S.W., Peripheral nerve injury and repair, *J. Am. Acad. Orthop. Surg*, 8, 243, 2000.
4. Ansselin, A.D. and Davey, D.F., Axonal regeneration through peripheral nerve grafts: The effect of proximo-distal orientation, *Microsurgery*, 9, 103, 1988.
5. Groves, M.L. et al., Axon regeneration in peripheral nerves is enhanced by proteoglycan degradation, *Exp Neurol*, 195, 278, 2005.
6. Nichols, C.M. et al., Effects of motor versus sensory nerve grafts on peripheral nerve regeneration, *Exp Neurol*, 190, 347, 2004.
7. Dodla, M.C. and Bellamkonda, R.V., Differences between the effect of anisotropic and isotropic laminin and nerve growth factor presenting scaffolds on nerve regeneration across long peripheral nerve gaps, *Biomaterials*, 29, 33, 2008.
8. Aravamudhan, S. and Bellamkonda, R.V., Toward a convergence of regenerative medicine, rehabilitation, and neuroprosthetics, *J Neurotrauma*, 28(11), 2329, 2011.
9. Lacour, S.P. et al., Long micro-channel electrode arrays: A novel type of regenerative peripheral nerve interface, *IEEE Trans Neural Syst Rehabil Eng*, 17(5), 454, 2009.
10. Navarro, X. et al., A critical review of interfaces with the peripheral nervous system for the control of neuroprostheses and hybrid bionic systems, *J Peripher Nerv Syst*, 10(3), 229, 2005.
11. Ramachandran, A. et al., Design, *in vitro* and *in vivo* assessment of a multi-channel sieve electrode with integrated multiplexer, *J Neural Eng*, 3(2), 114, 2006.
12. Branner, A. and Normann, R.A., A multielectrode array for intrafascicular recording and stimulation in sciatic nerve of cats, *Brain Res Bull*, 51(4), 293, 2000.
13. Rutten, W.L.C., van Wier H.J., and Put, J.M.H., Sensitivity and selectivity of intraneural stimulation using a silicon electrode array, *IEEE Trans Biomed Eng*, 38(2), 192, 1991.
14. Rutten, W.L.C. et al., 3D neuroelectronic interface devices for neuromuscular control, *Biosens Bioelectron*, 10(1–2), 141, 1995.
15. Malagodi, M.S., Horch, K.W., and Schoenberg, A.A., An intrafascicular electrode for recording of action potentials in peripheral nerves, *Ann Biomed Eng*, 17(4), 397, 1989.
16. Garde, K. et al., Early interfaced neural activity from chronic amputated nerves, *Front Neuroeng*, 2, 5, 2009.
17. Noble, J., et al., Analysis of upper and lower extremity peripheral nerve injuries in a population of patients with multiple injuries. *J Trauma*, 45(1), 116–22, 1998.
18. Kline, D.G. et al., Management and results of sciatic nerve injuries: A 24-year experience, *J Neurosurg*, 89(1), 13, 1998.
19. Nichols, C.M. et al., Effects of motor versus sensory nerve grafts on peripheral nerve regeneration, *Exp Neurol*, 190(2), 347, 2004.
20. Bini, T.B. et al., Peripheral nerve regeneration by microbraided poly(L-lactide-co-glycolide) biodegradable polymer fibers, *J Biomed Mater Res A*, 68(2), 286, 2004.
21. Itoh, S. et al., Evaluation of cross-linking procedures of collagen tubes used in peripheral nerve repair, *Biomaterials*, 23(23), 4475, 2002.
22. Millesi, H., Meissl, G., and Berger, A., The interfascicular nerve-grafting of the median and ulnar nerves. *J Bone Joint Surg Am*, 54(4), 727–50, 1972.
23. Taras, J.S., Nanavati, V., and Steelman, P., Nerve conduits, *J Hand Ther*, 18(2), 191, 2005.
24. FitzGerald, J.J. et al., Microchannel electrodes for recording and stimulation: *In vitro* evaluation, *IEEE Trans Biomed Eng*, 56(5), 1524, 2009.

25. Veraart, C., Grill, W.M., and Mortimer, J.T., Selective control of muscle activation with a multipolar nerve cuff electrode, *IEEE Trans Biomed Eng*, 40(7), 640, 1993.
26. Thorsen, R. A. et al., Functional electrical stimulation reinforced tenodesis effect controlled by myoelectric activity from wrist extensors, *J Rehabil Res Dev*, 43(2), 247, 2006.
27. Tuday, E.C., Olree, K.S., and Horch, K.W., Differential activation of nerve fibers with magnetic stimulation in humans, *BMC Neurosci*, 7, 58, 2006.
28. Peckham, P.H. and Knutson, J.S., Functional electrical stimulation for neuromuscular applications, *Annu Rev Biomed Eng*, 7, 327, 2005.
29. Dar, A. and Nathan, R. H., U.S. Patent No. 7, 149, 582, 2006.
30. Wallman, L. et al., The geometric design of micromachined silicon sieve electrodes influences functional nerve regeneration, *Biomaterials*, 22, 1187, 2001.
31. Bradley, R.M. et al., Long term chronic recordings from peripheral sensory fibers using a sieve electrode array, *J Neurosci Methods*, 73, 177, 1997.
32. Navarro, X. et al., Stimulation and recording from regenerated peripheral nerves through polyimide sieve electrodes, *J Peripher Nerv Syst*, 3, 91, 1998.
33. Lago, N. et al., Long term assessment of axonal regeneration through polyimide regenerative electrodes to interface the peripheral nerve, *Biomaterials*, 26, 2021, 2005.
34. Fitzgerald, J.J. et al., Microchannels as axonal amplifiers, *IEEE Trans Biomed Eng*, 55, 1136, 2008.
35. Lacour, S.P. et al., Polyimide micro-channel arrays for peripheral nerve regenerative implants, *Sens Actuators A*, 147, 456, 2008.
36. Srinivasan, A., Guo, L., and Bellamkonda, R.V., Regenerative microchannel electrode array for peripheral nerve interfacing, in *5th International IEEE/EMBS Conference of Neural Engineering*, IEEE, Piscataway, 2011, 253.
37. Starobin, J.M., Varadarajan, V., and Aravamudhan, S., High-density peripheral nerve resonant stimulation system, in: *Proceedings of BMES—Biomedical Engineering Society Annual Meeting*, Hartford, Connecticut, 2011.
38. Kim, Y.T. and Romero-Ortega, M.I., Material considerations for peripheral nerve interfacing, *MRS Bull*, 37(06), 573, 2012.
39. Lotfi, P. et al., Modality-specific axonal regeneration: Toward selective regenerative neural interfaces, *Front Neuroeng*, 4, 11, 2011.
40. Hess, A. and Young, J.Z., The nodes of Ranvier, *Proc R Soc Lond, Ser. B*, 140(900), 301, 1952.
41. Starobin, J.M. and Varadharajan, V., Entrainment of marginally stable excitation waves by spatially extended sub-threshold periodic forcing, *Nonlinear Biomed Phys*, 5, 8, 2011.
42. Starobin, J.M. and Varadarajan, V., Critical scale of propagation influences dynamics of waves in a model of excitable medium, *Nonlinear Biomed Phys*, 3, 4, 2009.
43. Chernyak, Y.B., Starobin, J.M., and Cohen, R.J., Where do dispersion curves end? A basic question in theory of excitable media, *Phys Rev E*, 58, 4108, 1998.
44. Chernyak, Y.B., Starobin, J.M., and Cohen, R.J., Class of exactly solvable models of excitable media, *Phys Rev Lett*, 80, 5675, 1998.
45. Richtmayer, R.D., *Difference Methods for Initial-Value Problems*, Interscience, New York, 1957.
46. Plonsey, R. and Barr, R.C., *Bioelectricity—A Quantitative Approach*, Kluwer Academic/Plenum Publishers, New York, 2000.
47. Quian, N. and Sejnowski, T.J., An electro-diffusion model for computing membrane potentials and ionic concentrations in branching dendrites, spines and axons, *Biol Cybern*, 62, 1, 1989.
48. Idriss, S.F., Feasibility of non-invasive determination of the stability-of-propagation reserve in patients, *Comput Cardiol*, 39, 353, 2012.
49. Press, W.H. et al., *Numerical Recipes in Fortran 77: The Art of Scientific Computing*, Cambridge University Press, New York, 2009.

Section II

Nanobio

5 Characterization of Biological and Condensed Matter at the Nanoscale

Adam R. Hall, Osama K. Zahid, Furat Sawafta, and Autumn T. Carlsen

CONTENTS

5.1 OVERVIEW

There is an undeniable drive toward understanding materials and systems at nanometer length scales. Electronic devices, of course, are continually pushed toward smaller and smaller dimensions; bulk material properties are largely determined by nanoscale characteristics; and the basic functions of biology are carried out by molecular machines. For these and many other reasons, the tools of nanometrology are of increasing importance.

In this chapter, we will describe the general operation of four types of microscopy tools used in nanoscience and nanoengineering research: atomic force microscopy (AFM), transmission electron microscopy (TEM), scanning electron microscopy (SEM), and helium ion microscopy (HIM). These different instruments offer specific advantages and disadvantages in characterizing nanomaterials. Throughout the chapter, these pros and cons will be highlighted as we discuss the central components of each instrument, their general mechanisms of operation, and some selected techniques that extend their applicability in the lab.

5.2 ATOMIC FORCE MICROSCOPY

5.2.1 INTRODUCTION AND COMPONENTS

Atomic force microscopy (AFM) is one of a class of microscopes collectively referred to as scanning probe microscopes (SPMs). This family of instruments has its origin in the 1982 work of Gerd Binning and Heinrich Rohrer of IBM Zurich, who first demonstrated the scanning tunneling microscope.[1] Like that earliest instrument, the majority of SPMs are able to achieve incredibly high-resolution images by operating in a manner that is very different from that of previously existing microscopes. SPMs rely on the accurate movement of a microfabricated probe across a sample surface and the collection of spatially confined data at each location. These data can then be stitched together to form a detailed image.

AFMs, like SPMs in general, accomplish their task through the use of four main components: a sharp probe, a planar substrate, scanners, and a feedback loop (Figure 5.1). Typically, the probe is fabricated using common silicon-processing techniques, resulting in a monolithic cantilever with a pyramidal or cone-shaped tip at its end. The size and shape of the cantilever can be customized for various applications and imaging modalities. The apex of the tip can be brought into intimate contact with the substrate of choice for imaging. The substrate itself must be relatively planar in order to reduce damage to both tip and sample. To move the tip (i) across the substrate and (ii) toward and away from the surface with a high degree of precision, piezoelectric scanners are typically used. Crystal lattice asymmetry produces a mechanical strain in piezoelectric materials when an electrical bias is applied. Scanners formed from such materials—which usually take the form of either stacks for one-dimensional actuation or cylinders for three-dimensional actuation—are able to produce tip movements with sub-nanometer accuracy. During the imaging process, the tip is scanned across the substrate surface, side to side and top to bottom, until each point is eventually probed.

The final component—the feedback loop—is a circuit that continually monitors the interaction between the probe tip and the surface and actively makes adjustments to maintain a constant level of interaction; a value known as the set point. Tip–sample interactions are recorded by reflecting a laser off the cantilever and into a quadrant photodiode. In the various modes of AFM operation (see next section), the relevant value of interaction (or set point) is dependent on tip–sample separation,

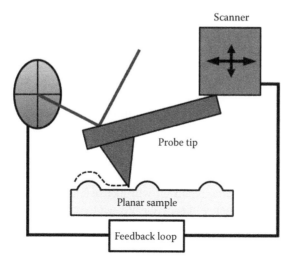

FIGURE 5.1   Schematic illustration of atomic force microscopy. A sharp probe tip (center) is scanned across a planar sample surface (bottom). At each point, the probe–sample interaction is determined and kept at a constant value ("set point") by feedback control of the vertical position of the tip by the scanner (left).

with the feedback loop actively adjusting the position of the piezoelectric scanner perpendicular to the sample surface (z-direction). Plotting the amount of z-extension or retraction necessary to maintain the set point at each point (pixel) along the surface produces a topographical image.

5.2.2   Mechanisms of Operation

The force experienced by a probe tip as it approaches a surface transitions from an attractive to a repulsive regime (Figure 5.2) is a consequence of the transition

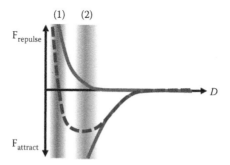

FIGURE 5.2   Tip–sample force regimes plot describing the interplay of attractive and repulsive inter-atomic forces acting on the probe tip as a function of tip–sample distance, D. The green line represents electron–nucleus attraction and the red line represents nucleus–nucleus repulsion. At small separation (1), the net force is repulsive, while at slightly larger separation (2), the net force is attractive.

from dominant long-range nucleus–electron interactions to dominant short-range nucleus–nucleus interactions. Using the piezoelectric scanner, the probe tip can be positioned in either of these regimes to perform imaging in one of three possible modes: contact, intermittent contact, and noncontact.

In contact mode, the tip apex is brought into intimate static contact with the surface such that repulsion dominates. As the tip approaches the substrate, these repulsive forces act to deflect the cantilever away from its equilibrium position, and this deflection is measured as a vertical shift of the reflected laser spot across the photodiode (Figure 5.3a). The position of the laser spot relative to its original (unperturbed) position acts as the set point value. If the apex encounters a feature that resides above the surface during the movement of the tip across the substrate (nominally in a plane parallel to the surface itself), repulsion will increase and thus further deflect the cantilever. The feedback loop will subsequently act to retract the cantilever position such that the set point constant is maintained (Figure 5.3b and c). The inverse reaction is true for features below the surface (i.e., gutters, Figure 5.3d and e). For every position of the tip, the scanner z-position is recorded, yielding a faithful topographic representation of the entire scanned surface.

This is the most straightforward method used in AFM, and is thus the simplest to set up. However, several disadvantages exist. Chief among them is the potential for damage, both to the probe tip and to the substrate itself. Because the probe tip stays in close contact with the sample surface at all times, the tip can easily become worn, with the altered shape reducing resolution or possibly causing artifacts.[2] Conversely, as the tip apex is scanned and comes in contact with sample features, the strain on the sample is temporarily high until the feedback loop adjusts the cantilever position. For this reason, sample features can undergo large strains. Thus, contact mode is generally not suitable for imaging soft matter, like biological molecules.

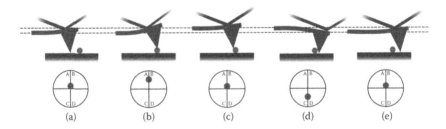

(a)   (b)   (c)   (d)   (e)

FIGURE 5.3   Set point maintenance. A diagram of feedback control of laser position on a quadrant photodiode in contact mode imaging. Below each state is a snapshot of laser position on the photodiode. (a) The tip is in contact with the sample surface, resulting in an initial laser position. (b) As the tip moves over a feature, the cantilever must deflect, resulting in a shift in laser position. (c) To compensate, the scanner retracts the tip slightly (to the upper dashed line), returning the laser to its initial position. (d) As the tip moves off of the feature (or into a depression on the sample surface), the cantilever deflects downward, resulting in laser position movement. (e) The scanner compensates by moving the tip down (to the lower dashed line), again returning the laser spot to its initial position. The figure is based on contact mode imaging.

Intermittent contact mode also utilizes the repulsive regime, but in a different manner. Here, the cantilever is driven near resonance such that the probe tip moves up and down in the z-direction in a continuous fashion. The amplitude of this motion can be monitored through the same laser signal described above. Geometry determines the oscillatory behavior of the cantilever and is thus an important consideration. For instance, the cantilever should be stiff enough to overcome adhesion forces during periods of contact. Otherwise, the resonant signal will be continuously perturbed. The probe tip is moved toward the substrate until the repulsive forces of close contact damp the oscillatory motion. The level of this damping—the reduction in measured amplitude compared to the amplitude in free-space—is then used as the set point value in the feedback loop. As in contact mode, intermittent-contact imaging maps the z-position needed for the scanners to account for changes in repulsive force as the probe tip is scanned across the surface. Unlike in contact mode, however, the oscillatory motion of the tip confines its interaction with the substrate mainly to the vertical direction; the tip "samples" the surface at one location and then moves to the next location discretely. Since shear forces on sample features are greatly reduced, intermittent contact mode is used widely to image both hard and soft nanomaterials (Figure 5.4). It should be noted, however, that some damage is still induced to both the tip and the sample, dependent on the level of damping.

Finally, noncontact mode is capable of imaging a sample using the attractive regime of atomic interaction between the probe tip and the substrate. Here, much like in intermittent-contact mode, the probe cantilever is driven near resonance (although at a comparably low amplitude). In noncontact mode, however, the cantilever is positioned at a point above the substrate such that attractive (van der Waals) forces are dominant. When the tip is scanned across the surface, these attractive forces perturb the measured amplitude signal in a separation-dependent manner, and the feedback loop again adjusts the probe z-position to maintain a constant value. The central advantage to noncontact mode imaging is the absence of damage to the tip or the substrate. This renders the method ideal for very delicate samples and can also result

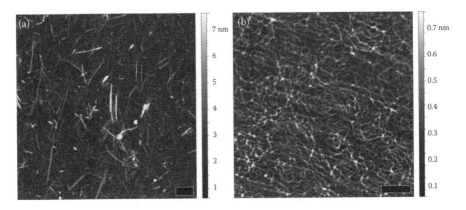

FIGURE 5.4   Intermittent contact images. (a) Polydisperse carbon nanotubes (scale bar is 1 μm). (b) A dense layer of double-strand DNA (scale bar is 500 nm). Both materials are deposited on a cleaved mica surface.

in exquisite resolution.[3] However, one disadvantage is that the tip can be perturbed by interactions with the surface or with a liquid meniscus on top of the surface. This can easily mask target features. Therefore, a pristine surface is vital in noncontact imaging and can make measurement of some biological materials challenging.

5.2.3   HIGH-SPEED AFM

The high-speed AFM (HS-AFM) technique is a novel approach to increase conventional AFM image-capture rates, allowing for visualization of unstained single biomolecule dynamics at nanometer spatial resolution and millisecond temporal resolution.[4] While a conventional AFM requires capture rates on the order of seconds to minutes for a single frame, HS-AFM is capable of capturing more than 10 frames per second.[5] The high-speed nature of this system allows for direct monitoring and recording of the behavior of various biomolecular processes, such as structural changes, and label-free molecular tracking during protein function.[4,5]

To observe the dynamics of a system directly, various components of the conventional AFM process are replaced by smaller, more robust versions of the same components. HS-AFM incorporates small cantilevers, ~2–10 µm in length, to allow for high resonance frequency as well as lower force noise which results in smaller acquisition times per pixel.[5] Additionally, high temporal resolution is achieved by way of a high-speed piezo scanner, which is designed to be stiff and compact via active damping and feedback controllers that allow for the maintenance of discrete tip–sample interaction forces.[4] Among the limitations that exist in the HS-AFM system is spatial resolution; large scan areas affect the quality of the resulting image due to increased tip velocity, which produces large forces that can cause damage when measuring biomolecule dynamics.[5] Other challenges include accurate enhancement of tip speed and the necessity for ultra-sharp cantilever tips.

HS-AFM has the capability to capture high-resolution videos of biomolecule activity. For instance, the structure and dynamic behavior of tail-truncated myosin V molecules (a two-headed cargo transporter in cells) has been monitored during its motion along a protein filament.[6] The movement of the myosin V molecules was described as a "stomping-like" attachment followed by detachment of one head after another, which produced intermolecular tension and conformational switching within the motor function of the molecule throughout biomolecular processes.[6] This study was crucial to understanding both the previously expected motor movements of myosin as well as previously unknown behavior, such as the discrete movement of the uncoiling of the tail ends of these two-headed myosin molecules.

5.2.4   SCANNING CONDUCTANCE MICROSCOPY

Scanning conductance microscopy (SCM) is a technique that measures the conductance of samples at nanometer spatial resolution without the use of electrical contacts. As such, the structural and electrical properties of low- or nonconducting materials can be measured.

This dual-pass technique tracks the phase shift of a voltage-biased cantilever as a function of tip position over a sample placed above an insulating substrate and ground plane. An AFM tip passes over each line of the sample twice, first with inter-mittent mode AFM to obtain a topography profile, and second with noncontact mode at a fixed tip-surface height (typically 30–120 nm), where the topographic informa-tion from the first pass is used to maintain a constant height separation. A potential is applied between the tip and the sample during the second-pass scan, and the phase shift determined by the total capacitance at the tip provides the dielectric properties of the scanned sample as a function of height, quality factor, spring constant, and tip radius.[7] The electrostatic forces that arise from the tip and substrate dictate the direction of the phase shift.[8]

Applications for this technique include imaging single-wall carbon nanotubes (CNTs), distinguishing between conducting and insulating polyaniline/poly(ethylene oxide) nanofibers, and demonstrating the insulating properties of λ-DNA.[7] Characterization of these samples has shown that carbon nanotubes always show a negative phase shift while doped poly(ethylene oxide) nanofibers show an increas-ing positive shift with increasing fiber diameter.[7] A phase shift occurs when the tip approaches the sample and both the capacitive force and an attractive inter-atomic force act on the tip. The magnitude of these forces then determines the direction of the phase shift for a given sample size. Short carbon nanotubes, for example, experience smaller phase shifts due to decreased capacitance with respect to the ground plane than do longer nanotubes, a difference which can be explained by the division of volt-age between the tip and the carbon nanotube.[8] Properties of λ-DNA were determined using carbon nanotube SCM measurements as a control on the same sample. While CNTs provide a clear signal in the sample measurements, the λ-DNA does not and is therefore verified as an electrical insulator that exhibits low conductivity.[8]

5.2.5 SCANNING MICROWAVE MICROSCOPY

Scanning microwave microscopy (SMM) is a measurement method that is capable of monitoring local material properties during conventional AFM imaging of a sample. SMM operates by use of a network analyzer that sends a microwave electromag-netic signal (typically 1–6 GHz) through a resonant circuit to a solid-metal or metal-coated probe tip. This tip acts as both a transmitter and receiver and can therefore measure any microwave signal that is reflected back from the sample. The electrical properties of the substrate material that is in contact with the probe tip during this measurement determines the amount of reflection that occurs. Thus, comparison of the amplitude and phase of the reflected signal to that of the generated signal can be used in conjunction with a model or calibration standard to arrive at a quantification of local dielectric constant, capacitance, impedance, and dopant density.

A measurement of this kind is performed at each point in the imaged area during the collection of topographic data using contact-mode imaging as described above. The result is simultaneous structural and electrical measurement of the surface. By way of example, Figure 5.5 shows both topography and contact-capacitance spec-troscopy obtained simultaneously from a calibration standard in which sample oxide thickness is decreased in a stepwise manner from top to bottom. As the thickness decreases, capacitance is found to increase due to the smaller separation between the probe and the underlying sheet layer.

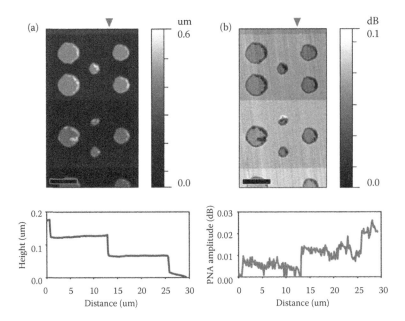

FIGURE 5.5 SMM imaging. Topographic (a) and capacitive (b) images of a calibration standard collected simultaneously with SMM. Below each image is a vertical cross-section, measured at the location of the arrow. PNA amplitude is the output of the SMM hardware and scales with capacitance. Scale bar is 5 μm.

5.3 TRANSMISSION ELECTRON MICROSCOPY

5.3.1 INTRODUCTION AND COMPONENTS

Transmission electron microscopy (TEM) is a technique that can produce a projected image of a thin sample with a magnification up to about 1 million times. This instrument is among the most accurate tools for visualization at the nanoscale, reaching atomic resolution under certain conditions. In this section, we will describe the function of TEM by addressing each major component and its function in the ultimate production of an image.

At the top of the instrument column (Figure 5.6), a metal source produces a coherent beam of electrons. Generally, electrons are freed from this source in one of three ways: thermionic emission, in which increased temperature is used; field emission, in which electric field is used; or Schottky emission, in which a combination of the two is used. Each type of emission has advantages and disadvantages; for instance, while a thermionic emitter has high temporal stability, the beam it produces is typically less bright and has lower phase consistency (coherence) than other emitter types. The effect of these factors must be taken into account for specific applications. After being freed from the emitter, electrons are next accelerated away from the source and down the column by a large positive voltage (typically 100–400 kV), which is applied to a solid disk electrode with an opening (the anode). The beam that emerges from the anode aperture is divergent.

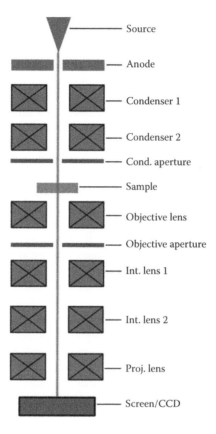

FIGURE 5.6 Schematic of a transmission electron microscope. Cross-sectional illustration of the components in TEM. See text for descriptions.

Next, the beam encounters a set of electromagnetic condenser lenses. The first of these is used to converge the beam to a cross-over, creating a virtual image of the source. The second lens is used to control the angle at which the beam converges toward the sample below and thus the focal plane. These lenses (in conjunction with the following component) determine the "spot size" or the diameter of the beam that interacts with the sample. A smaller spot size will result in higher resolution, but at the expense of lowered brightness. After the condenser lenses, the beam reaches the condenser aperture. This is a small opening in a solid plate that is used to prevent high-angle electrons from passing. It is used to limit the brightness of the beam and its size can typically be determined by the user.

Following the condenser aperture, the beam interacts with the sample itself, where it is diffracted by the material as it passes through. The most severe limitation that must be taken into account with regard to the sample is its thickness; since the TEM operates via projection through a sample, obtaining an image requires that a measurable amount of the original beam passes through the sample without undergoing total internal scattering. In general, this requirement for electron transparency means that a sample must be less than about 100 nm thick to be compatible with TEM imaging. For many nanomaterials deposited from powder or liquid suspensions, this thickness is easily obtainable. However, bulk samples may also be studied by creating a thin section of the material to image. This is commonly achieved through the use of an ultramicrotome, a device composed of a sharp blade and a thermal or piezoelectric actuator. The sample material is attached to the actuator above the cutting edge and is extended a sufficiently small distance between cuts to result in an electron transparent cross section that can be introduced to a grid or membrane for analysis. More recently, a focused beam of Ga ions has been used to sculpt thin sections of a sample that can then be transferred to a grid with micromanipulators.[9,10] This approach has the advantage of being able to target specific small areas of a sample. This section sculpting can be performed on both solid-state materials and biological materials, including whole cells (Figure 5.7). With ultramicrotome, the latter are often embedded in a resin to create a bulk sample. While solid-state materials usually do not require additional treatments prior to imaging, the scattering density of many biological materials is rather low and uniform, providing insufficient contrast for direct imaging. Therefore, contrast-enhancing agents like ruthenium or osmium tetroxide are often employed to label specific components[11,12] (cf. Figure 5.7).

Below the sample is the objective lens, which re-focuses the beam to a cross-over point and ultimately results in a diverging and inverted image of the transmitted beam. Importantly, the beam image at the cross-over (i.e., the back focal plane) is the diffraction pattern that results from the sample. This pattern can be used to achieve local crystallographic information about the material. The pattern can also be exploited to result in different imaging modes; by placing a small (select area) aperture at the back focal plane, the user can select either the undeflected beam (bright field) or any of the diffraction spots or areas (dark field). The latter results in

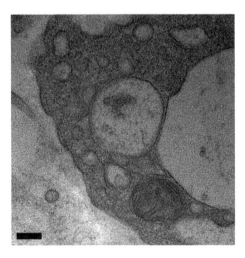

FIGURE 5.7 TEM image of biological material. Image of a rat kidney cell, sectioned by ultramicrotome and stained with osmium tetroxide. Scale bar is 200 nm.

an image formed predominantly from a given crystal orientation, revealing selective structural information about the sample.

After passing through the objective aperture, the beam is expanded using a set of intermediate lenses and a projector lens which collectively magnify the resulting image up to about 1 million times in a typical TEM. Finally, the projected beam is converted to a visible image. This is accomplished with a phosphorescent screen for normal usage and a CCD camera for digital image acquisition.

Together, the components of a TEM result in some of the highest resolution microscopy available in the laboratory. As applied to solid-state nanomaterials, the instrument is capable of imaging atomic rearrangement at material edges,[13,14] the identification of crystalline domains in single-layer graphene,[15] and even dynamic measurements of inter-shell coupling in concentrically wrapped carbon nanotubes.[16] In the area of bionanoscience, TEM has found utility in the study of conformations of individual DNA–protein complexes[17] and has recently emerged as a tool for ultra-high resolution imaging of samples in solution.[18]

5.3.2 ELECTRON ENERGY LOSS SPECTROSCOPY

When an electron beam of known energy strikes a sample, a percentage of the incident electrons scatter inelastically. These deflected electrons lose energy through a variety of phenomena, including band transitions, plasmon excitations, and photon excitations.[19] Using an electron spectrometer, the energies of electrons transmitted through the sample can be analyzed. This process, known as electron energy loss spectroscopy (EELS), provides a wide spectrum of electron energy loss with energy resolution on the order of a few meV.[20] Careful interpretation of the spectrum reveals information like the type and number of atoms and the local atomic thickness of the sample. This method yields information averaged over the entire field of electron incidence simultaneously. To obtain spatial information, either analysis must be performed at discrete points across the sample, or the beam itself must be filtered to allow only select energies to contribute to the image (see the following section).

5.3.3 CONTRAST TUNING

As mentioned above, the interaction of the primary electron beam with the sample results in both electrons that have been inelastically scattered and electrons that have transmitted without scattering. Due to the myriad different interactions that occur during inelastic scattering, the emerging electron energies can vary over a wide range (Figure 5.8a), especially for thick samples. Commonly in TEM, the objective lens is not able to properly focus all energies present in the emerging beam (a distortion called chromatic aberration), resulting in a reduction in image contrast. While the objective aperture (see Section 5.3.1) is able to account for this to a certain extent, the opening is generally of a fixed size and thus does not have high selectivity.

To address this lack of selectivity, a recent advance has allowed for precise energy filtering by first expanding the beam energy spatially and then using an adjustable aperture to select a given energy range. The former can be accomplished through electron optics that perform prismatic expansion of the beam. In a typical case, this requires the

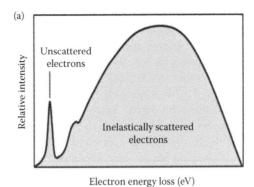

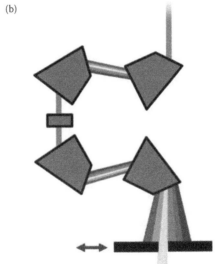

FIGURE 5.8 (**See color insert.**) Energy filtering TEM for contrast tuning. (a) Plot (not real data) of electron energy loss spectrum, showing unscattered and inelastically scattered populations. (b) "Omega filter" schematic, showing prismatic effect on electron beam in line with the optical axis. The aperture (bottom) can be moved to select for a given energy range (i.e., a given vertical slice through the plot shown in (a)).

beam to be directed around a 90° angle, with higher energy electrons being deflected more than those with lower energy. However, it is often not ideal to divert the beam path so drastically. Therefore, an alternative approach utilizes an "omega filter,"[21] so called because it diverts the beam path in the shape of the Greek letter, ultimately bringing the expanded beam back to its original path (Figure 5.8b). The adjustable aperture can then be placed in the beam path to allow a narrow band of energies (down to 10–20 eV wide) to pass. Importantly, this aperture can select for either unscattered electrons or any range of inelastically scattered electrons, resulting in high-contrast bright-field imaging or enhancement of small or weak constituents, respectively. This approach can

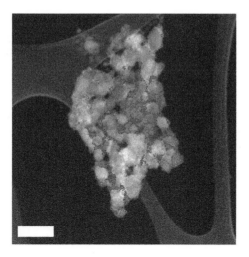

FIGURE 5.9   (See color insert.) EELS imaging. Dark-field TEM image of nanoparticles on a lacy carbon grid. Red overlay shows the location of Fe as determined by energy-filtered EELS imaging. Scale bar is 200 nm.

be utilized in the imaging of unstained biological samples, removing the effects of the staining process on structures of interest.[22] It can also be used to isolate narrow energy bands within an EELS spectrum to allow for elemental mapping (Figure 5.9).

5.3.4   TEM TOMOGRAPHY

In TEM tomography, an electron beam passes through a sample that is incrementally rotated around a central axis. The data collected are then reassembled to form a three-dimensional image of sample composition and morphology (Figure 5.10).

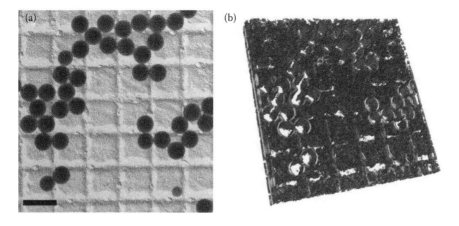

FIGURE 5.10   TEM tomography. (a) Planar bright-field TEM image of 50 nm beads supported by a grid. Scale bar is 100 nm. (b) Tomography of the same sample, computed from multiple images taken at various axial tilt angles relative to the beam.

These 3D images, or tomograms, can reveal detailed information about the organization within both crystalline and noncrystalline structures. Tomography has been used to study a wide variety of biological and inorganic specimens, ranging from viruses and cells to carbon nanotubes and nanoparticle superlattices.[23] Typically, resolution is on the scale of 5–30 nm, although 2.4 Å resolution using high-speed lasers has been reported.[24]

While equilibrium structures have traditionally been the focus of tomograms, time-resolved scanning has recently been reported.[25,26] These studies have shown that morphological and mechanical motion of a structure can be reconstructed from timed snapshots collected over a series of tilt angles.

5.4 SCANNING ELECTRON MICROSCOPY

5.4.1 INTRODUCTION AND COMPONENTS

The scanning electron microscope (SEM) has been a workhorse of nanoscale characterization for decades. While the resolution of SEM is significantly reduced compared to TEM (1–20 nm in most practical applications), the sample requirements are not as stringent with regards to thickness or vacuum environment; while TEM must operate in the high vacuum range (below 10^{-4} Pa), some SEM systems can operate near ambient conditions.[27]

Many of the components of an SEM are very similar to those of a TEM: the electron source operates in the same manner (although usually at a lower accelerating voltage of 0.1–100 kV); the condenser lenses converge the beam and then form a steep-angle, narrow spot; and apertures are used to remove higher angle electrons. However, the order of these pieces is altered and there are additional components as well. Below, we will detail these differences.

A diagram of a typical SEM is shown in Figure 5.11. As the beam passes through the condenser lenses, you will note that it passes through two apertures as well; the condenser aperture positioned between the lenses and the objective aperture after. Unlike in the TEM, the objective aperture is not moveable and thus does not select different areas of the beam. Both apertures are used to maximize coherency by allowing only the central (focused) region of the beam to pass.

Unlike in TEM, which uses a static electron beam to project through a sample, SEM operates by analyzing the products of beam–sample interaction at each discrete point across a sample surface. This necessitates that the beam position be actively adjustable. Therefore, following the objective aperture, the beam approaches a set of coils that can be used to deflect it. Two stacked sets of coils are needed to accomplish the task—while a single coil set can deflect (tilt) the beam, the resulting spot is elongated by the process and would result in a loss in resolution for areas away from the optical axis (Figure 5.12b). Two sets working in tandem are able to tilt and shift the beam to new locations, effectively moving the virtual location of the electron source laterally (Figure 5.12c). This maintains the spot shape and size.

Next the beam passes through an array of alternating (N–S) pole pieces, known as stigmators, which are arranged radially around the beam axis. These are used to control the eccentricity of the beam. When the beam spot is far from circular,

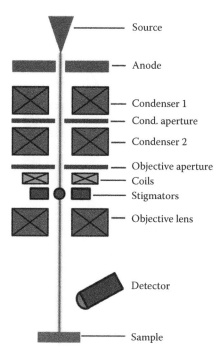

FIGURE 5.11   Schematic of an SEM. Cross-sectional illustration of the components in SEM. See text for descriptions.

resulting images can become streaked, and focus may become unachievable. This is because the size and shape of the beam determines the size and shape of pixels in the image, and eccentricity causes overlap along the major axis. In its simplest form (known as the quadrupole geometry), four elements are positioned at right angles to one another around the optical axis. Controlling the strength and asymmetry of the net field created by the stigmators can produce a circular beam cross section. It should be noted that aberrant eccentricity can result from a number of sources, including geometrical imperfections in the shapes of the lenses, inhomogeneities in the lens material, contamination on the poles, and external magnetic fields. This variety of sources means that changes in beam eccentricity are time variable, and therefore must be continually corrected.

Once the beam shape is optimized, it passes through the objective lens. As in TEM, this component will form a virtual image of the source. The strength of the lens determines the focal plane of the beam and thus the size of the beam spot that interacts with the sample directly below. Minimization of the spot size results in a focused image.

As described above, the beam is rastered across the sample surface in SEM and the interaction between the primary beam and the sample material is quantified at each point by a detector, thereby determining pixel brightness. While the beam–sample interaction produces a large number of particles and photons (some of which are described in later sections), the products most commonly used for image

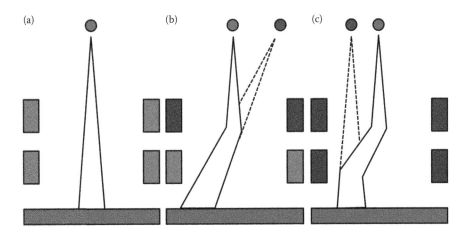

FIGURE 5.12 Beam scanning with coils. (a) The electron beam path through nonenergized deflection coils (light circle represents actual beam source). (b) Using a single set of coils moves the beam to a new lateral position, resulting in a virtual source (dark circle), but also a tilted beam path. Notice the elongated beam at the surface below. (c) Two coil sets in tandem can be used to maintain spot shape while moving the beam to a new location with a beam path parallel to the original.

formation in an SEM are secondary electrons and backscattered electrons. Each has a different origin and is detected in a different way.

Secondary electrons (SE) are electrons ejected from inner shells of the sample material through inelastic scattering of the primary beam. They are typically low energy (~50 eV), having received a fraction of the primary electron energy, and can only escape from depths on the order of 1–10 nm.[28] One common type of detector[29] utilizes a scintillator/photomultiplier apparatus to quantify SE production at each pixel and is positioned near the sample, but away from the optical path. At the head of this detector is a collector grid or cup that is held at a moderate positive voltage (100–300 V) to attract produced electrons. This low voltage is selective toward SE due to their low energy and, thus, does not affect the primary beam.

Backscattered electrons (BSE) are primary electrons that scatter elastically after interactions with nuclei. These tend to lose very little energy (<1 eV) and are directed back along the optical path. Because of their high energy, BSE cannot be selectively steered toward a detector without affecting the primary beam, and so BSE detectors are typically annular detectors positioned directly around the beam path. These tend to be semiconductor materials in which electron–hole pairs are created by impinging BSE, resulting in a quantifiable current, much like in a charge-coupled device.

SEM is used widely in nanotechnology and is capable of imaging materials on the order of 1 nm (Figure 5.13). However, some limitations exist. Among the most serious is sample charging[30]: for poorly conducting materials, an electron charge is unable to dissipate from the sample properly and eventually floods the SE signal with a constant high count at the detector. This effect is especially prevalent with polymers and biological samples, making imaging of these materials challenging. Typically, two methods are used to compensate for sample charging. First,

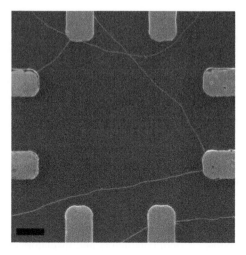

FIGURE 5.13   SEM image of nanomaterials. SEM image of single-wall carbon nanotubes (bright strings, diameter ~1 nm) pinned under lithographically defined metal electrodes on a silicon oxide substrate. Scale bar is 5 μm.

reducing the accelerating voltage of the primary beam can reduce the rate of charging. However, this also typically reduces the lateral resolution. Second, the sample can be coated with a thin conductive layer. This is often done by sputtering metals like AuPd, which has a small grain size of around 5 nm. However, this thin coating can mask small features on the surface.

5.4.2   ENERGY-DISPERSIVE X-RAY IMAGING

Aside from SE and BSE, the primary electron beam generates various other types of radiation as it interacts with the sample. These radiative products include characteristic x-rays. Such x-rays are generated as a by-product of SE production: when a primary electron knocks out an inner shell electron in an atom, an outer shell electron can replace it, resulting in the emission of a photon that corresponds to the energy difference between the two electron orbitals involved. There are a number of orbitals within an atom and the energy transition can occur between several of them, meaning that a single element under scrutiny releases a range of x-rays with discrete energy. This results in a spectral "fingerprint" for a given element and allows for fast determination of the chemical composition of the sample. Since analysis is performed at each pixel, it also allows for the spatial mapping of elements within the sample through a process known as energy-dispersive x-ray (EDX) imaging.[31] An example of such analysis is shown in Figure 5.14.

5.4.3   CATHODOLUMINESCENCE

Cathodoluminescence (CL) occurs when the primary electron beam is focused on samples made of certain materials, resulting in emission of photons with

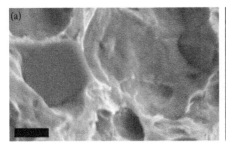 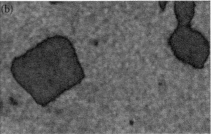

FIGURE 5.14 (**See color insert.**) Energy-dispersive x-ray mapping. (a) SEM image of an alloy material. Scale bar is 2 μm. (b) EDX map of the same field of view, showing the location of Al (green) and Mg (blue).

characteristic wavelengths.[32,33] Semiconductors and insulators exhibit this phenomenon due to the presence of an intermediate gap between the conduction and the valence band, unlike conductors that do not have a gap between the conduction and the valence band. The primary beam can transfer energy to an electron in the valence band, which allows it to jump across the band gap into the conduction band. When such an electron relaxes back into the ground state conduction band, it can do so directly or it can be trapped temporarily by intrinsic traps (structural defects) or extrinsic traps (impurities) between the two bands. In either case, the relaxation results in the release of radiation with discrete energy in the form of photons. Since each element that constitutes the sample will luminesce with a unique wavelength, the sample composition as well as its impurities and defects can be determined with the spatial resolution of SEM.

CL has been used to characterize nanostructures such as nanorods and nanowires. For example, ZnO nanostructures formed under various conditions have been investigated.[34] Structural variations in the nanomaterials were identified by analysis of CL spectra and intensity.

5.5 HELIUM ION MICROSCOPY

5.5.1 INTRODUCTION AND COMPONENTS

The helium ion microscope (HIM) is a relatively new instrument for the characterization of nanomaterials.[35] The optics (Figure 5.15) and mechanism of image formation are similar to those found in SEM with one central difference: rather than an electron beam, the HIM uses a coherent beam of He[+] ions. By the de Broglie relation, the wavelength of He[+] ions is much smaller than that of electrons, and would therefore be expected to yield an increased resolution according to the Rayleigh criterion. While a higher resolution is indeed achieved (down to 0.3 nm edge resolution), it is more a result of practical advantages, as discussed below. Regardless, HIM is capable of imaging a variety of materials with nanoscale features (Figure 5.16).

The beam originates at a field ion source[36] that is generated to come to an atomically small point (Figure 5.15, inset). A voltage applied to the metal tip first polarizes

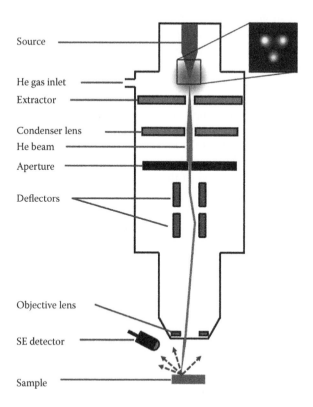

FIGURE 5.15 (**See color insert.**) Helium ion microscope. A simplified cross-sectional illustration of the components in HIM. Top inset: view of three atoms forming the atomically small source tip.

and then ionizes trace amounts of He gas injected around it. The enhanced field gradient at the sharpest point results in the highest ionization, and thus a large ion current is produced from the atoms at the apex. A select area aperture allows the beam produced at a single source atom to proceed through subsequent optics. Due to the large mass of He relative to electrons, the electromagnetic lenses used in SEM must

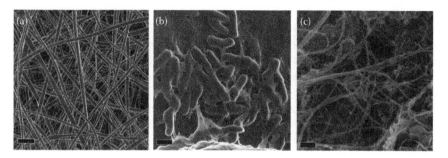

FIGURE 5.16 HIM images. (a) Electrospun polymeric fibers. Scale bar is 5 µm. (b) Uncoated bacteria. Scale bar is 1 µm. (c) Single-wall carbon nanotube. Scale bar is 100 nm.

be replaced with stronger electrostatic lenses in HIM. However, the interaction of the ion beam with the sample yields SE and backscattered ions that can be detected through means similar to those described previously.

An important aspect of the interaction is that He$^+$ ions scatter significantly less as they penetrate the sample material. This offers two specific advantages over SEM. Because the primary electrons in SEM scatter significantly, they generate SE emission from a lateral area of the sample surface that is large compared to the spot size of the beam. The low scattering of He$^+$ ions means that SE are produced only from an area that is roughly the size of the focal spot. This is central explanation for the incredible resolution of HIM. Second, because the He$^+$ beam penetrates deep into the sample, the build-up of charge at the surface is slow, even for poorly conducting samples. For this reason, imaging of polymers and biological material can be performed even without a conductive layer (cf. Figure 5.16), thereby avoiding the possibility of surface structure modifications caused by metallization. An additional electron beam can also be used to counter the accumulation of positive charge, further reducing the effect.

5.5.2 APPLICATIONS TO NANOFABRICATION

The momentum of He$^+$ ions impinging on the sample can cause significant structural damage. While this is certainly a disadvantage for the imaging of small and delicate features (especially at high magnification and thus high ion density), the effect can also be exploited as a means to modify materials at the nanometer scale controllably. An early demonstration of this ability showed that graphene sheets could be patterned with high precision,[37] including the fabrication of electrical devices.[38] HIM modification has also been applied to thin, free-standing silicon nitride membranes. At relatively low ion exposure, a membrane can be controllably thinned to a desired dimension.[39] At a higher point exposure, through holes can be fabricated with diameters below 4 nm.[40] These nanopores can subsequently be used as devices with which to characterize biological molecules electrically one at a time.

5.6 CONCLUSIONS

In this chapter, we have provided a brief introduction to a range of characterization tools used in both biological and condensed matter nanotechnology. The atomic force microscope is an instrument in which a sharp tip is scanned across a surface and probes its topography with nanometer accuracy. We discussed the various modes of AFM operation as well as two variations—high-speed AFM and scanning conductance microscopy—that add to the utility of the technique. The transmission electron microscope uses a coherent beam of electrons passed through a thin sample to form a projected image. It is compatible with both solid-state and biological materials to varying extents and offers extraordinary resolution, reaching the atomic scale. We further described electron energy loss spectroscopy, contrast tuning, and tomography as additional capabilities. The SEM rasters an electron beam across a sample and quantifies the interaction of the primary beam with the surface material at every discrete point to form an image of the entire sample. Normal imaging is accomplished by detecting secondary electrons or backscattered electrons, but other types of radiation produced through the interaction can also be used. For instance, we describe the analysis of resultant photons both in the x-ray range (EDX imaging) and in other ranges (cathodoluminescence). Finally, we described the new technology of helium ion microscopy, in which a coherent beam of He$^+$ ions is used to perform scanning charged particle microscopy in the same fashion as SEM. This instrument achieves high-resolution imaging and can be used for nanoscale fabrication as well.

REFERENCES

1. Binning, G., Rohrer, H., Gerber, C., and Weibel, E. Surface studies by scanning tunneling microscopy. *Physical Review Letters* 49, 57–61, 1982.
2. Grutter, P., Zimmermannedling, W., and Brodbeck, D. Tip artifacts of microfabricated force sensors for atomic force microscopy. *Applied Physics Letters* 60, 2741–2743, 1992.
3. Gross, L., Mohn, F., Moll, N., Schuler, B., Criado, A., Guitian, E., Pena, D., Gourdon, A., and Meyer, G. Bond-order discrimination by atomic force microscopy. *Science* 337, 1326–1329, 2012.
4. Ando, T., Uchihashi, T., Kodera, N., Yamamoto, D., Miyagi, A., Taniguchi, M., and Yamashita, H. High-speed AFM and nano-visualization of biomolecular processes. *Pflugers Archiv-European Journal of Physiology* 456, 211–225, 2008.
5. Katan, A.J. and Dekker, C. High-speed AFM reveals the dynamics of single biomolecules at the nanometer scale. *Cell* 147, 979–982, 2011.
6. Kodera, N., Yamamoto, D., Ishikawa, R., and Ando, T. Video imaging of walking myosin V by high-speed atomic force microscopy. *Nature* 468, 72–76, 2010.
7. Staii, C., Johnson, A.T., and Pinto, N.J. Quantitative analysis of scanning conductance microscopy. *Nano Letters* 4, 859–862, 2004.
8. Bockrath, M., Markovic, N., Shepard, A., Tinkham, M., Gurevich, L., Kouwenhoven, L.P., Wu, M.S.W., and Sohn, L.L. Scanned conductance microscopy of carbon nanotubes and lambda-DNA. *Nano Letters* 2, 187–190, 2002.
9. Giannuzzi, L.A., Drown, J.L., Brown, S.R., Irwin, R.B., and Stevie, F. Applications of the FIB lift-out technique for TEM specimen preparation. *Microscopy Research and Technique* 41, 285–290, 1998.
10. Giannuzzi, L.A. and Stevie, F.A. A review of focused ion beam milling techniques for TEM specimen preparation. *Micron* 30, 197–204, 1999.
11. Angermuller, S. and Fahimi, H.D. Imidazole-buffered osmium-tetroxide—An excellent stain for visualization of lipids in transmission electron-microscopy. *Histochemical Journal* 14, 823–835, 1982.
12. van den Bergh, B.A.I., Swartzendruber, D.C., Bosvander Geest, A., Hoogstraate, J.J., Schrijvers, A., Bodde, H.E., Junginger, H.E., and Bouwstra, J.A. Development of an optimal protocol for the ultrastructural examination of skin by transmission electron microscopy. *Journal of Microscopy-Oxford* 187, 125–133, 1997.
13. Fischbein, M.D. and Drndic, M. Electron beam nanosculpting of suspended graphene sheets. *Applied Physics Letters* 93, 113107, 2008.
14. Song, B., Schneider, G.F., Xu, Q., Pandraud, G., Dekker, C., and Zandbergen, H. Atomic-scale electron-beam sculpting of near-defect-free graphene nanostructures. *Nano Letters* 11, 2247–2250, 2011.
15. Huang, P.Y., Ruiz-Vargas, C.S., van der Zande, A.M., Whitney, W.S., Levendorf, M.P., Kevek, J.W., Garg, S. et al. Grains and grain boundaries in single-layer graphene atomic patchwork quilts. *Nature* 469, 389–392, 2011.

16. Lin, L.T., Cui, T.R., Qin, L.C., and Washburn, S. Direct measurement of the friction between and shear moduli of shells of carbon nanotubes. *Physical Review Letters* 107, 206101, 2011.

17. Mastrangelo, I.A., Courey, A.J., Wall, J.S., Jackson, S.P., and Hough, P.V.C. Dna looping and Sp1 multimer links—A mechanism for transcriptional synergism and enhancement. *Proceedings of the National Academy of Sciences of the United States of America* 88, 5670–5674, 1991.

18. de Jonge, N. and Ross, F.M. Electron microscopy of specimens in liquid. *Nature Nanotechnology* 6, 695–704, 2011.

19. Scholl, J.A., Koh, A.L., and Dionne, J.A. Quantum plasmon resonances of individual metallic nanoparticles. *Nature* 483, 421–427, 2012.

20. Egerton, R.F. Electron energy-loss spectroscopy in the TEM. *Reports on Progress in Physics* 72, 016502, 2009.

21. Probst, W., Benner, G., Bihr, J., and Weimer, E. An omega energy filtering TEM—Principles and applications. *Advanced Materials* 5, 297–300, 1993.

22. Debruijn, W.C., Sorber, C.W.J., Gelsema, E.S., Beckers, A.L.D., and Jongkind, J.F. Energy-filtering transmission electron-microscopy of biological specimens. *Scanning Microscopy* 7, 693–709, 1993.

23. Midgley, P.A. and Dunin-Borkowski, R.E. Electron tomography and holography in materials science. *Nature Materials* 8, 271–280, 2009.

24. Scott, M.C., Chen, C.C., Mecklenburg, M., Zhu, C., Xu, R., Ercius, P., Dahmen, U., Regan, B.C., and Miao, J.W. Electron tomography at 2.4-angstrom resolution. *Nature* 483, 444–447, 2012.

25. Kwon, O.H. and Zewail, A.H. 4D Electron tomography. *Science* 328, 1668–1673, 2010.

26. Su, D.S. Electron tomography: From 3D statics to 4D dynamics. *Angewandte Chemie-International Edition* 49, 9569–9571, 2010.

27. Stokes, D.J. Recent advances in electron imaging, image interpretation and applications: Environmental scanning electron microscopy. *Philosophical Transactions of the Royal Society of London Series A—Mathematical Physical and Engineering Sciences* 361, 2771–2787, 2003.

28. Seiler, H. Secondary-electron emission in the scanning electron-microscope. *Journal of Applied Physics* 54, R1–R18, 1983.

29. Everhart, T.E. and Thornley, R.F.M. Wide-band detector for micro-microampere low-energy electron currents. *Journal of Scientific Instruments* 37, 246–248, 1960.

30. Fitting, H.-J., Touzin, M., Cornet, N., Goeuriot, D., Juvé, D., and Guerret-Piécourt, C. Non-conductive sample charging in SEM and ESEM. *Microscopy and Microanalysis* 13, 76–77, 2007.

31. Anderhalt, R. X-ray microanalysis in nanomaterials. In: *Scanning Microscopy for Nanotechnology: Techniques and Applications* (eds. Zhou, W. and Wang, Z.), Springer, New York, 2006.

32. Bresse, J.F., Remond, G., and Akamatsu, B. Cathodoluminescence microscopy and spectroscopy of semiconductors and wide bandgap insulating materials. *Mikrochimica Acta* 13, 135–166, 1996.

33. Remond, G., Cesbron, F., Chapoulie, R., Ohnenstetter, D., Roquescarmes, C., and Schvoerer, M. Cathodoluminescence applied to the microcharacterization of mineral materials—A present status in experimentation and interpretation. *Scanning Microscopy* 6, 23–68, 1992.

34. Grym, J., Fernandez, P., and Piqueras, J. Growth and spatially resolved luminescence of low dimensional structures in sintered ZnO. *Nanotechnology* 16, 931–935, 2005.

35. Scipioni, L., Stern, L.A., Notte, J., Sijbrandij, S., and Griffin, B. Helium ion microscope. *Advanced Materials and Processes* 166, 27–30, 2008.

36. Scipioni, L., Alkemade, P., Sidorkin, V., Chen, P., Maas, D., and van Veldhoven, E. The helium ion microscope: Advances in technology and applications. *American Laboratory* 41, 26–28, 2009.

37. Bell, D.C., Lemme, M.C., Stern, L.A., Rwilliams, J., and Marcus, C.M. Precision cutting and patterning of graphene with helium ions. *Nanotechnology* 20, 455301, 2009.

38. Lemme, M.C., Bell, D.C., Williams, J.R., Stern, L.A., Baugher, B.W.H., Jarillo-Herrero, P., and Marcus, C.M. Etching of graphene devices with a helium ion beam. *Acs Nano* 3, 2674–2676, 2009.

39. Marshall, M.M., Yang, J., and Hall, A.R. Direct and transmission milling of suspended silicon nitride membranes with a focused helium ion beam. *Scanning* 34, 101–106, 2012.

40. Yang, J., Ferranti, D.C., Stern, L.A., Sanford, C.A., Huang, J., Ren, Z., Qin, L.-C., and Hall, A.R. Rapid and precise scanning helium ion microscope milling of solid-state nanopores for biomolecule detection. *Nanotechnology* 22, 285310, 2011.

6 Biological Nanostructures of Insect Cuticles

Formation, Function, and Potential Applications

Dennis LaJeunesse, Adam Boseman, Kyle Nowlin, and Alan Covell

CONTENTS

6.1 INTRODUCTION

This may not be a surprise, but at the cellular and molecular level, biology operates at the nanoscale. Integrin-based adhesion between a cell and its underlying substrate has specific dimensional parameters: integrin proteins must be properly spaced and organized within the plasma membrane at nanoscale tolerances to facilitate the intracellular bundling of actin filaments and attachment to the extracellular matrix [1–4]. Stacks of nanoscale intracellular lamellar discs must be spaced 12–15 nm within the outer segment of the photoreceptor cell to display the photoreactive rhodopsin complexes for excitation by light [5–7]. The mechanisms that underlie the nanoscale nature of biology are built into the molecules that make up life itself. Biological molecules such as proteins, phospholipids, and nucleic acids rapidly form elaborate structures spontaneously as dictated by constrains provided by their composition [8–11]. Cytoskeletal structures such as eukaryotic microfilaments, tubulin-based microtubules, or even actin-like filamentous structure in bacteria assemble to regulate linear nanostructures that play many critical roles in the cell such as vesicle transport, segregation of genetic information, and cell shape and structure [12–15].

Advances in scanning particle beam microscope and atomic force microscope technologies have opened up an entire new world in surface science at the nanometer scale [16–19]. While most light microscopes have diffraction-limited resolution down to a several hundred nanometers, particle beam microscopes (i.e., electron and ion) have nanoscale resolution, with helium ion microscopy having near subnanometer resolution [16,17,20]. Many of the newer particle beam microscopic techniques (e.g., low electron voltage excitation and helium ion beam microscopy) preclude the need for the addition of a conductive metal surface. This has allowed exploration of the biological system's surface in an unprecedented manner [17,19,21,22].

Insect cuticles have been a particularly attractive subject for microscopic study, as many of the spectacular colors and patterns that exist at the macroscale are the product of surface properties at the micro- and nanoscale. In this chapter, we will detail recent work, involving the biological nanostructures of the insect cuticle. Insect cuticles provide a unique paradigm for the study of biological nanostructures that will be applicable for understanding other biological nanoscale structures. The cuticle is the natural composite material that is the product of a living cell, and being biological in origin, the formation of the cuticle and the nanostructures associated with it will contain the aspect of self-assembly that many biological systems possess. Insect cuticles are remarkable materials and their study will lead to the development of novel biomaterials and bioinspired processes for the formation of these materials.

6.2 INSECT CUTICLE COMPOSITION AND FORMATION

The insect cuticle is a nonliving composite material that is secreted extracellular from an underlying epithelium [23]. The insect cuticle functions as a protective barrier that prevents desiccation and physical damage and assumes a variety of morphologies and configurations to serve in multiple mechanical functions [23–30]. Depending on the cuticle and its location in the organism, the insect cuticle displays an extraordinarily broad range of physical and mechanical properties and represents some of the strongest, toughest, and hardest materials in the biological world [24,27,30–34]. For instance, the mandibles of larval jewel beetles *(Pseudotaenia frenchi)* are extremely hard, rivaling the hardness of stainless steel [33]. The cuticle is a composite of protein and the polysaccharide chitin along with smaller amounts of polyphenols and lipids [23]. However, the cuticle also serves as a site of mineral and metal deposits that have pronounced effects on the structure and properties of the cuticle [34]. Although the mechanisms involved in the organization of chitin are not well described, in all insects cuticles, chitin fiber bundles are arranged in layers forming a helicoidal pattern that resists stress from all directions (Figure 6.1) [23,30–32]. Cuticular components are secreted from the underlying epithelial cell layer as a liquid crystal matrix and are deposited into an assembly zone where chitin nanofibers and specific proteins self-assemble into regularly arranged fibrils [35–37]. In some cases, such as in the bristles of the fruit fly *Drosophila melanogaster*, there appears to be localized intracellular chitin deposition zones that direct chitin secretion into the assembly zone [38]. Within this assembly zone, nanofibers of the chitin polysaccharide, β-1,4-linked acetylated glucosamine are arranged into highly crystalline bundles that are characterized by an antiparalleled hydrogen bonding

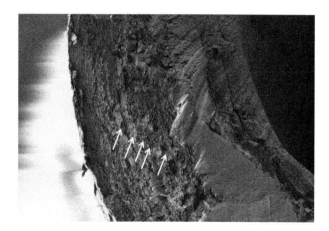

FIGURE 6.1 Laminar organization of the insect cuticle. A cross-sectional SEM (scanning electron) micrograph of a cicada wing shows the layered structure of the cicada wing and the arrows point out each layer. The outer surface contains the conical nanostructures. All layers are secreted from a now-collapsed cellular epithelium.

between the sugar chains [23,39]. Almost all cuticles have two distinct layers, a thin waxy water-resistant epicuticle and an inner bilaminate procuticle that is further subdivided into an outer exocuticle and an inner endocuticle [23,28,29,31]. The exocuticle is hard and pigmented, while the endocuticle contains a higher water content and is more flexible [32]. The balance of the thickness, arrangement, and specific composition of the cuticular layers determines the properties of a given cuticle. For instance, the soft cuticle found within the articulated joints of appendages has a thinner, less-sclerotized exocuticle, while the exocuticle in cuticles of mandibles and carapaces is more developed and harder [23,28,30,31].

The interaction of specific chitin-binding proteins with chitin determines whether the properties of a given layer within the cuticle are endocuticle or exocuticle [23,28,31, 40–43]. Perhaps, the best known chitin-binding protein is the rubber-like protein RESILIN that is the cuticular component of the tendons of certain insects' legs and is responsible for storing mechanical energy needed for jumping and flight in various insect species [42,44–47]. Several hundred putative chitin-binding proteins have been identified by genome analysis of insects and other arthropods [48]. However, the manner in which this binding domain interacts with chitin nanofibers and/or chitin crystallites—the nature of protein–protein interactions within the cuticle, as well as the manner in which chitin nanofibers are organized to confer specific functions—remains unclear [28,31,47,49,50]. Some chitin-binding proteins appear to organize chitin into elaborate structures; the chitin containing the peritrophic membrane of the insect midgut has a hexagonal lattice laced with ~10-nm nanopores that filter food particles [51,52].

6.3 INSECT NANOSCALE STRUCTURE

Over the past 450 million years, insects have evolved countless adaptations for survival in extreme conditions. Insects occupy every continent; they have colonized extreme environments [53–55]. Many of the interactions between insects and their environment are mediated via their skin or cuticle and it is reasonable to hypothesize that this epicuticle surface has evolved unique adaptations at the micro- and nanoscale to accommodate these challenges. At the microscale, insect cuticles display a diverse set of structures, hairs, pores, bristles, and scales (Figure 6.2). These microscale structures have long been studied; some such as the hairs on the crane fly (*Nephrotoma australasiae*) mediate the shedding of water from the insect's wings and legs [56]. In certain water insects, the presence of dense arrays of microscale hairs called microtrichia trap air on the surface of water insects for up to 3 months even when the insect is dead [57].

At the nanoscale, an impressive and diverse set of structures has been discovered. Some nanostructured surfaces, such as the conical structures of cicada wings and moth eyes, function as stand-alone features and many of the insect circular microstructures (e.g., bristle, scales, and hairs) are also decorated with nanoscale textures and structures (Figures 6.2 and 6.3). The nanoscale structures often endow that surface with new properties [58–60]. The nanoridges of the abdomen of the oriental hornet *Vespa orientalis* act as a solar-powered, metabolic starter engine [61]. Nanotextured scales of certain moths function as sonic-dampening systems as a protection from bat predation [62]. The bristles of the fruit fly, *D. melanogaster*, are decorated with transverse 55-nm-wide nanoribs along the length of the bristle (Figure 6.1) [22]. While the role that these structures give to the bristles is unknown, is based on the role that these bristles play in the cleaning behavior of the fruit fly [63] and/or in providing sensory information during flight.

While, the formation of these nanoribs is a mystery, they must involve a process similar to the formation of the cuticle. Insect bristles and bristle-like structures such

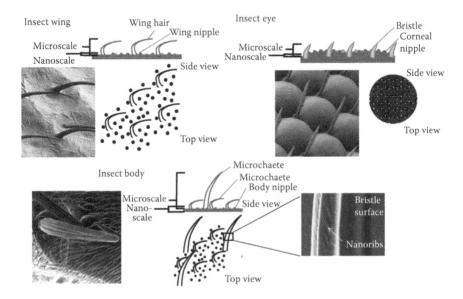

FIGURE 6.2 Generalized hierarchy of microscale and nanoscale structures on insect cuticles.

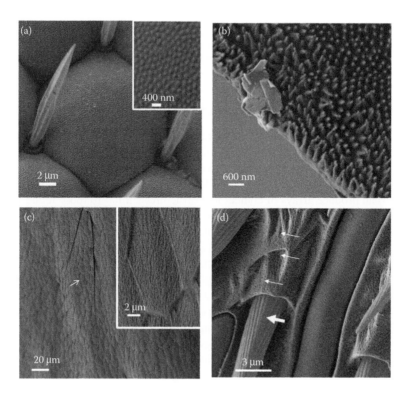

FIGURE 6.3 **(See color insert.)** Examples of insect epicuticle nanostructures. (a) Nano-nipples cover the surface (arrow) on the corneal surface of the eye of the fruit fly, *D. mela-nogaster*. The insert shows greater detail on the corneal nipple array. (b) Wing surface of the dog day cicadae, *Tibieins* ssp, is decorated with tall conical nanonipples (arrow); scale bar 600 nm. (c) Scaled surface (arrow) of the rostral plate of the European hornet, *Vespa crabro*. The insert shows that each has a vermiform, irregular nanopatterned surface. (d) The mouth-part of the European hornet, *V. crabro*, exhibits bristles with nanogrooves (thick arrows) that are surrounded by a epicuticle covered with nanopores (thin arrows).

as the scales found on moths, flies, and mosquitos function as mechanical or chemical sensory organs and are derived from cellular extensions [38,60,64,65]. In the fruit fly *D. melanogaster*, primordial bristles are formed through complicated cytoskeletal dynamics, involving filamentous actin and microtubules [64,66,67]. In the adult thoracic macrochaetes, the scaffold of bundled filamentous actin is manifested as longitudinal ridges that span the length of the bristle [65]. The microfilament/actin cytoskeleton is an intracellular nanoscale component with each fiber that is 5 nm wide, which is involved in cell motility, cell division, cell shape, and in vesicle trafficking [68,69]. Many actin-binding proteins bundle and organize actin in the cell creating elaborate networks of filamentous actin; for instance, proteins of the FASCIN family have been shown to cross-link and self-organize actin into regular nanoscale structures *in vitro* [70]. One potential mechanism for the formation of nanoribs on the bristles of the fruit fly and other nanoscale structures found on butterfly and mosquito scales may involve some form of subcellular scaffolding of the forming cuticle via cytoskeletal network by either directly influencing the structure or the deposition of the material. In support of this idea, recently extracellular chitin deposition has been shown in fly bristles to be mediated by interactions of the secretory machinery with the cytoskeleton [38].

6.4   EPICUTICLE WAXES AND NANOSTRUCTURE FORMATION

Although the mechanisms behind the formation of specific cuticular nanostructured surfaces in insects remain unresolved, synthetic means of forming similar structures are well known. Block copolymer and polymer-demixing techniques have been used to generate nanostructured surfaces [71,72]. Far ultraviolet (UV) photolithographic methods can produce arrays of conical nanostructures similar to those found in the cuticle of cicada wings [73]. The formation of native biological nanostructured surfaces has also evolved in the cuticle of plants. Many plants exhibit a hierarchical surface structure analogous to those found in many insect cuticles and these types of structured surfaces provide a variety of different properties to these surfaces (Figure 6.5d) [74,75]. Many of the plant nanostructures are composed of secreted epicuticular waxes that are readily observed on fruits such as grapes and plums as an opaque blush [75]. Plant epicuticular waxes are crystalline and form a variety of nanoscale structures depending on the chemical composition of the wax. Common structures such as tubules, flakes, granules, and rings range in size from 200 nm to 100 μm. The type and size of the structure is dependent on the type and amount of epicuticular wax and, in some cases, uses the underlying cuticle structure as a scaffold. In other cases, however, these waxes can self-assemble in two-dimensional (2D) and three-dimensional (3D) crystalline nanostructures. The wax nonacosan-10-ol, which is found on many leaves, is the major component of self-assembling tubular crystal nanostructures [74]. Insects have a waxy outer epicuticle and secrete hydrocarbon and waxes through nanopores in this cuticle [76,77]. Recently, nanostructured waxy tubes that are involved in antiwetting and color formation have been reported in damsel flies [78]. The elaborate structure of these is species specific and appears to be essential for the viability of the organism; however, the type of wax and whether these waxes exhibited self-assembly remains uncharacterized.

6.5   OPTICAL PROPERTIES: ANTIREFLECTION AND
STRUCTURAL COLOR

Certain insects have hexagonally packed, conical nanostructure arrays covering the cornea surfaces of their compound eyes and wings (Figure 6.4a–c) [79]. While all have similar optical properties, the size, shape, aspect ratio, and distribution of these conical nanostructures varies from species to species and dictates the extent and nature of these optical properties [73]. While these nanoscale protrusions are essentially transparent, they reduce reflection and increase transmission of incident light by an optical impedance mechanism, which involves the gradual transitioning of light from air into surface-by-surface geometry alone as the conical shape of these nanostructures results in a partitioning of the refraction index of the surface material (Figure 6.4e) [79–81]. The conical nanostructure arrays have been shown to improve

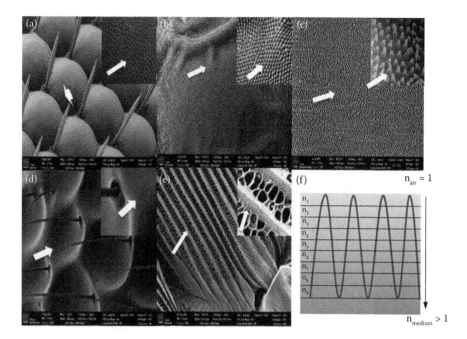

FIGURE 6.4 (a) SEM image of *D. melanogaster* (fruit fly) compounds the eye showing several ommatidia with insert showing protrusions that are (nipples) 40 nm wide. (b) SEM image of mosquito *Ommatildia* with insert showing nipples about 100 nm wide. (c) Helium ion microscope (HIM) image of cicada wing surface showing dense nipple coverage with insert showing conical-like structure of individual nipples 100 nm wide at the base and 300 nm tall. (d) SEM image of orchard bee thorax showing large dimple-like features with insert showing 200-nm-diameter dimples/pores. (e) SEM image of butterfly wing scales showing major ridges spaced 2 μm apart along with the insert showing minor ridges spaces that are 150–200 nm apart. (f) Carton cross section of cicada wing nipples demonstrating impedance matching for antireflection function.

vision of the insects by increasing light transmission into the eye and reduce glare for countering predation while they are less active during certain times of the day [19,82–84]. Arrays of conical nanostructures are being applied to surfaces to reduce reflection and increase transmission such as in car windshields, the screens of electronic devices, camera lenses, and photovoltaic solar panels [73].

Cuticular nanostructures also play significant roles in color production. The bright colors displayed on the cuticle of beetles, bees and wasps, and the wings of many moths and butterflies are the product of the nanoscale-structured surfaces (Figure 6.5d and e) [85–87]. Structural colors result from the way light is reflected from multilayer thin films, diffraction gratings, and/or 2D or 3D photonic crystals. Single-layer thin films strongly reveal constituent colors of white light that constructively interfere after reflections from successive interfaces. To do this, they must emerge in phase, otherwise known as the Bragg condition [88]. This is dependent on the layer thickness, index of refraction, and the angle of incidence. Many insect wings demonstrate this phenomenon as different colors become apparent from their transparent

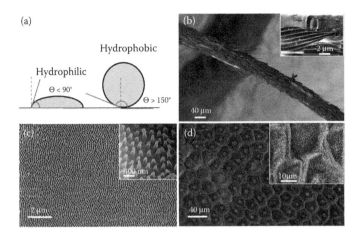

FIGURE 6.5 (a) Schematic demonstrating static contact angle measurement and generalized droplet shapes for hydrophilic/hydrophobic surfaces, (b) SEM micrograph of the mosquito (*A. aegypti*) leg covered with scales (arrow) and insert showing details of the scales with their porous grid pattern (arrow), (c) helium ion microscope micrograph of the surface of a cicada wing (*Tibicen* ssp) showing the array of conical nanostructures and the insert shows details of the conical nanostructures 100 nm wide and 300 nm tall; and (d) SEM of the surface of a *Colocasia esculenta* (elephant ear) leaf showing hierarchical roughness with 20–30-μm-diameter papillae and, the insert shows nanometer-sized waxy tubules on this surface.

wings when seen from various angles [89]. Some insects make use of multilayer reflectors, a more rich and complex extension of the single-film reflectors; some beetles have metallic iridescent colors due to this effect [17,88–90]. One-dimensional diffraction gratings are also common in nature and consist of periodically spaced adjacent chitin-based structures that create a variation in the refraction index on the order of the light wavelength [88]. The chitinous photonic crystals found in some butterflies are essentially an extension on the one-dimensional diffraction grating into two and three dimensions, analogous to a combination of the stacking form of a multilayer thin film and the adjacent form of a diffraction grating [85,91]. The application of natural photonic crystals will be important for developing novel security-labeling methods and structural, perhaps even tunable, paints [91,92].

6.6 WETTING: HYDROPHOBICITY AND SUPERHYDROPHOBOCITY

The interaction of a liquid with a surface is referred to as wetting; surfaces that are resistant to wetting are referred to as hydrophobic and those that promote wetting are hydrophilic [21]. This property is measured by determining the contact angle measurement of the liquid with the surface and this measurement corresponds to the area of contact that a drop of liquid has with a surface (Figure 6.5a) [21,93]. In the most exaggerated, hydrophobic state, which is termed superhydrophobicity, an aqueous liquid drop will form a perfect sphere and contact the surface at a single point, while in the most "wettable" state, a liquid drop will spread completely onto a surface [56]. Generally, surfaces that present a water droplet with a contact angle >150° are called superhydrophobic, while hydrophilic surfaces have a contact angle of <90° (Figure 6.5a). Wettability or contact angle depends on the multiple variable, including the cohesive forces and surface tensions within the liquid, the adhesive forces of the surface and surface chemistry, the roughness of the surface both at the micro- and nanoscale, and properties of the surrounding atmosphere [21,93].

In the context of the insect cuticle, hydrophobicity is a feature that must have been selected for evolution. Wet surfaces add weight and for flying insects in which every microgram counts, hydrophobic or better yet superhydrophobic surfaces facilitate flight by preventing the accumulation of water droplets. Almost all insect cuticles contain a hydrophobic waxy layer; however, the most striking aspect of the wetting properties of insect cuticles is the elaboration of a surface structure that augments the surface chemistry. These structures can be a combination of microscale and nanoscale structures. For instance, the scales found on the legs of the mosquito (*Aedes aegypti*) that have nanoscale grids (Figure 6.5b) provide extreme hydrophobicity to the insect [94]. Nanostructured surfaces such as the arrays of conical nanostructures on the cicada wing (Figure 6.5c) also generate highly water-repellant surfaces [95]. Biologically, the native self-cleaning mechanisms of the cuticle require these hydrophobic and superhydrophobic properties; rolling beads of water collect and remove dirt, dust, and other particulate materials such as fungal spores and bacteria. Recently, the superhydrophobic surface of the cicada wing has been shown to have a unique self-cleaning "jumping water" mechanism, whereby a dirty superhydrophobic surface ejects the dirt off the surface through the release of surface energy that is generated by the coalescence of water around the dirt particle [96]. Superhydrophobic surfaces have many applications. Self-cleaning is a hot topic for such surfaces because as the water beads up and rolls off the surface, it can collect particulate matter and take it with it. This is very simple to demonstrate with some soil, water, and a lotus leaf or other superhydrophobic surfaces [73,81].

6.7 INSECT TARSAL ADHESION AND APPLICATION

Insects walk and run freely on smooth, vertical, and even inverted surfaces [97–99]. Many survival advantages are gained by sticking to a surface, especially when dealing with the incredible forces small insects must face to survive; something as simple as raindrops landing on the surface of a hanging leaf will be enough to trampoline a small ant onto the ground if it was not able to stick to the surface. The adhesion is generated by elaborate pads on the feet and is remarkably similar between insects and some animals such as the geckos (Figure 6.6). These footpads are derived from a specialized cuticle found at the tarsal terminal and consist of fine micro- and nanoscale structures that effectively increase the surface contact area through pillars with millions of tiny hair-like structures (Figure 6.6a and b) or smooth ridges (Figure 6.6c) [97,98]. The mechanisms behind this phenomenon can be broken down into two classes: dry adhesion and wet adhesion [17,97,100]. Dry adhesion involves the interaction of the dry footpad with the surfaces through van der Waals interactions; the shape and orientation of participating molecules and

FIGURE 6.6   Gecko and insect-adhesive pads. (a) Setae and spatulae on Tokay gecko (*G. gecko*) foot; spatulae are shown in the insert. (Images courtesy Dr. Kellar Autumn, Dr. Jonathan Puthoff, and Dr. Jijin Yan.) (b) Hairy-type of setae or footpads on the fruit fly tarsus (*D. melanogaster*). (c) Smooth type of setae on hornet (*V. crabro*).

structures are important for these interactions to occur [101–103]. In addition to dry adhesion, many insects use a different set of interactions to adhere to surfaces called wet adhesion, which requires a thin film of liquid between the surface and footpad. Compression of the pad or capillary suction by contact with the surface causes a small release of fluid, while detaching the pad causes a decompressed fluid and much of the fluid is reabsorbed into the pad [104–106]. In contrast to dry adhesion, there is a greater diversity of structures that use a wet adhesion [98,107,108]. Polydimethylsiloxane (PDMS) replica versions of an insect footpad reproduce the physical adhesive properties demonstrated by the native insect and show that these adhesive properties do not require biological material components [109]. Using these same design principles, others fabricate a synthetic hand-sized adhesive pad that could hold around 2950 N, or 660 lbs and could be effortlessly peeled off the surface when it was no longer needed [110].

6.8   BIOLOGICAL NANOSTRUCTURES AND EVOLUTION

To understand the mechanisms behind the formation of naturally occurring biological nanostructures, as well as understand the properties they possess, we must recognize that the fundamental force guiding the formation of these structures and the processes that led to their formation is evolution. Since all biological nanostructures are derived from living creatures, they are products of eons of evolution. But with this, there is a caveat; although the formation and/or function of biological nanostructures may have potentially arisen from natural selection, not all biological nanostructures need to be the products of natural selection. The implications are that over the course of evolution, there was some selective force (e.g., disease, mating, and environmental change) and that the presence of these nanostructures provided an advantage to an organism having this nanostructure (e.g., disease resistance, greater mating success, and surviving environmental changes). This logic would incorrectly suggest that all biological nanostructures have specific properties that provide a selective advantage and this is probably not always the case.

The structure/function relationship is not always clear and in many cases, such relationships simply may not exist. Just as is the case with some microscale or macroscale biological structures, some biological nanostructures may have no discernable properties and may have simply arisen as a by-product of some other process. Similar arguments have been made about architectural features [111,112]. Spandrels are architectural features associated with the domed arches that are found in many medieval European churches and are often ornately decorated. However, the formation of these spaces was a by-product of the construction of domed arches. If one looked at the churches, one might incorrectly assume that the spaces were purposely created for ornamentation. When considering the function of biological nanostructures, we must consider this paradigm. Not all biological nanostructures will have evolutionary and/or biological relevance. Double-stranded DNA (deoxyribonucleic acid) can been manipulated to form complex 2D and 3D structures such as smiley faces and treasure chests in the technique known as DNA origami [10]. The self-assembly properties of DNA make this process efficient and remarkably reproducible, but DNA in the living cell never assumes these structures. The conical nanostructured surfaces of the cicada wing have an antimicrobial property of killing Gram-negative bacteria [113–115]. Do cicada wings need an anti-Gram-negative bacterial surface? It will take more work to evaluate whether this was a selected feature of the surface or simply a by-product of an elaborate array of conical nanostructures. Cicada shows a great deal of variability in the size and aspect ratio of the conical nanostructures in its wings [95], some of which may significantly alter its wetting and optical properties. Arguments could be constructed to determine whether aspects of the life histories of these different species provide selective forces for the presence/or absence of these features. Some biological nanostructures may have functions that simply have not been described because we have not identified or understood the selective force that shaped them. In this manner, the relevance of any biological nanostructure may not be fully understood.

In the Arctic, there are two strikingly distinct design traditions, one from those of the eastern Pacific Aleutian Island natives and another design tradition that is found in kayaks of those Native Americans living in the western Atlantic/Greenland area. Aleutian-style kayaks are long, flat boats with little bow (front-to-back) curvature or rocker, while Greenland-style kayaks are long boats with significant bow curvature or rocker. Both kayaks are fast, seaworthy boats that were an integral part of these people subsistence hunting and fishing cultures. Based solely on boat speed, a bow curvature appears to be a simple flourish, that is, something that produces a different style of kayak, but does not change the performance of the craft itself. However, this is not the case when one considers other aspects of their development and use. For a greater part of the year, the seas around Greenland are filled with heavy flows of sea ice that often results in capsizing, while the seas around the Aleutian Islands are virtually ice free and, therefore, pose less of a risk for capsizing. Owing to their bow curvature, paddlers in Greenland-style kayaks can easily perform a self-righting rescuing technique called rolling, while the flat, rocker-less Aleutian-style kayaks cannot be rolled. In the evolution of long, skinny seaworthy hunting boats, the presence of a new selective force (i.e., sea ice) resulted in an additional trait (i.e., the need to self-right a capsized kayak) within the development of a seacraft with a slight rocker. In a similar manner, biological nanostructures can also be selected for more than one trait and in many instances this duality is apparent [81]. Furthermore, this duality often creates internal restraints on the system. For instance, using our kayak analogy once more, adding too much rocker to a craft will result in a loss of seaworthiness, but does create a situation for considerably easier rolling. In the context of biological nanostructures, this implies that in some cases, extremes will not be present in the native structure, because of evolutionarily-based optimization, but may be a source for new structures when applied to real-world solutions.

REFERENCES

1. Khang D., Choi J., Im Y.M., Kim Y.J., Jang J.H., Kang S.S., Nam T.H., Song J., Park J.W.: Role of subnano-, nano- and submicron-surface features on osteoblast differentiation of bone marrow mesenchymal stem cells. *Biomaterials* 2012, 33(26):5997–6007.
2. Yu C.H., Law J.B., Suryana M., Low H.Y., Sheetz M.P.: Early integrin binding to arg-gly-asp peptide activates actin polymerization and contractile movement that stimulates outward translocation. *Proceedings of the National Academy of Sciences of the United States of America* 2011, 108(51):20585–20590.
3. Vignaud T., Galland R., Tseng Q., Blanchoin L., Colombelli J., Thery M.: Reprogramming cell shape with laser nano-patterning. *Journal of Cell Science* 2012, 125(Pt 9): 2134–2140.
4. Pennisi C.P., Dolatshahi-Pirouz A., Foss M., Chevallier J., Fink T., Zachar V., Besenbacher F., Yoshida K.: Nanoscale topography reduces fibroblast growth, focal adhesion size and migration-related gene expression on platinum surfaces. *Colloids and surfaces B, Biointerfaces* 2011, 85(2):189–197.
5. Calvert P.D., Strissel K.J., Schiesser W.E., Pugh E.N., Arshavsky V.Y.: Light-driven translocation of signaling proteins in vertebrate photoreceptors. *Trends in Cell Biology* 2006, 16(11):560–568.
6. Januschka M.M., Burkhardt D.A., Erlandsen S.L., Purple R.L.: The ultrastructure of cones in the walleye retina. *Vision Research* 1987, 27(3):327–341.
7. Makino C.L., Wen X-H., Michaud N.A., Covington H.I., DiBenedetto E., Hamm H.E., Lem J., Caruso G.: Rhodopsin expression level affects rod outer segment morphology and photoresponse kinetics. *PLoS One* 2012, 7(5):e37832.
8. Horwich A.L.: Protein folding in the cell: An inside story. *Nature Medicine* 2011, 17(10):1211–1216.
9. Birkedal V., Dong M.D., Golas M.M., Sander B., Andersen E.S., Gothelf K.V., Besenbacher F., Kjems J.: Single molecule microscopy methods for the study of DNA origami structures. *Microscopy Research and Technique* 2011, 74(7):688–698.
10. Saaem I., Labean T.H.: Overview of DNA origami for molecular self-assembly. *Wiley Interdisciplinary Reviews Nanomedicine and Nanobiotechnology* 2013, 5(2): 150–162.
11. Kinoshita M.: Importance of translational entropy of water in biological self-assembly processes like protein folding. *International Journal of Molecular Sciences* 2009, 10(3):1064–1080.
12. Wang S., Shaevitz J.W.: The mechanics of shape in prokaryotes. *Frontiers in Bioscience* 2013, 5:564–574.

13. Kondo T., Hayashi S.: Mitotic cell rounding accelerates epithelial invagination. *Nature* 2013, 494(7435):125–129.
14. Tolbert C.E., Burridge K., Campbell S.L.: Vinculin regulation of F-actin bundle formation: What does it mean for the cell? *Cell Adhesion and Migration* 2013, 7(2):219–225.
15. Subramanian R., Kapoor T.M.: Building complexity: Insights into self-organized assembly of microtubule-based architectures. *Development Cell* 2012, 23(5):874–885.
16. Vanden Berg-Foels W.S., Scipioni L., Huynh C., Wen X.: Helium ion microscopy for high-resolution visualization of the articular cartilage collagen network. *Journal of Microscopy* 2012, 246(2):168–176.
17. Boden S.A., Asadollahbaik A., Rutt H.N., Bagnall D.M.: Helium ion microscopy of lepidoptera scales. *Scanning* 2012, 34(2):107–120.
18. Hlawacek G., Veligura V., Lorbek S., Mocking T.F., George A., van Gastel R., Zandvliet H.J., Poelsema B.: Imaging ultra thin layers with helium ion microscopy: Utilizing the channeling contrast mechanism. *Beilstein Journal of Nanotechnology* 2012, 3:507–512.
19. Kryuchkov M., Katanaev V.L., Enin G.A., Sergeev A., Timchenko A.A., Serdyuk I.N.: Analysis of micro- and nano-structures of the corneal surface of *Drosophila* and its mutants by atomic force microscopy and optical diffraction. *PloS One* 2011, 6(7):e22237.
20. Joy D.C., Griffin B.J.: Is microanalysis possible in the helium ion microscope? Microscopy and microanalysis. *The Official Journal of Microscopy Society of America, Microbeam Analysis Society, Microscopical Society of Canada* 2011, 17(4):643–649.
21. Koch K., Barthlott W.: Superhydrophobic and superhydrophilic plant surfaces: An inspiration for biomimetic materials. *Philosophical Transactions Series A, Mathematical, Physical, and Engineering Sciences* 2009, 367(1893):1487–1509.
22. Boseman A., Nowlin K., Ashraf S., Yang J., Lajeunesse D.: Ultrastructural analysis of wild type and mutant *Drosophila melanogaster* using helium ion microscopy. *Micron* 2013, 51:26–35.
23. Vincent J.F., Wegst U.G.: Design and mechanical properties of insect cuticle. *Arthropod Structure and Development* 2004, 33(3):187–199.
24. Andersen S.O.: Regional differences in degree of resilin cross-linking in the desert locust, *Schistocerca gregaria*. *Insect Biochemistry Molecular Biology* 2004, 34(5):459–466.
25. Beament J.W.: The water relations of insect cuticle. *Biology Review Cambridge Philosophical Society* 1961, 36:281–320.
26. Galbreath R.A.: Water balance across the cuticle of a soil insect. *Journal of Experimental Biology* 1975, 62(1):115–120.
27. Neumann D., Woermann D.: Physical conditions for trapping air by a microtrichia-covered insect cuticle during temporary submersion. *Naturwissenschaften* 2009, 96(8):933–941.
28. Willis J.H., Muthukrishnan S.: Insect cuticle. Foreword. *Insect Biochemistry Molecular Biology* 2010, 40(3):165.
29. Wigglesworth V.B.: The insect cuticle. *Biology Review Cambridge Philosophical Society* 1948, 23(4):408–451.
30. Muller M., Olek M., Giersig M., Schmitz H.: Micromechanical properties of consecutive layers in specialized insect cuticle: The gula of *Pachnoda marginata* (*Coleoptera, Scarabaeidae*) and the infrared sensilla of *Melanophila acuminata* (*Coleoptera, Buprestidae*). *Journal of Experimental Biology* 2008, 211(Pt 16):2576–2583.
31. Hamodrakas S.J., Willis J.H., Iconomidou V.A.: A structural model of the chitin-binding domain of cuticle proteins. *Insect Biochemistry Molecular Biology* 2002, 32(11):1577–1583.
32. Klocke D., Schmitz H.: Water as a major modulator of the mechanical properties of insect cuticle. *Acta Biomaterialia* 2011, 7(7):2935–2942.

33. Cribb B.W., Lin C.L., Rintoul L., Rasch R., Hasenpusch J., Huang H.: Hardness in arthropod exoskeletons in the absence of transition metals. *Acta Biomaterialia* 2010, 6(8):3152–3156.
34. Schofield R.M., Nesson M.H., Richardson K.A., Wyeth P.: Zinc is incorporated into cuticular "tools" after ecdysis: The time course of the zinc distribution in "tools" and whole bodies of an ant and a scorpion. *Journal of Insect Physiology* 2003, 49(1):31–44.
35. Wolfgang W.J., Fristrom D., Fristrom J.W.: An assembly zone antigen of the insect cuticle. *Tissue Cell* 1987, 19(6):827–838.
36. Yoon C.S., Hirosawa K., Suzuki E.: Corneal lens secretion in newly emerged *Drosophila melanogaster* examined by electron microscope autoradiography. *Journal of Electron Microscopy (Tokyo)* 1997, 46(3):243–246.
37. Gangishetti U., Breitenbach S., Zander M., Saheb S.K., Muller U., Schwarz H., Moussian B.: Effects of benzoylphenylurea on chitin synthesis and orientation in the cuticle of the *Drosophila* larva. *European Journal of Cell Biology* 2009, 88(3):167–180.
38. Nagaraj R., Adler P.N.: Dusky-like functions as a Rab11 effector for the deposition of cuticle during *Drosophila* bristle development. *Development* 2012, 139(5):906–916.
39. Neville A.C., Parry D.A., Woodhead-Galloway J.: The chitin crystallite in arthropod cuticle. *Journal of Cell Science* 1976, 21(1):73–82.
40. Charles J.P.: The regulation of expression of insect cuticle protein genes. *Insect Biochemistry Molecular Biology* 2010, 40(3):205–213.
41. Papandreou N.C., Iconomidou V.A., Willis J.H., Hamodrakas S.J.: A possible structural model of members of the CPF family of cuticular proteins implicating binding to components other than chitin. *Journal of Insect Physiology* 2010, 56(10):1420–1426.
42. Donoughe S., Crall J.D., Merz R.A., Combes S.A.: Resilin in dragonfly and damselfly wings and its implications for wing flexibility. *Journal of Morphology* 2011, 272(12):1409–1421.
43. Li L., Teller S., Clifton R.J., Jia X., Kiick K.L.: Tunable mechanical stability and deformation response of a resilin-based elastomer. *Biomacromolecules* 2011, 12(6):2302–2310.
44. Appel E., Gorb S.N.: Resilin-bearing wing vein joints in the dragonfly *Epiophlebia superstes*. *Bioinspiration and Biomimetics* 2011, 6(4):046006.
45. Andersen S.O.: Studies on resilin-like gene products in insects. *Insect Biochemistry Molecular Biology* 2010, 40(7):541–551.
46. Charati M.B., Ifkovits J.L., Burdick J.A., Linhardt J.G., Kiick K.L.: Hydrophilic elastomeric biomaterials based on resilin-like polypeptides. *Soft Matter* 2009, 5(18):3412–3416.
47. Iconomidou V.A., Willis J.H., Hamodrakas S.J.: Unique features of the structural model of "hard" cuticle proteins: Implications for chitin–protein interactions and cross-linking in cuticle. *Insect Biochemistry Molecular Biology* 2005, 35(6):553–560.
48. Willis J.H.: Structural cuticular proteins from arthropods: Annotation, nomenclature, and sequence characteristics in the genomics era. *Insect Biochemistry Molecular Biology* 2010, 40(3):189–204.
49. Cornman R.S., Willis J.H.: Annotation and analysis of low-complexity protein families of *Anopheles gambiae* that are associated with cuticle. *Insect Molecular Biology* 2009, 18(5):607–622.
50. Cornman R.S., Togawa T., Dunn W.A., He N., Emmons A.C., Willis J.H.: Annotation and analysis of a large cuticular protein family with the R&R consensus in *Anopheles gambiae*. *BMC Genomics* 2008, 9:22.
51. Wang P., Granados R.R.: Molecular structure of the peritrophic membrane (PM): Identification of potential PM target sites for insect control. *Archives of Insect Biochemistry Physiology* 2001, 47(2):110–118.
52. Hegedus D., Erlandson M., Gillott C., Toprak U.: New insights into peritrophic matrix synthesis, architecture, and function. *Annual Review Entomology* 2009, 54:285–302.

53. Teets N.M., Peyton J.T., Colinet H., Renault D., Kelley J.L., Kawarasaki Y., Lee R.E., Denlinger D.L.: Gene expression changes governing extreme dehydration tolerance in an Antarctic insect. *Proceedings of the National Academy of Sciences of the United States of America* 2012, 109(50):20744–20749.

54. Clark M.S., Thorne M.A.S., Purac J., Burns G., Hillyard G., Popovic Z.D., Grubor-Lajsic G., Worland M.R.: Surviving the cold: Molecular analyses of insect cryoprotective dehydration in the Arctic springtail *Megaphorura arctica (Tullberg)*. *BMC Genomics* 2009, 10:1–19.

55. Clark T.M., Flis B.J., Remold S.K.: pH tolerances and regulatory abilities of freshwater and euryhaline *Aedine* mosquito larvae. *Journal of Experimental Biology* 2004, 207(13):2297–2304.

56. Hu H.M.S., Watson G.S., Cribb B.W., Watson J.A.: Non-wetting wings and legs of the cranefly aided by fine structures of the cuticle. *Journal of Experimental Biology* 2011, 214(6):915–920.

57. Balmert A., Bohn H.F., Ditsche-Kuru P., Barthlott W.: Dry under water: Comparative morphology and functional aspects of air-retaining insect surfaces. *Journal of Morphology* 2011, 272(4):442–451.

58. Dickerson A.K., Shankles P.G., Madhavan N.M., Hu D.L.: Mosquitoes survive raindrop collisions by virtue of their low mass. *Proceedings of the National Academy of Sciences of the United States of America* 2012, 109(25):9822–9827.

59. Watson J.A., Cribb B.W., Hu H.M., Watson G.S.: A dual layer hair array of the brown lacewing: Repelling water at different length scales. *Biophysical Journal* 2011, 100(4):1149–1155.

60. Cho E.H., Nijhout H.F.: Development of polyploidy of scale-building cells in the wings of *Manduca sexta*. *Arthropod Structure and Development* 2013, 42(1):37–46.

61. Plotkin M., Hod I., Zaban A., Boden S.A., Bagnall D.M., Galushko D., Bergman D.J.: Solar energy harvesting in the epicuticle of the oriental hornet (*Vespa orientalis*). *Naturwissenschaften* 2010, 97(12):1067–1076.

62. Zeng J., Xiang N., Jiang L., Jones G., Zheng Y., Liu B., Zhang S.: Moth wing scales slightly increase the absorbance of bat echolocation calls. *PloS One* 2011, 6(11):e27190.

63. Usui-Ishihara A., Ghysen A., Kimura K.: Peripheral axonal pathway and cleaning behavior are correlated in *Drosophila microchaetes*. *Developmental Biology* 1995, 167(1):398–401.

64. Fei X., He B., Adler P.N.: The growth of *Drosophila* bristles and laterals is not restricted to the tip or base. *Journal of Cell Science* 2002, 115(Pt 19):3797–3806.

65. Guild G.M., Connelly P.S., Ruggiero L., Vranich K.A., Tilney L.G.: Long continuous actin bundles in *Drosophila* bristles are constructed by overlapping short filaments. *The Journal of Cell Biology* 2003, 162(6):1069–1077.

66. Tilney L.G., Connelly P.S., Ruggiero L., Vranich K.A., Guild G.M.: Actin filament turnover regulated by cross-linking accounts for the size, shape, location, and number of actin bundles in *Drosophila* bristles. *Molecular Biology of the Cell* 2003, 14(10):3953–3966.

67. Tilney L.G., Connelly P., Smith S., Guild G.M.: F-actin bundles in *Drosophila* bristles are assembled from modules composed of short filaments. *The Journal of Cell Biology* 1996, 135(5):1291–1308.

68. Kim M.C., Kim C., Wood L., Neal D., Kamm R.D., Asada H.H.: Integrating focal adhesion dynamics, cytoskeleton remodeling, and actin motor activity for predicting cell migration on 3D curved surfaces of the extracellular matrix. *Integrative Biology-UK* 2012, 4(11):1386–1397.

69. Schwarz U.S., Gardel M.L.: United we stand—Integrating the actin cytoskeleton and cell-matrix adhesions in cellular mechanotransduction. *Journal of Cell Science* 2012, 125(13):3051–3060.

70. Cant K., Knowles B.A., Mooseker M.S., Cooley L.: *Drosophila* singed, a fascin homolog, is required for actin bundle formation during oogenesis and bristle extension. *The Journal of Cell Biology* 1994, 125(2):369–380.

71. Tang C., Lennon E.M., Fredrickson G.H., Kramer E.J., Hawker C.J.: Evolution of block copolymer lithography to highly ordered square arrays. *Science* 2008, 322(5900):429–432.

72. Dalby M.J., Giannaras D., Riehle M.O., Gadegaard N., Affrossman S., Curtis A.S.: Rapid fibroblast adhesion to 27 nm high polymer demixed nano-topography. *Biomaterials* 2004, 25(1):77–83.

73. Park K.C., Choi H.J., Chang C.H., Cohen R.E., McKinley G.H., Barbastathis G.: Nanotextured silica surfaces with robust superhydrophobicity and omnidirectional broadband supertransmissivity. *Acs Nano* 2012, 6(5):3789–3799.

74. Bhushan B., Jung Y.C., Niemietz A., Koch K.: Lotus-like biomimetic hierarchical structures developed by the self-assembly of tubular plant waxes. *Langmuir* 2009, 25(3):1659–1666.

75. Koch K., Ensikat H.J.: The hydrophobic coatings of plant surfaces: Epicuticular wax crystals and their morphologies, crystallinity and molecular self-assembly. *Micron* 2008, 39(7):759–772.

76. Locke M.: Pore canals and related structures in insect cuticle. *Journal of Biophysics Biochemistry Cytology* 1961, 10(4):589–618.

77. Jarau S., Zacek P., Sobotnik J., Vrkoslav V., Hadravova R., Coppee A., Vasickova S., Jiros P., Valterova I.: Leg tendon glands in male bumblebees (*Bombus terrestris*): Structure, secretion chemistry, and possible functions. *Naturwissenschaften* 2012, 99(12):1039–1049.

78. Kuitunen K., Gorb S.N.: Effects of cuticle structure and crystalline wax coverage on the coloration in young and old males of *Calopteryx splendens* and *Calopteryx virgo*. *Zoology* 2011, 114(3):129–139.

79. Bernhard C.G., Gemne G., Moller A.R.: Modification of specular reflexion and light transmission by biological surface structures. *Quarterly Reviews of Biophysics* 1968, 1(1):89–105.

80. Watson G.S., Myhra S., Cribb B.W., Watson J.A.: Putative functions and functional efficiency of ordered cuticular nanoarrays on insect wings. *Biophysical Journal* 2008, 94(8):3352–3360.

81. Park K.C., Choi H.J., Chang C.H., Cohen R.E., McKinley G.H., Barbastathis G.: Nanotextured silica surfaces with robust superhydrophobicity and omnidirectional broadband supertransmissivity. *Acs Nano* 2012, 6(5):3789–3799.

82. Stavenga D.G., Foletti S., Palasantzas G., Arikawa K.: Light on the moth-eye corneal nipple array of butterflies. *Proceedings of the Biological Sciences/The Royal Society* 2006, 273(1587):661–667.

83. Bernhard C.G., Miller W.H., Moller A.R.: Function of the corneal nipples in the compound eyes of insects. *Acta Physiologica Scandinavica* 1963, 58:381–382.

84. Gemne G.: Ultrastructural ontogenesis of cornea and corneal nipples in the compound eye of insects. *Acta Physiologica Scandinavica* 1966, 66(4):511–512.

85. Ding Y., Xu S., Wang Z.L.: Structural colors from *Morpho peleides* butterfly wing scales. *Journal of Applied Physics* 2009, 106(7), 074702:1–6.

86. Saranathan V., Osuji C.O., Mochrie S.G., Noh H., Narayanan S., Sandy A., Dufresne E.R., Prum R.O.: Structure, function, and self-assembly of single network gyroid (I4132) photonic crystals in butterfly wing scales. *Proceedings of the National Academy of Sciences of the United States of America* 2010, 107(26):11676–11681.

87. Mark G.I., Vertesy Z., Kertesz K., Balint Z., Biro L.P.: Order–disorder effects in structure and color relation of photonic-crystal-type nanostructures in butterfly wing

scales. *Physical Review E, Statistical, Nonlinear, and Soft Matter Physics* 2009, 80(5 Pt 1):051903.

88. Parker A.R.: 515 million years of structural colour. *Journal of Optical a-Pure Applied Optics* 2000, 2(6):R15–R28.

89. Shevtsova E., Hansson C., Janzen D.H., Kjaerandsen J.: Stable structural color patterns displayed on transparent insect wings. *Proceedings of the National Academy of Sciences of the United States of America* 2011, 108(2):668–673.

90. Simonis P., Vigneron J.P.: Structural color produced by a three-dimensional photonic polycrystal in the scales of a longhorn beetle: *Pseudomyagrus waterhousei* (*Coleoptera*: *Cerambicidae*). *Physical Review E, Statistical, Nonlinear, and Soft Matter Physics* 2011, 83(1 Pt 1):011908.

91. Kolle M., Salgard-Cunha P.M., Scherer M.R., Huang F., Vukusic P., Mahajan S., Baumberg J.J., Steiner U.: Mimicking the colourful wing scale structure of the *Papilio blumei* butterfly. *Nature Nanotechnology* 2010, 5(7):511–515.

92. Wiesendanger S., Zilk M., Pertsch T., Rockstuhl C., Lederer F.: Combining randomly textured surfaces and photonic crystals for the photon management in thin film microcrystalline silicon solar cells. *Optics Express* 2013, 21 (Suppl 3):A450–A459.

93. Murphy B.P., Cuddy H., Harewood F.J., Connolley T., McHugh P.E.: The influence of grain size on the ductility of micro-scale stainless steel stent struts. *Journal of Materials Science Materials in Medicine* 2006, 17(1):1–6.

94. Wu C.W., Kong X.Q., Diane W.: Micronanostructures of the scales on a mosquito's legs and their role in weight support. *Physics Review E* 2007, 76(1):017301.

95. Sun M., Watson G.S., Zheng Y., Watson J.A., Liang A.: Wetting properties on nanostructured surfaces of cicada wings. *Journal of Experimental Biology* 2009, 212(19):3148–3155.

96. Wisdom K.M., Watson J.A., Qu X., Liu F., Watson G.S., Chen C.H.: Self-cleaning of superhydrophobic surfaces by self-propelled jumping condensate. *Proceedings of the National Academy of Sciences of the United States of America* 2013, 110(20):7992–7997.

97. Bullock J.M.R., Drechsler P., Federle W.: Comparison of smooth and hairy attachment pads in insects: Friction, adhesion and mechanisms for direction-dependence. *Journal of Experimental Biology* 2008, 211(20):3333–3343.

98. Federle W.: Why are so many adhesive pads hairy? *Journal of Experimental Biology* 2006, 209(14):2611–2621.

99. Aristotle Balme D.M.: *History of Animals*. Books VII–X. Cambridge, M.A.: Harvard University Press; 1991.

100. Autumn K., Sitti M., Liang Y.A., Peattie A.M., Hansen W.R., Sponberg S., Kenny T.W., Fearing R., Israelachvili J.N., Full R.J.: Evidence for van der Waals adhesion in gecko setae. *Proceedings of the National Academy of Sciences of the United States of America* 2002, 99(19):12252–12256.

101. Hornyak G.L.: *Introduction to Nanoscience and Nanotechnology*. Boca Raton: CRC Press; 2009.

102. Autumn K., Liang Y.A., Hsieh S.T., Zesch W., Chan W.P., Kenny T.W., Fearing R., Full R.J.: Adhesive force of a single gecko foot-hair. *Nature* 2000, 405(6787):681–685.

103. Guo C., Sun J., Ge Y., Wang W., Wang D., Dai Z.: Biomechanism of adhesion in gecko setae. *Science China Life Sciences* 2012, 55(2):181–187.

104. Dirks J.H., Federle W.: Mechanisms of fluid production in smooth adhesive pads of insects. *Journal of the Royal Society Interface* 2011, 8(60):952–960.

105. Federle W., Riehle M., Curtis A.S., Full R.J.: An integrative study of insect adhesion: Mechanics and wet adhesion of pretarsal pads in ants. *Integrative Computer Biology* 2002, 42(6):1100–1106.

106. Kesel A.B., Martin A., Seidl T.: Getting a grip on spider attachment: An AFM approach to microstructure adhesion in arthropods. *Smart Material Structure* 2004, 13(3): 512–518.

107. Kovalev A.E., Filippov A.E., Gorb S.N.: Insect wet steps: Loss of fluid from insect feet adhering to a substrate. *Journal of the Royal Society, Interface/The Royal Society* 2013, 10(78):20120639.

108. Bullock J.M., Federle W.: Beetle adhesive hairs differ in stiffness and stickiness: *In vivo* adhesion measurements on individual setae. *Naturwissenschaften* 2011, 98(5):381–387.

109. Yao H., Della Rocca G., Guduru P.R., Gao H.: Adhesion and sliding response of a biologically inspired fibrillar surface: Experimental observations. *Journal of the Royal Society Interface* 2008, 5(24):723–733.

110. Bartlett M.D., Croll A.B., King D.R., Paret B.M., Irschick D.J., Crosby A.J.: Looking beyond fibrillar features to scale gecko-like adhesion. *Advanced Materials* 2012, 24(8):1078–1083.

111. Gould S.J., Lewontin R.C.: Spandrels of San-Marco and the panglossian paradigm—A critique of the adaptationist program. *Proceedings of the Royal Society B—Biology Science* 1979, 205(1161):581–598.

112. Gould S.J.: The narthex of San Marco and the pangenetic paradigm. *Nature History* 2000, 109(6):24–37.

113. Ivanova E.P., Hasan J., Webb H.K., Truong V.K., Watson G.S., Watson J.A., Baulin V.A. et al.: Natural bactericidal surfaces: Mechanical rupture of *Pseudomonas aeruginosa* cells by cicada wings. *Small* 2012, 8(16):2489–2494.

114. Hasan J., Webb H.K., Truong V.K., Pogodin S., Baulin V.A., Watson G.S., Watson J.A., Crawford R.J., Ivanova E.P.: Selective bactericidal activity of nanopatterned superhydrophobic cicada *Psaltoda claripennis* wing surfaces. *Applied Microbiology and Biotechnology* 2013, 97(20):9257–9262.

115. Pogodin S., Hasan J., Baulin V.A., Webb H.K., Truong V.K., Phong Nguyen T.H., Boshkovikj V. et al.: Biophysical model of bacterial cell interactions with nanopatterned cicada wing surfaces. *Biophysical Journal* 2013, 104(4):835–840.

Section III

Nano Medicine

7 Current Nanodelivery Systems for Imaging and Therapeutics

Effat Zeidan, Stephen Vance,
and Marinella G. Sandros

CONTENTS

7.1 INTRODUCTION

Expanding on recent progress in control of matter at the nanoscale, nanoparticles have impinged tremendous advancement on their application as delivery vehicles for therapeutic and imaging agents. More specifically, nanodelivery vehicles have offered selectivity to targeted organ, tissue, and specific cell type. Besides harboring a payload of therapeutic drugs, functionalized nanoparticles facilitate targeting only diseased tissue and minimizing exposure to healthy tissue. Strategies aimed at further improving nanotherapeutic carriers' accumulation within diseased tissue are under intensive investigation and could enable, in a not too distant future, the realization of the long sought after treatment of cancer, cardiovascular, and neurological diseases with or without reduced side effects.

In vivo imaging provides real-time tracking of nanodelivery vehicles; however, the complexities presented in *in vivo* systems impose several challenges such as penetration across the vascular endothelium, control on their bioaccumulation, and excretion. The former challenge impedes effectiveness of nanoparticle delivery from the blood stream into targeted tissue and the latter on the efficacy of their toxicity. Thus far, the greatest breakthrough has been the ability of nanocarriers to cross the blood–brain barrier, a tightly packed layer of endothelial cells surrounding the brain, offering common neurological disorders, such as stroke, tumors, and Alzheimer's means for treatment.

7.2 NANOPARTICLES

The application of nanoparticles in the field of medicine offered and continues to offer promising improvements to the current health care system. It provides pertinent information on biological processes to improve diagnosis and therapy. Early diagnosis of most diseases has been shown to increase the survival rate and enhance the therapeutic effect of the patient. In therapeutics, nanocarriers are considered to be promising vehicles for improving the efficiency of drugs due to their exceptional properties that allow for both controlled and site-specific delivery of the drug. Compared to bulk material, nanoparticles have a high size-to-volume ratio offering a decrease in the dosage of the drug and the toxicity level to healthy tissue. In regards to optimal characteristics of nanoparticles, the size and surface charge were studied in order to depict the effect of these two factors on the uptake by tumor tissue and permeability through the blood–brain barrier. And the results showed that nanoparticles of size 100 nm or larger were subject to opsonization; whereas, hydrophilic nanoparticles with a particle size smaller than 100 nm offered a higher biological distribution, improved site-specific targeting and effect of the drug [1]. Additionally, the zeta potential of the nanoparticles should be addressed because it represents the surface charge and the repulsive forces existing between the nanoparticles. Therefore, as the repulsive interactions increase between the nanoparticles, the nanosuspension will become more stable and a more uniform interaction with the cells *in vivo* is attained; hence, increasing the circulation time at the site of the disease.

Traditional therapies in the fight against cancer have been limited in effectiveness mainly due to the nonspecific mode of action and the reduced bioavailability of the drug inside tumor tissue. In this mode of drug delivery, the anticancer drugs induce tumor cell death; however, exhibit high levels of toxicity to the healthy tissue as well. These major drawbacks have instigated the integration of nanoparticles with drugs to provide more effective cancer treatment.

Nanotechnology has also presented solutions to challenges residing in neurotherapeutics; for example, controlling drug release at the target site and facilitating permeability through the blood–brain barrier. By large, neurodegenerative diseases include disorders characterized by the deterioration of neurons, cells in the brain and spinal cord. This damage leads to altered physical and mental behavior such as dementia, uncoordinated movement, and impaired cognitive abilities. The brain possesses a self-protective mechanism due to two major barriers along the linings of the cerebral blood capillaries. These two barriers, the blood–brain barrier and the blood cerebrospinal fluid barrier, protect the central nervous system against microscopic foreign material, selectively allowing nutrients and only small hydrophobic molecules into the brain tissue. In fact, the permeability of the blood–brain barrier depends on the lipophilicity and the molecular weight of the drug. Nonionized and low-molecular weight drugs are allowed to diffuse through the barrier; whereas, other compounds require special carriers for delivery [2]. Hence, inefficiency in the conventional drug delivery strategies has led to extensive research for alternative delivery mechanisms.

Scientists utilize nanotechnology in the aim of achieving a better understanding of the molecular stages of the disease and a more desirable therapeutic result. Nanoparticles constructed to deliver antitumor drugs such as paclitaxel were successful in enhancing the therapeutic effect of the drug due to their small size that can permeate through the vasculatures in cancerous tissue. In fact, this advantage is due mostly to a phenomenon known as enhanced permeability and retention (EPR), defined by an increased permeability of the vasculatures in cancer tissue that strive for nutrients and oxygen [3]. A major advantage of a nanoparticle is that it has been shown capable of crossing the blood–brain barrier *in vivo* and *in vitro* through openings in the tight junctions existing between the endothelial cells. The promising nanosystems include nonmetal nanocarriers such as liposomes, polymeric micelles, dendrimers, polymer–drug conjugate nanoparticles, and metal-based such as superparamagnetic iron oxide nanoparticles, quantum dots, gold, and upconverting nanoparticles.

In this chapter, we will review current metal and nonmetal-based nanocarriers, namely, compare the various synthetic techniques and conjugation strategies and survey their application as therapeutics for cardiovascular diseases, cancer, and neurological disorders.

7.3 METALLIC NANOPARTICLES

Metal nanoparticle systems form an attractive area of research in both diagnosis and treatment of cancerous tissue. Due to their exceptional optical properties and surface chemistry, quantum dots, iron oxide, rare earth metal, and gold nanoparticles demonstrate tissue-selective targeting strategies for a more controlled therapeutic application.

7.4 QUANTUM DOTS

Quantum dots are a special class of semiconductors that are smaller in size than the Bohr radius of the particle and as a result their properties differ significantly from bulk semiconductors [4]. To describe its properties, the linear combination

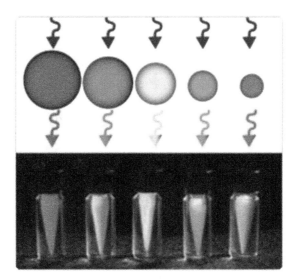

FIGURE 7.1 **(See color insert.)** This figure illustrates the tunability of quantum dots. Five different nanocrystal solutions are shown excited with the same long-wavelength UV lamp; the size of the nanocrystal determines the color. (Reproduced from Life Technologies. www. invitrogen.com/site/us/en/home/brands/Molecular-Probes/Key-Molecular-Probes-Products/ Qdot/Technology-Overview.html. With permission.)

of atomic orbitals–molecular orbitals (LCAO–MO) must be used since the particle is between the size of bulk materials and individual molecules [5]. This infers that as the particle size decreases its energy bands become discrete and the band gap increases [4]. Different size quantum dots can be excited with UV light and will fluoresce from near infrared to blue depending on the size of the particle and composition (Figure 7.1) [6]. Quantum dots can be synthesized by either the conventional hot injection technique or the more recently developed microwave synthesis.

7.4.1 Quantum Dot Synthesis

Hot injection nucleation occurs by a rapid injection of metal organics into a hot solvent that raises the reactants concentration above a nucleation threshold. This super saturation of reactants is relieved by nucleation of the metals [7]. Additionally, these nucleation sites are used to grow larger nanoparticles during which some of the smaller nanoparticles dissolve back into solution and are reformed as part of the larger ones [8]. This growth results in larger particles with a narrow spread of size [9].

The standard method of producing trioctylphosphine oxide capped CdSe quantum dots is performed by using the hot injection method. The first method involves using trioctylphosphine oxide as the solvent that is heated under an inert atmosphere to 300°C. Dimethylcadmium and selenium powder are dissolved in trioctylphosphine and injected into the hot trioctylphosphine oxide [8]. The solution is then heated slowly such that the nucleation site growth will be uniform. The longer the solution is heated the larger the quantum dots will grow and the longer the emission wavelength will be. The second method which is similar involves using tributylphosphine instead of trioctylphosphine to dissolve the powders [10]. Implications of these methods limit the nucleation of nanoparticles to occur only under an extremely high temperature and inert atmosphere. In addition, controlling the monodispersity of the product is a challenge [11]. To alleviate some of the latter challenges, microwave-assisted synthesis is employed as an alternative method.

Microwave heating allows for faster and better control of nanoparticle sizes than conventional heating. Microwave radiation is concentrated at the center of a round-bottom flask and the heating occurs by dielectric polarization and by conduction [12]. Dielectric polarization is an effect that occurs when the microwave radiation oscillates and the molecules rotate to align their dipole with the oscillating field. This heating is considered to be superheating, where certain locations of the solution are able to go above its boiling point. The heating rate in microwave synthesis is much quicker than conventional heating as it heats by two methods and it is done directly onto the solution as opposed to heating the glass, which then transfers heat to the solution. As a result of the dielectric polarization, the molecules rotate and collide leading to an increase in the kinetics of the reaction. This method also requires less solvent and has direct control over the heating so that after the reaction is complete the heat can be reduced at a much higher rate, which inhibits potential unwanted reactions from occurring [12].

Trioctylphosphine oxide capped CdSe quantum dots were accomplished by microwave-assisted synthesis through dissolving cadmium acetate in trioctylphosphine oxide and hexadecylamine and separately dissolving selenium powder in trioctylphosphine. The two solutions are mixed together and then irradiated in a microwave apparatus. The reaction time takes no longer than a few minutes depending on the size of the quantum dot to be produced [11].

For biological applications, quantum dots need to be hydrophilic. CdTe nanoparticles with high quantum yield were prepared successfully using microwave-assisted technology by reacting $CdCl_2$ dissolved in 3-mercaptopropionic acid with NaHTe. The period of microwave irradiation controlled the production of nanoparticles with different sizes [13]. With microwave synthesis, it is possible to increase the particle size by altering heating time, heating temperature, or by increasing wattage without increasing temperature [14].

7.4.2 Conjugation Strategies

For nanoparticles to be used in *in vitro* or *in vivo* imaging, the ligands must be exchanged to render the nanoparticles water soluble. This is done by replacing non-polar capping agents with polar or ionic capping agents (Figure 7.2). Typically, for example, it is performed by reacting the trioctylphosphine oxide capped quantum dots with mercaptohexadecanoic acid and then the reactants are dissolved in dimethylformamide and reduced by potassium-t-butoxide [15]. Another method involves coating the quantum dot with a ZnS shell and capping it with mercaptopropionic acid [11].

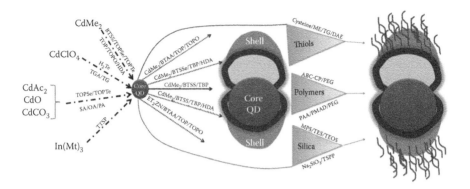

FIGURE 7.2 (See color insert.) A schematic illustration of the synthesis of core QDs, coating with materials to form a shell, and overlaying the core/shell QDs with additional layers to facilitate additional conjugation and improve biocompatibility. (SA, stearic acid; OA, oleic acid; TOP, trioctylphosphine; TOPO, trioctylphosphine oxide; BTSS, bis(trimethylsilyl)sulfide; TGA, thioglycolic acid; TG, thioglycerol; TBP, tributyl phosphine; HAD, hexadecyl amine; MPA, methiopropamine; DHLA, dihydrolipoic acid; PAA, polyacrylic acid; PMAD, poly(maleic anhydride alt-1-tetradecane); PEG, polyethylene glycol; MPS, 3-mercaptopropyl trimethoxysilane; TES, tetraethoxysilane; TEOS, tetra ethyl orthosilicate.) (Adapted from Biju, V., Itoh, T., and Ishikawa, M., *Chemical Society Reviews*, 39, 3031, 2010.)

7.4.3 APPLICATION OF QUANTUM DOTS

Quantum dots exhibit superior properties for serving as biological imaging tools and drug delivery vehicles. They are characterized by long fluorescent times, resistance to chemical degradation, and an absorption spectra ranging from the UV to the near infrared. One application employs the use of polyethylene glycol-poly (lactic acid) as a coating for quantum dots, which are further functionalized with wheat germ agglutinin rendering them water soluble and highly penetrable through the blood–brain barrier [16]. Similarly, cerium oxide nanoparticles exhibit interesting abilities enabling the inhibition of superoxide and hydroxyl radicals production, preventing neuronal death as well as blocking the radical formation by the amyloid beta and many more in the brain [17].

Heavy metal-based quantum dots have been extensively studied for biological imaging applications; however, they have been limited to *in vitro* applications mainly due to the toxicity of the incorporated elements like cadmium. To overcome this drawback, quantum dots that consist of a nontoxic element that can be easily cleared from the body such as silicon were synthesized, characterized, and examined for possible uptake by tumor cells. In a recent study, CdTe and CdSe quantum dots coated with a silica shell proved to be proper agents for photothermal therapy of cancer. In specific, red and brown quantum dots were examined and shown to convert laser radiation into heat both *in vivo* and *in vitro* (Figure 7.3). This resulted in the inhibition of melanoma growth in mouse models [18].

In another study, the silicon quantum dots were functionalized with biological molecules such as antimesothelin and folate that have been shown to be successful in targeting cancer cells. The proper conjugation with proteins and antibodies

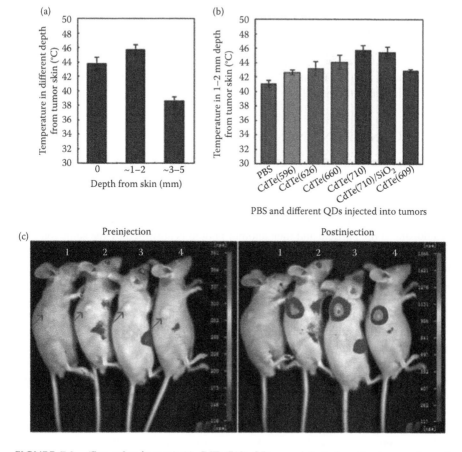

FIGURE 7.3   (See color insert.) (a) CdTe(710) QDs was injected on the tumor site and temperature at different depths from the irradiated surface are displayed after a 20 min irradiation. (b) A plot representation after 20 min irradiation of the temperature of the tumors injected with PBS, PBS-dispersed CdTe, CdTe/SiO₂ and CdSe QDs. (c) *In vivo* images of tumors (see arrows) pre- and postinjection of (1) CdTe(626), (2) CdTe(660), (3) CdTe(710) and (4) CdTe(710)/SiO₂ QDs. (Reproduced from Chu, M. et al., *Biomaterials*, 33, 7071, 2012. With permission.)

was achieved by a nontoxic carboxylic acid, undecylenic acid, through carbodiimide chemistry. The study resulted in the effective uptake of silicon quantum dots into pancreatic cancer cells by confocal microscopy emphasizing their potential as imaging tools [19].

Understanding the cardiac contractions at the molecular level is mainly dependent on knowing the average length of the sarcomere. Many studies have been done for calculating the average of that length and its variations in a single myocyte, which determines the influences on the mechanical properties of the sarcomere [20]. Due to their high efficiency, brightness, and fluorescing properties in imaging applications, quantum dots were examined as potential markers for determining the length of the sarcomere with a resolution of 30 nm. These quantum dots were functionalized with anti-α-actinin antibody that would allow it to bind to the Z disks of the sarcomere in a rat's ventricular muscles. The results also showed the advantages of using quantum dots as novel imaging agents under varying physiological conditions in a myocyte [21]. In their most recent study, their research went further to add additional insight to the dynamics of the sarcomere in a single intact or skinned myocyte under different physiological conditions [22].

7.5 UPCONVERTING NANOPARTICLES

7.5.1 SYNTHESIS OF UPCONVERTING NANOPARTICLES

Upconverting nanoparticles unlike quantum dots emit light with a higher energy than the excitation light. This antistokes emission is possible due to the rare earth metals in the particle. When an ion is excited by a photon the energy is transferred to a dopant in the particle and then another photon is absorbed by the same ion and energy is again transferred [23]. The dopant will then emit a higher energy photon. The dopant can be altered to allow for different emission. Typical dopants are erbium and thulium that result in 650 and 800 nm emission, respectively [24].

Cubic $NaYF_4$ nanoparticles doped with either Er^{3+}/Yb^{3+} or Tm^{3+}/Yb^{3+} have been synthesized through a thermal decomposition method consisting of trifluoroacetate precursors within a mixture of octadecene and oleic acid [25]. However, this approach involves long reaction times (1–2 days) in a pressurized reactor (i.e., autoclave). Another common chemical synthetic approach for UCNs is coprecipitation [26], which is less labor intensive than the thermal decomposition method. However, this technique requires postsynthetic treatments to enhance the crystallinity of the final product.

To have better control on UCNs crystallinity, reduce the length of production and monodispersity, microwave-assisted synthesis was found to be effective by using trifluoroacetate reactants (Figure 7.4) [24]. These reactants follow a classical La Mer mechanism [7] since they are polar, absorb microwave radiation efficiently, and undergo thermal decomposition with release of fluoride [24]. As a result the reaction can be controlled, readily allowing for high reproducibility and size control.

Many synthetic procedures [27–29] are available to exchange the oleic acid capping ligand on UCNs to render them water soluble. A specific example involves mixing UCNs with Igepal CO-520 in the presence of ammonia forming a clear emulsion that was then introduced to tetraethyl orthosilicate [30]. This procedure rendered the UCNs highly soluble in water and amenable for further surface functionalization. It is worth noting that the silica coating had no effect on the luminescence of the UCNs.

The UCNs have a benefit when used for biological samples in that the excitation wavelength is near infrared which is not absorbed by cells or tissue. As a result biological fluorophores are not excited and the excitation light can penetrate the sample deeper [23].

7.5.2 UPCONVERTING NANOPARTICLES INFLUENCE ON BIOIMAGING

Upconverting nanoparticles as previously described have been shown to be great candidates as noninvasive deep-tissue imaging agents. A recent study incorporated

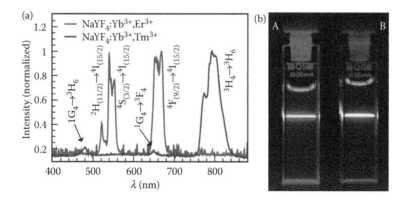

FIGURE 7.4 (**See color insert.**) (a) Emission spectrum of (red) $NaYF_4$:Yb^{3+},Er^{3+} nanocrystals and (green) $NaYF_4$:Yb^{3+},Tm^{3+} nanocrystals excited by a 980 nm laser. (b) A digital image of the upconversion luminescence of the rare-earth co-doped $NaYF_4$ nanocrystals in chloroform after excitation with a 980 nm laser diode (A) Luminescence of $NaYF_4$:Yb^{3+},Er^{3+} nanocrystals; (B) luminescence of $NaYF_4$:Yb^{3+},Tm^{3+} nanocrystals. (Reproduced from Wang, H.Q. and Nann, T., *ACS Nano*, 3, 3804, 2009. With permission.)

upconverting nanoparticles to a photosensitizer forming a nanotransducer [31], which displays an enhancement effect in photodynamic therapy of tumor tissue. This device is capable of overcoming the main obstacle facing the effectiveness of this therapy, which is deep-tissue accessibility by radiation. Mainly, the upconverting nanoparticles convert the near-infrared radiation to energy in the visible spectrum, which in turn excites the photosensitizer to convert molecular oxygen into toxic radicals. This important work emphasizes on near-infrared radiation, which is the most biocompatible form in imaging applications.

7.6 IRON OXIDE NANOPARTICLES

7.6.1 SYNTHESIS OF IRON OXIDE NANOPARTICLES

Iron oxide (Fe_2O_3) nanoparticles can be used as MRI contrast agents due to their ability to alter relaxation times of surrounding tissues [32]. Typically, iron oxide nanoparticles are coated with organic material that will provide biocompatibility and allow binding to drugs, proteins, and other molecules [33,34]. These particles are also designed to increase saturation magnetism values to allow for high MRI signal [35]. This value depends on the particle size, spacing between the particles, and the crystalline structure of the iron oxide [32]. Typically nanoparticles of 6–20 nm have the highest electromagnetic unit per gram (emu/g). In addition, enhancing their stability and minimizing their surface defects can also influence this value [32,36].

Fe_2O_3 nanoparticles can be synthesized by dissolving ferrous and ferric salts in a basic aqueous medium forming magnetite nanoparticles. The nanoparticles size and shape can be adjusted by altering pH, ion strength, temperature, the chemical composition of the salts, and ratio of Fe^{II} and Fe^{III} [37]. Higher pH and ionic strength can lead to smaller core size nanoparticles with a narrow size distribution [38,39]. Sun et al. [40] reported a mediated growth method to have a better control on size through seeding a small nanoparticle (>8 nm) in order to generate nanoparticles up to 16 nm.

7.6.2 Fe₂O₃ NANOPARTICLES AS THERANOSTIC AGENTS

Superparamagnetic iron oxide nanoparticles have attracted much attention as multifunctional theranostic agents in the field of cancer research. These promising vehicles are good contrast imaging agents and carry specific functional groups and therapeutic drugs that allow for a guided and controlled therapy of the tumor. SPION, which include mainly two forms, magnetite or maghemite, display significant magnetic moments in the presence of an external magnetic field. This property grants these nanoparticles the ability to be good MRI imaging contrast agents. However, magnetic nanoparticles are still facing a challenge as they require a large external magnetic gradient supplied by nearby large magnets to generate the sufficient force. The recent work was aimed at overcoming this drawback and improving the targeting of the magnetic nanoparticles by incorporating nickel micromeshes, which generate a large magnetic gradient once magnetized. Along with the micromesh, the superparamagnetic iron oxide nanoparticle is coated with a siliceous shell rendering it biocompatible and functionalized with an organic fluorophore for direct imaging of the vehicles *in vivo*. In conclusion, the study shows this system as a potential approach for early detection of cancer as well as guided therapy [41].

7.7   GOLD NANOPARTICLES

7.7.1 SYNTHESIS OF GOLD NANOPARTICLES

Gold nanoparticles are typically synthesized by using a two-phase reaction [42]. An aqueous solution of chloroauric acid and tetraethylammonium bromide dissolved in toluene was mixed together. The mixture is reacted with sodium borohydride and then suspended in toluene. A thiolated polyethylene glycol (PEG) derivative is then mixed into the solution, which allows for suspension in water [43]. The functional groups on PEG allow for the many binding affinities discussed earlier. Another method for synthesizing gold nanoparticles is by mixing chloroauric acid with hexadecyltrimethylammonium bromide (CTAB) and adding sodium borohydride. These nanoparticles can then be used in the production of gold nanorods. The seed solution is added to a solution of CTAB, silver nitrate, ascorbic acid, and chloroauric acid [44]. Finally, unlike their bulk and molecular counterparts, gold nanoparticles exhibit vivid colors that depend on the size and shape of the particle.

7.7.2 GOLD NANOPARTICLES FOR BIOSENSING AND THERMAL THERAPY

The sizes of the particles cause preferential accumulation at tumor sites, which allows for detection of cancerous cells by looking at the Raman scattering. When cancerous cells are incubated in the presence of gold nanorods, a surface-enhanced

Raman scattering occurs and the Raman peaks are intensified [45]. Gold nanoparticles also exhibit a two-photon luminescence that converts near-infrared into visible light, allowing for high spatial resolution without the chance of autofluorescence of cells [46,47].

Additionally, it is important to note the use of gold nanoparticles in killing tumor tissue through a number of strategies. Gold nanoshells are capable of destroying the tumor cells by photothermal therapy through excitation with near-infrared radiation [48].

7.8 NONMETAL-BASED NANOPARTICLES

7.8.1 LIPOSOMES

Liposomal formulations for antitumor drugs were one of the first nanosystems to be approved for clinical use; such as Doxil® which is a PEGylated liposome encapsulating doxorubicin. Further functionalization of Doxil® with monoclonal antibodies is currently in phase I clinical trial, and has proved to show more potential in targeting breast, ovarian, and multiple myeloma cancer cells. Moreover, liposomes are being employed in a relatively new approach for cancer targeting that designs antitumor drugs specific to the surface of tumor vasculature. Particularly, PEGylated liposomal formulation carrying the antimicrotubule agent paclitaxel, that has been previously mentioned, is again synthesized with specific functionalization and proved to reduce tumor growth both *in vivo* and *in vitro*. The formulation allowed for enhanced solubility of the drug as well as a more effective targeted delivery due to further functionalization of the liposome with alpha-integrins, which are overexpressed in tumor vessels [49].

Liposomes, artificial vesicles made of uni- or multilamellar lipid bilayers, are one of the first types of nanomaterials used for drug delivery. Hydrophilic drug molecules can be incorporated inside the aqueous structure in large amounts; however, liposomes are prone to clearance from circulation by the rectinoendothelial system. To overcome that problem and enhance circulation, liposomes can be coated with PEG and further functionalized with monoclonal antibodies for target-specific delivery in the brain. For instance, Citicoline, a drug designed for protecting the nervous system cells, showed an improved therapeutic effect after encapsulation within the liposomal cavity [16]. Multifunctional nanoparticles have attracted much attention for allowing better understanding of the different stages of a disease as well as guiding the procedure for effective delivery of the therapeutic agent. In this regard, gadolinium-loaded liposomes bound to adenoviral vectors have been shown to allow the tracking of the adenoviral vectors to three areas in the primate brain [2].

Thrombo-occlusion diseases and reduced blood circulation to the vital organs are the main consequences of vascular diseases and causes for organ damage. To date, the different forms of traditional medical treatments for revascularization have been successful; however, they have posed serious side effects on the patient. To reduce the side effects and allow for a specific targeted delivery of agents responsible for dissolving blood clots, conjugated liposomal carriers of thrombolytics have been formulated in a few studies. In this particularly interesting work, the liposome entrapped streptokinase in the aqueous cavity, coated with PEG and functionalized with RGD and P-selectin-targeted motifs, shows potential in protecting the streptokinase from degradation in the circulatory system and in targeting thrombo-occlusion sites. More importantly, it addresses how varying encapsulation techniques can affect the percent of drug encapsulation and opens the door for further examination of optimizing the efficiency of these nanocarriers [50]. Another application of liposomal vehicles is the encapsulation of hemoglobin; hence, improving cerebral artery thrombosis [16].

7.8.2 POLYMERS

A second class of nonmetal nanoparticles includes polymeric nanoparticles, which are synthesized from biodegradable natural polymers such as chitosan, albumin, or collagen and synthetic polymers such as PEG, polyglycolic acid (PGA), or polyaspartate (PAA).

Abraxane® designated for the treatment of ovarian, breast, head, and neck cancer is a formulation of nanoparticle-paclitaxel bound to albumin which renders the drug soluble and less toxic. Moreover, cyclodextrin-based nanoparticles are currently being examined for delivering small interference RNA into the nucleus of the tumor cells in the aim of silencing certain oncogenes involved in tumor growth and resistance to common therapies [51]. Similarly and employing the same posttranscriptional silencing mechanism, small interference RNA, specialized for inhibiting the expression of the oncoprotein Murine double Minute 1, were functionalized with a series of biocompatible agents for an enhanced delivery into the nucleus. This oncoprotein is expressed in high rates in breast tumor tissue and plays a role in rendering the carcinomas resistance to radiotherapy. Interestingly, the delivery vehicles incorporated an imaging component for tracking the fate of the therapeutic agents. The nanoplex system consisted of Streptavidin functionalized near-infrared quantum dots bound to the biotinylated small interference RNA which are further covered with N-acetyl histilated glycol chitosan polymer. The coating offers the vehicle biocompatibility and a more enhanced permeability into the cell. Additionally, the novelty of this vehicle is not solely due to the multifunctional properties of the nanoplex, but also to the chemical properties of the imidazole group in histidine, which once in the endosome gets protonated leading to the proton sponge effect [52] (Figure 7.5). As a result, this effect leads to the swelling and bursting of the endosomal cavity releasing the nanoplex into the cytosol which then permeate through the nuclear pores to the site of delivery [53].

Moreover, one interesting strategy developed by You et al. examined a nanodelivery system composed of a pH-sensitive paclitaxel release from poly (N,N-dimethylaminoethyl methacrylate (DMAEMA)/2-hydroxyethyl methacrylate (HEMA)) nanoparticles to the relatively acidic brain tumor microenvironment [54]. This technique employs hydrophilic surfactants that use a sensing mechanism in the target's microenvironment for specific delivery of the drug.

7.8.3 DENDRIMERS

Lewy body dementia is a particular neurodegenerative disorder characterized by the existence of cytoplasmic inclusions containing alpha-synuclein protein accumulations

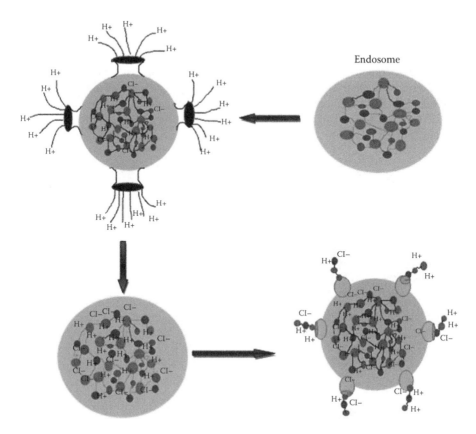

FIGURE 7.5 **(See color insert.)** An illustration of the proton sponge effect that enhances delivery of payloads (polyamines) into the cytosol of the cell trapped inside the endosomes by buffering H^+ and subsequent accumulation of Cl^-. (Adapted from Biju, V., Itoh, T., and Ishikawa, M., *Chemical Society Reviews*, 39, 3031, 2010.)

in the affected neurons. Patients diagnosed with this disorder have symptoms combining dementia from Parkinson's disease and Lewy bodies. Unfortunately, effective treatment of this disease has not been fully accomplished; however, current work involving nanocarrier systems present a potentially successful therapy. One example involves polyamidoamine dendrimers for the purpose of inhibiting the fibrillation of alpha-synuclein [55].

7.8.4 MICELLES

Another nonmetal-based vehicle used for the same purpose is comprised of active micelles attached to an anticoagulant drug and a fluorophore. These micelles bind to fibrin which is expressed on plaque and deliver a high concentration of the drug, Hirulog, resulting in a more controlled release of the drug to prevent clog formation [56]. Recently, Bastakoti et al. [57] engineered a core–shell–corona-type multifunctional polymeric micelles using asymmetric triblock copolymer (poly(styrene-acrylic acid ethylene glycol), PS-PAA-PEG) and imbedded a fluorescent probe (Nile Red) inside the core and an anticancer drug (cisplatin) inside the PAA shell. *In vitro* investigation using a cancer cell line showed that the drug is taken up into the cell and inhibited cell proliferation. Also multifunctional micelles have been widely used *in vivo* to deliver intranasally into the lungs MRI contrast agents together with a drug and genetic material [58] offering potential use in diagnosing and treating lung cancer. Using a mouse model, OX26-conjugated micelles were able to cross the blood–brain barrier [59] presenting a promising vehicle to treat a wide range of neurological diseases.

7.9   CONCLUSION

The application of metal and nonmetal-based nanoparticles has thus far made a huge impact in medicine enough to generate a new field itself known as "nanomedicine." This chapter surveyed a series of nanomaterials. The commonality between them involves the ability to overcome barriers with conventional therapy by providing direct transport to diseased areas and avoiding healthy tissue, controlled drug release, and an enhanced mode of administration. These improved multifunctional capabilities featured within nanocarriers are attributed to charge, increase in surface area, size, and composition. We look forward to their potential application in the clinical phase. Perhaps, not all vehicles will make it into clinical trials as they could pose some toxicity or inability to clear the body safely. Nevertheless, they could still serve as excellent and sophisticated tools to promote a better understanding of diseases.

REFERENCES

1. Sahni, J.K. et al., Neurotherapeutic applications of nanoparticles in Alzheimer's disease, *Journal of Controlled Release*, 152, 208, 2011.
2. Bhaskar, S. et al., Multifunctional nanocarriers for diagnostics, drug delivery and targeted treatment across blood–brain barrier: Perspectives on tracking and neuroimaging, *Part Fibre Toxicology*, 7, 3, 2010.
3. Chakraborty, M., Jain, S., and Rani, V., Nanotechnology: Emerging tool for diagnostics and therapeutics, *Applied Biochemistry and Biotechnology*, 165, 1178, 2011.
4. Wang, Y. and Herron, N., Nanometer-sized semiconductor clusters: Materials synthesis, quantum size effects, and photophysical properties, *The Journal of Physical Chemistry*, 95, 525, 1991.
5. Burdett, J.K., From bonds to bands and molecules to solids, *Progress in Solid State Chemistry*, 15, 173, 1984.
6. Life Technologies. www.invitrogen.com/site/us/en/home/brands/Molecular-Probes-Key-Molecular-Probes-Products/Qdot/Technology-Overview.html
7. LaMer, V.K. and Dinegar, R.H., Theory, production and mechanism of formation of monodispersed hydrosols, *Journal of the American Chemical Society*, 72, 4847, 1950.
8. Murray, C.B., Norris, D.J., and Bawendi, M.G., Synthesis and characterization of nearly monodisperse CdE (E = sulfur, selenium, tellurium) semiconductor nanocrystallites, *Journal of the American Chemical Society*, 115, 8706, 1993.
9. Murray, C.B., Kagan, C.R., and Bawendi, M.G., Synthesis and characterization of monodisperse nanocrystals and close-packed nanocrystal assemblies, *Annual Review of Materials Science*, 30, 545, 2000.

10. Katari, J.E.B., Colvin, V.L., and Alivisatos, A.P., X-ray photoelectron spectroscopy of CdSe nanocrystals with applications to studies of the nanocrystal surface, *The Journal of Physical Chemistry*, 98, 4109, 1994.
11. Roy, M.D. et al., Emission-tunable microwave synthesis of highly luminescent water soluble CdSe/ZnS quantum dots, *Chemical Communications*, 0, 2106, 2008.
12. Patel, D. and Patel, B., Microwave assisted organic synthesis: An overview, *Journal of Pharmacy Research*, 4, 2090, 2011.
13. Li, L., Qian, H., and Ren, J., Rapid synthesis of highly luminescent CdTe nanocrystals in the aqueous phase by microwave irradiation with controllable temperature, *Chemical Communications*, 0, 528, 2005.
14. Gerbec, J.A. et al., Microwave-enhanced reaction rates for nanoparticle synthesis, *Journal of the American Chemical Society*, 127, 15791, 2005.
15. Sandros, M.G. et al., InGaP A versatile fluorescent probe for deep-tissue imaging, *Advanced Functional Materials*, 17, 3724, 2007.
16. Nair, S., Dileep, A., and Rajanikant, G., Nanotechnology based diagnostic and therapeutic strategies for neuroscience with special emphasis on ischemic stroke, *Current Medicinal Chemistry*, 19, 744, 2012.
17. Spuch, C., Saida, O., and Navarro, C., Advances in the treatment of neurodegenerative disorders employing nanoparticles, *Recent Patents on Drug Delivery & Formulation*, 6, 2, 2012.
18. Chu, M. et al., The therapeutic efficacy of CdTe and CdSe quantum dots for photothermal cancer therapy, *Biomaterials*, 33, 7071, 2012.
19. Erogbogbo, F. et al., Bioconjugation of luminescent silicon quantum dots for selective uptake by cancer cells, *Bioconjugate Chemistry*, 22, 1081, 2011.
20. Bub, G. et al., Measurement and analysis of sarcomere length in rat cardiomyocytes in situ and *in vitro*, *American Journal of Physiology-Heart and Circulatory Physiology*, 298, H1616, 2010.
21. Serizawa, T. et al., Real-time measurement of the length of a single sarcomere in rat ventricular myocytes: A novel analysis with quantum dots, *American Journal of Physiology Cell Physiology*, 301, C1116, 2011.
22. Kobirumaki-Shimozawa, F. et al. Sarcomere imaging by quantum dots for the study of cardiac muscle physiology, *Journal of Biomedicine and Biotechnology*, 2012, 7, 2012.
23. Wang, H.Q. and Nann, T., *Upconverting Nanoparticles*, Springer, Berlin Heidelberg, 2011.
24. Wang, H.Q. and Nann, T., Monodisperse upconverting nanocrystals by microwave-assisted synthesis, *ACS Nano*, 3, 3804, 2009.
25. Boyer, J.C., Cuccia, L.A. and Capobianco, J.A., Synthesis of colloidal upconverting NaYF4: Er3+/Yb3+ and Tm3+/Yb3+ monodisperse nanocrystals, *Nano Letters*, 7, 847, 2007.
26. Yi, G. et al., Synthesis, characterization, and biological application of size-controlled nanocrystalline NaYF4:Yb,Er infrared-to-visible up-conversion phosphors, *Nano Letters*, 4, 2191, 2004.
27. Boyer, J.C. et al., Surface modification of upconverting NaYF4 nanoparticles with PEG – phosphate ligands for NIR (800 nm) biolabeling within the biological window, *Langmuir*, 26, 1157, 2009.
28. Jiang, G. et al., An effective polymer cross-linking strategy to obtain stable dispersions of upconverting NaYF4 nanoparticles in buffers and biological growth media for biolabeling applications, *Langmuir*, 28, 3239, 2012.
29. Zhang, Q. et al., Hexanedioic acid mediated surface–ligand-exchange process for transferring NaYF4:Yb/Er (or Yb/Tm) up-converting nanoparticles from hydrophobic to hydrophilic, *Journal of Colloid and Interface Science*, 336, 171, 2009.
30. Wilhelm, S. et al., Multicolor upconversion nanoparticles for protein conjugation, *Theranostic*, 3(4), 239, 2012.

31. Chatterjee, D.K. and Yong, Z., Upconverting nanoparticles as nanotransducers for photodynamic therapy in cancer cells, *Nanomedicine (Lond)*, 3, 73, 2008.
32. Lodhia, J. et al., Development and use of iron oxide nanoparticles (Part 1): Synthesis of iron oxide nanoparticles for MRI, *Biomedical Imaging & Intervention Journal*, 6, 1, 2010.
33. Gupta, A.K. and Gupta, M., Synthesis and surface engineering of iron oxide nanoparticles for biomedical applications, *Biomaterials*, 26, 3995, 2005.
34. Jun, Y.W., Lee, J.H., and Cheon, J., *Nanoparticle Contrast Agents for Molecular Magnetic Resonance Imaging*, Wiley-VCH Verlag GmbH & Co. KGaA, Weinheim, Germany, 2007.
35. Wang, Y.-X., Hussain, S., and Krestin, G., Superparamagnetic iron oxide contrast agents: Physicochemical characteristics and applications in MR imaging, *European Radiology*, 11, 2319, 2001.
36. Jun, Y.W. et al., Nanoscale size effect of magnetic nanocrystals and their utilization for cancer diagnosis via magnetic resonance imaging, *Journal of the American Chemical Society*, 127, 5732, 2005.
37. Laurent, S. et al., Magnetic iron oxide nanoparticles: Synthesis, stabilization, vectorization, physicochemical characterizations, and biological applications, *Chemical Reviews*, 108, 2064, 2008.
38. Vayssieres, L. et al., Size tailoring of magnetite particles formed by aqueous precipitation: An example of thermodynamic stability of nanometric oxide particles, *Journal of Colloid and Interface Science*, 205, 205, 1998.
39. Jiang, W. et al., Preparation and properties of superparamagnetic nanoparticles with narrow size distribution and biocompatible, *Journal of Magnetism and Magnetic Materials*, 283, 210, 2004.
40. Sun, S. and Zeng, H., Size-controlled synthesis of magnetite nanoparticles, *Journal of the American Chemical Society*, 124, 8204, 2002.
41. Fu, A. et al., Fluorescent magnetic nanoparticles for magnetically enhanced cancer imaging and targeting in living subjects, *ACS Nano*, 6(8), 6862, 2012.
42. Tshikhudo, T.R., Wang, Z., and Brust, M., Biocompatible gold nanoparticles, *Materials Science & Technology*, 20, 980, 2004.
43. Templeton, A.C., Wuelfing, W.P., and Murray, R.W., Monolayer-protected cluster molecules, *Accounts of Chemical Research*, 33, 27, 1999.
44. Nikoobakht, B. and El-Sayed, M.A., Preparation and growth mechanism of gold nanorods (NRs) using seed-mediated growth method, *Chemistry of Materials*, 15, 1957, 2003.
45. Dreaden, E.C. et al., The golden age: Gold nanoparticles for biomedicine, *Chemical Society Reviews*, 41, 2740, 2012.
46. Boyd, G.T., Yu, Z.H., and Shen, Y.R., Photoinduced luminescence from the noble metals and its enhancement on roughened surfaces, *Physical Review B*, 33, 7923, 1986.
47. Zhou, Y. et al., A comparison study of detecting gold nanorods in living cells with confocal reflectance microscopy and two-photon fluorescence microscopy, *Journal of Microscopy*, 237, 200, 2010.
48. Huttunen, K.M. et al., Cytochrome P450-activated prodrugs: Targeted drug delivery, *Current Medicinal Chemistry*, 15, 2346, 2008.
49. Meng, S. et al., Integrin-targeted paclitaxel nanoliposomes for tumor therapy, *Medical Oncology*, 28, 1180, 2011.
50. Holt, B. and Gupta, A.S., Streptokinase loading in liposomes for vascular targeted nanomedicine applications: Encapsulation efficiency and effects of processing, *Journal of Biomaterials Applications*, 26, 509, 2012.
51. Egusquiaguirre, S. et al., Nanoparticle delivery systems for cancer therapy: Advances in clinical and preclinical research, *Clinical and Translational Oncology*, 14, 83, 2012.

52. Biju, V., Itoh, T., and Ishikawa, M., Delivering quantum dots to cells: Bioconjugated quantum dots for targeted and nonspecific extracellular and intracellular imaging, *Chemical Society Reviews*, 39, 3031, 2010.
53. Azari, F., Sandros, M.G., and Tabrizian, M., Self-assembled multifunctional nanoplexes for gene inhibitory therapy, *Nanomedicine*, 6, 669, 2011.
54. Wang, C.X. et al., Antitumor effects of polysorbate-80 coated gemcitabine polybutyl-cyanoacrylate nanoparticles *in vitro* and its pharmacodynamics *in vivo* on C6 glioma cells of a brain tumor model, *Brain Research*, 1261, 91, 2009.
55. Milowska, K., Malachowska, M., and Gabryelak, T., PAMAM G4 dendrimers affect the aggregation of α-synuclein, *International Journal of Biological Macromolecules*, 48, 742, 2011.
56. Peters, D. et al., Targeting atherosclerosis by using modular, multifunctional micelles, *Proceedings of the National Academy of Sciences*, 106, 9815, 2009.
57. Bastakoti, B.P. et al., Multifunctional core-shell-corona-type polymeric micelles for anticancer drug-delivery and imaging, *Chemistry—A European Journal*, 19(15), 4812, 2013.
58. Howell, M. et al., Manganese-loaded lipid-micellar theranostics for simultaneous drug and gene delivery to lungs, *Journal of Controlled Release*, 167(2), 210, 2013.
59. Yue, J. et al., Fluorescence-labeled immunomicelles: Preparation, *in vivo* biodistribution, and ability to cross the blood–brain barrier, *Macromolecular Bioscience*, 12, 1209, 2012.

8 Nanodevices and Systems for Clinical Diagnostics

Smith Woosley, Jun Yan, and Shyam Aravamudhan

CONTENTS

8.1 INTRODUCTION

Conventional clinical diagnostic technologies aim to detect the symptoms of a disease. These methods include measurements of a particular antibody produced by the body in response to foreign body infection or observation of a specific bacterium known to cause a disease. Such methods are considered to be slow and inefficient as they involve identifying or measuring a disease after the person has contracted it. In addition, most current clinical analyzers are dedicated only to a single class of analytes and are hampered by bulky, expensive, and laboratory-sized instrumentation, which is not amenable to use in portable applications or in remote locations. Recent advances in nanotechnology-based diagnostics could potentially alleviate some of the current limitations in clinical diagnosis and subsequently in therapeutic systems. This chapter will discuss methods in point-of-care (PoC) diagnostics, the current status, and future trends in nanodevices and systems for clinical diagnostics. Nanodevices and systems, when implemented within the current diagnostic technology, have the potential to analyze the entire genome in minutes instead of hours. For example, based on which DNA (deoxyribonucleic acid) sequence is deviated from the normal sequence, the doctor will be able to determine a person's predisposition to a specific disease. This chapter will also review the current research in microfluidic, lab-on-chip (LoC), and biochips or bio-nano-electro-mechanical systems (bio-NEMS), where a number of complex diagnostic procedures, such as sample collection, sample pretreatment, sample preparation, target detection, postprocessing, and waste disposal, are integrated into one platform so that PoC diagnosis is implemented. Finally, we will discuss some of the challenges and safety aspects of nanodiagnostic systems.

8.2 CLINICAL DIAGNOSTICS

Clinical diagnosis is the identification of a disease based on a study of its signs and symptoms. These studies are known as diagnostic tests, and they pose questions that aim to provide medical information about the condition of a patient. These diagnostic tests may include a combination of an assessment of the patient's medical history, a physical examination, radiological imaging, laboratory investigations, and/or surgical operations [1]. One example of a familiar clinical diagnostic technique is the use of imaging methods to search for the cancerous tissue in a patient. This process usually begins when a patient experiences symptoms and seeks medical assistance. After a review of the patient's medical history and a physical examination, the doctor may suspect cancer as the cause for the symptoms and recommend an imaging scan to search for signs of the disease. If signs are found, a surgical biopsy of the tissue may be performed for verification and to help doctors decide on treatment options. Unfortunately, such methods rely on technologies that detect the symptoms of the disease and, therefore, rely on advanced stages of the disease (cancer) for accurate diagnosis. By this time, the patient has begun to feel the negative effects of the disease as its severity worsens; the cancer/disease would have spread to the other parts of the body (metastasis). In addition, conventional diagnostic techniques can often lead to a variety of diagnoses. Because of this, medical experts often rely on a method known as differential diagnosis. This method is performed by comparing and contrasting clinical findings in a process of elimination to narrow down possibilities to the diagnosis that best fits the test results. However, this method can require an excessive amount of tests, which adds to the total health-care cost and increases time from initial symptoms to the beginning of the treatment. Furthermore, these methods can be invasive, thus adding surgical complications to an already at-risk patient.

8.2.1 BIOMARKERS

An ideal diagnostic technique should detect the disease noninvasively in its early stages, before it presents any significant harm to the patient. Recent advances in *omics* methods such as genomics, proteomics, metabolomics, and glycomics have increased the information on the identification and study of disease-related biomarkers. The National Cancer Institute (NCI) defines a biomarker as "a biological molecule found in blood, other body fluids, or tissues that is a sign of a normal or abnormal process, or of a condition or disease." By identifying biomarkers that are present in conjunction with certain diseases, researchers can develop diagnostic tests to monitor and quantify levels of the markers, leading to accurate clinical diagnoses. Biomarkers also have many other potential applications including risk assessment, screening, differential diagnosis, determination of prognosis, prediction of the response to treatment, and monitoring of progression of the disease [2]. Even though a large number of biomarker "discoveries" (20,000 cancer and 6000 cardiac) have been reported, approximately one biomarker per year had received U.S. FDA (Food and Drug Administration) approval between 1995 and 2005 [3]. The biomarkers come in a wide variety of forms, including proteins, nucleic acids, antibodies, metabolites, peptides, human cells, and microbes [2]. To provide accurate diagnoses, the appropriate biomarker must be matched with its corresponding and specific disease. Toward this goal, researchers have been developing a library of known disease biomarkers [4–6]. This biomarker library aims to provide a catalog of markers that researchers may use to develop diagnostic tests. The important biomarker characteristics in this library include their type (e.g., protein and DNA), location (e.g., blood, saliva, and urine), and concentration. Cancer is one disease where biomarkers have been studied extensively. A wide range of the types and causes of cancer combined with the potential for late onset of symptoms from the disease makes biomarkers a highly desirable option for clinical diagnosis. Cancer biomarker diagnosis has the potential to identify specific types of cancer in their early stages; thus, it can prevent disease metastasis and can avoid additional challenges caused by late-stage detection. Table 8.1 presents a variety of applications for cancer biomarkers in diagnostic testing and a small subset of the cancer biomarker library [2]. In addition, biomarker diagnostics can even determine the risk of developing certain cancers before they occur. For example, mutations of the BRCA1 gene in women indicate an increased likelihood of developing breast or ovarian cancer [7]. At-risk women can undergo a diagnostic test to detect the mutation. If this biomarker is present, then

TABLE 8.1
Cancer Biomarker and Its Application

Application	Example of a Biomarker
To estimate risk of developing cancer	BRCA1 germline mutation (breast and ovarian cancer)
To screen	PSA (prostate cancer)
For differential diagnosis	Immunohistochemistry to determine the tissue of origin
To determine disease prognosis	21 gene recurrence score (breast cancer)
To predict response to therapy	KRAS mutation/anti-EGFR antibody (colorectal cancer)
	HER2 expression/anti-Her2 therapy (breast/gastric cancer)
	Estrogen receptor expression (breast cancer)
To monitor for disease recurrence	CEA (colorectal cancer)
	AFP, LDH, and βHCG (germ cell tumor)
To monitor response or progression in metastatic disease	CA15-3 and CEA (breast cancer)

Source: Adapted from Henry, N.L. and Hayes, D.F., *Mol. Oncol.*, 6, 140, 2012.

these women have an option to reduce their risk of developing the cancer through intensive screening or chemoprevention [2].

8.2.2 Detection of Biomarkers

There are a number of techniques currently available to detect biomarkers. Examples include southern and western blots [8], enzyme immunoassays (EIA) and enzyme-linked immunosorbent assays (ELISA) [9], mass spectrometry [10], and radioimmunoassays [11]. These methods have become common practice for the biomarker detection in modern laboratories. For example, western blot and ELISA tests are used to analyze the patient's blood for antibodies of a specific disease or virus. However, these techniques are burdened with a number of limitations. These include complicated procedures requiring a well-appointed centralized laboratory with trained personnel using specialized equipment to carry out multistep processes. For instance, a southern blot requires restriction enzyme treatment, gel electrophoresis, and *in situ* denaturation, each of which requires specific laboratory equipment and operational expertise [8]. This leads to delays from testing to diagnosis and limited access to the test in low-resource settings and remote areas. Cost is another issue with the current biomarker detection techniques. The necessary reagents and equipment add significant health-care overhead to these diagnostic processes. For example, the cost of one ILDR1 ELISA kit (which detects the biomarker for cancer progression) is $1165, excluding the equipment and specialized personnel to run the test. To alleviate some of these drawbacks, researchers have pursued a technology known as biosensors.

IUPAC defines a biosensor as "a device that uses specific biochemical reactions mediated by isolated enzymes, immunosystems, tissues, organelles, or whole cells to detect chemical compounds usually by electrical, thermal, or optical signals" [12]. A biosensor works in methods similar to traditional detection techniques, but carries out all functions in a single-device platform rather than using the infrastructure of a modern laboratory. By integrating the capabilities of biosensors into a diagnostic test, traditional testing methods drawbacks can be overcome. Some biosensors have already been successfully commercialized for clinical applications such as the electrochemical blood glucose sensor [13]. An ideal diagnostic biosensor needs to satisfy a number of requirements, including specificity, which is the ability to selectively choose a particular biomarker from a variety of other interferences. Next, sensitivity should be high, as these molecules are often in ultralow concentrations (<pg/L) in the human body. High sensitivity will ensure that there will be no missed diagnoses resulting from the inability to detect the biomarker. A fast response time is also necessary. Traditional testing methods have had long time constants and delays between screening and treatment. Lastly, an ideal diagnostic biosensor should enable quantification of the desired biomarker. This characteristic will ensure that the disease severity is known to the physician determining the treatment options. Molecular biosensors are generally preferred as a clinical diagnosis compared to other methods because of their capability for real-time measurements, rapid diagnosis, multitarget analyses, automation, and reduced costs [14].

8.3 POINT-OF-CARE DIAGNOSTIC SYSTEMS

An ideal PoC device, along with its diagnostic biosensor and other ancillary components is shown in Figure 8.1. A PoC device incorporates components, such as microfluidic sample preparation and handling elements, a biosensor, signal processor, power source, and a data display, into one handheld or desktop platform. PoC diagnostics have been extensively reviewed in recent literature, both from the point of view of application and development [3,15,16]. In this chapter, we will briefly review some of the advantages of PoC diagnostics, involving micro total analysis systems (μTAS) and nanodiagnostic systems. Gubala et al. [16] described three main advantages with PoC diagnostics: (1) time, (2) patient responsibility and compliance, and (3) cost. PoC tests return results quickly, thus minimizing the time between testing, diagnosis, and treatment. By empowering individuals to do their own tests, patient compliance (adherence to diagnosis and treatment regimens) can be improved. A recent study of the cost-effectiveness of PoC testing reveals significant increase in testing regularity and adherence to prescribed medications, along with improvements in clinical outcomes [16]. Lastly, as both reagent and test sample usage is efficient in PoC, the total cost of the diagnostic test is reduced. However, as stated earlier, there are several challenges and requirements that need to be addressed. The biosensor must be adequately sensitive and selective while maintaining a small footprint to enable integration within the portable device. A PoC device also requires intricate systems such as microfluidics to provide all desired transport and manipulation features. Table 8.2 summarizes some of the key factors and attributes of a successful PoC or microfluidic device implementation [17]. In recent years, advances in microfluidics and nanotechnology-based biosensors have enabled progress to meet some of these requirements and attributes. Aravamudhan et al. [18] showed that nanoscale biosensors can enable selective quantification of ultra-low-concentration biomarkers within a portable PoC device. The integration of nanobiosensors with microfluidics

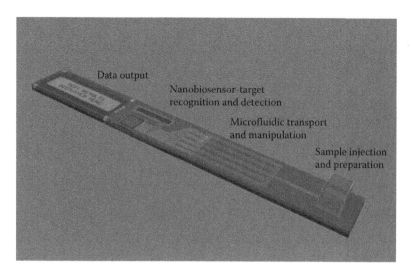

FIGURE 8.1 (**See color insert.**) Schematic of a portable PoC diagnostic system.

TABLE 8.2
POC-Enabling Factors

Key Factors	Key Attributes
Technology cost	Low cost
Degree of accuracy	High degree
Quality control	Reproducible performance
Level of training required	Interface with little training
Time to result	Relatively short time
Performance	Stable and low-power consumption
Performance under variable conditions	Reproducible operation and ruggedness

Source: Adapted from Yager, P. et al. *Nature,* 442(7101), 412–418, 2006.

and LoC technology will lead to the development of accurate and cost-effective PoC diagnostic tests, while eliminating the drawbacks of traditional clinical methods.

8.3.1 Microfluidics and Lab-on-Chip Technology

Microfluidic technology, the field of miniaturized fluid flow and control devices, will be one of the main enabling factors for the development of PoC diagnostics. Microfluidic devices have been developed and used for a wide variety of biomedical and analytical applications [19–22]. This technology not only enables a scaledown of macro-sized laboratory functions to a miniaturized single-platform format but also integration and automation of various unit biochemical operations [23,24]. The development of microfluidic technology started in the 1950s, when efforts were made to dispense ultralow volumes of liquids for use in ink-jet printers [25]. In the following years, various single-flow and multiflow unit operations have been developed including fluid transport, metering, mixing, valving, switching, and separation of samples (Figure 8.2). During the last decade, microfluidic devices have been

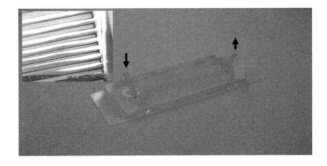

FIGURE 8.2   (**See color insert.**) The image shows a passive microfluidic sample preparation section with inset showing the polydimethylsiloxane (PDMS) microchannels.

incorporated for clinical diagnostics (for blood-based and body fluid-based assays) by including on-chip optical or electrochemical sensors. Microfluidic systems easily surmount many shortcomings associated with bulky and complex conventional bioanalytical methods in terms of consuming ultralow sample or reagent volumes, shorter assay times, and the potential for portable operation. This technology also raises the possibility of developing PoC systems particularly for molecular diagnosis of diseases. Microfluidic technology has also enabled the development of LoC systems.

LoC is a device that can perform specific laboratory functions using microfluidic components created on a chip [26]. This chip is usually millimeters to a few centimeters in size. LoC systems are also the basis for PoC diagnostics. For example, Vo-Dinh et al. [27] developed a multifunctional biosensor–chip that combined two different types of bioreceptors–nucleic acid probes and antibody probes on a single platform to measure DNA probes specific to gene fragments of *Bacillus anthracis* and antibody probes targeted to *Escherichia coli*. In other examples, portable microfluidic systems have been developed to measure physiological or metabolic parameters such as cholesterol, glucose, and lactate [18,28]. However, one of the significant challenges in large-scale implementation of high-throughput LoC systems has been the lack of reliable fluid manipulation and integration strategies [26]. Toward this goal, various researchers have explored active fluidic control such as microvalves, micromixers, and micropumps to move and manipulate fluids [29,30]. However, active systems remain difficult to implement because of high cost, complex fabrication procedures, and the difficulty of integrating with complementary technologies. However, passive microfluidics such as capillary-driven fluid manipulation is very attractive for portable PoC applications because of no external power requirements. For example, the ready-to-use test strips (for diabetes and pregnancy testing) use capillary-driven mechanisms for fluid manipulation [23]. Recently, Whitesides' group used paper-based substrates to realize capillary-driven microfluidic systems. This system promises to be low cost, easy to use, and portable [31]. The advantages of passive microfluidics make them highly viable for developing LoC platforms for biochemical analysis. The other important factor for an LoC is the choice of substrate material. Recently, there has been considerable effort to employ substrates other than silicon or glass. Particularly, paper, plastic, or polymer-based substrates [32,33] due to their low cost and ease of fabrication are very attractive. The other advantages include a range of surface modifications that can be carried out to address various manipulation and compatibility requirements of the PoC detection system [34]. Digital microfluidic–biosensor systems have also been developed to analyze biomarkers that are characteristic of liver injury, soft-tissue injury, and abdominal trauma. All these examples illustrate the robustness and potential for LoC systems in PoC molecular diagnostics with its excellent properties, such as miniaturization, integration, and automation. In summary, LoC systems create new opportunities for real-time and PoC detection technologies. However, effective system integration strategy that is both low cost and broadly applicable is essential [35]. Several promising system integration approaches for LoC integration have been proposed, including capillary-driven microfluidics, multilayer soft lithography, multiphase microfluidics, electrowetting-on-dielectric, electrokinetics, and centrifugal microfluidics.

8.4 NANOTECHNOLOGY IN DIAGNOSTICS

PoC diagnostic systems require small, fast, low cost, and easy-to-operate compo-nents. Nanotechnology can provide a means to meet these requirements through the implementation of microfluidics, nanobiosensors, and other system integra-tion components. PoC devices can consist of several subsystems, including sample preparation, biosensors, signal processing, and possible on-chip power source. The following section will discuss the current trends in nanotechnology for all these subsystems.

8.4.1 NANOTECHNOLOGY-BASED SAMPLE PREPARATION

PoC systems often deal with unrefined samples directly from the patient, such as blood, urine, or saliva. It is important to filter and/or concentrate these samples to enable thorough downstream analysis. Sample preparation is normally carried out using centrifuges and filtration systems. This equipment is generally not available on site, limiting the use of PoC devices. Nanotechnology can play an important role in sample preparation to enable PoC diagnostics. One example of nanotechnol-ogy in sample preparation is the use of nanofiltration. An important task in many diagnostic tests is extracting DNA from a sample. Kim and Gale [36] integrated a nanoporous aluminum oxide membrane in a microfluidic device to extract human genomic DNA from whole blood. The study used flow and ionic disruption to elute human genomic DNA from the porous membrane under different combinations of lysis solution concentration, nanopore size, and elution solution. The optimized com-bination was a low-salt lysis solution, 100 nm pore size, and a cationic elution buffer. These extracted human genomic DNA samples were then amplified using poly-merase chain reaction (PCR) and no inhibition effects were found. Another example is the use of vertically aligned nanowires for their large surface-to-volume ratio (large binding surface at small volume). Krivitsky et al. [37] developed an on-chip protein sample preparation platform that integrates filtering, selective separating, desalting, and preconcentrating. In this study, nanosphere lithography and metal-assisted chemical etching (MACE) were used to fabricate Si nanowires, which were later functionalized by antibody receptors. When the whole blood flow passed over the functionalized nanowires, the targeted protein bound to the receptors. It is evi-dent that nanotechnology can make on-chip sample preparation simpler and faster compared to the lab-based methods.

8.4.2 NANOTECHNOLOGY-BASED TRANSDUCTION AND BIOSENSORS

Nanobiosensors and their transduction mechanisms are the key technologies that will ultimately enable PoC diagnostic systems. Nanotechnology methods can aid PCR units, which most existing molecular diagnostic systems still need to amplify nucleic acid signals [38]. This is because the DNA/RNA is normally at low concen-trations in clinical samples. However, PCR adds complication and cost to the PoC system. Proteins, on the other hand cannot be amplified, which makes detection of low concentration of proteins difficult. Nanotechnology methods can help lower this limit of detection and increase the detection sensitivity, which may ultimately negate the need for amplification. Nanobiosensors for PoC systems can also advance trans-duction technology. Instead of using conventional optics and electronic systems, which are power hungry, nanotechnology methods can enable low-power visual and electrical transduction systems.

8.4.2.1 Nanoparticle-Based Transduction and Biosensors

Nanoparticles are particles around 1–100 nm in diameter. Their small sizes confer a unique set of electrical and optical properties and phenomenon. Nanoparticles can be engineered from a number of materials, including metals, semiconductors, and biomaterials and in a variety of shapes and forms, such as rods and spheres. In this section, we will review two widely used nanoparticles for diagnostics, namely gold and magnetic nanoparticles (MNPs).

One implementation method for gold nanoparticles (AuNPs) is to functionalize them with DNA. When another single-strand DNA is added to this solution, different degrees of aggregation can be expected, which may result in different interparticle distance and aggregate size. This in turn can lead to differences in color because the absorption frequency of the surface plasmon band of metal nanoparticles is a function of interparticle distance and aggregate size. Elghanian et al. [39] exploited this unique phenomenon to develop a colorimetric detection method, with detection of 10 fmol for an oligonucleotide. Storhoff et al. [40] demonstrated the differen-tiation of single-base mismatch using "melting point" measurements. Recently, by using AuNPs, Kalidasan et al. [41] directly detected *Salmonella* genomic DNA in femtomolar range with just a change in color. This method requires only minimum resources and room-temperature operation, a highly desirable feature for usage in limited-resource settings. Mancuso et al. [42] implemented a multiplexed colorimet-ric detection system, which detected Kaposi's sarcoma-associated herpesvirus and Bartonella's DNA (with nM resolution).

MNPs also have the ability to bind to proteins or DNA with suitable function-alization scheme. Chung et al. [43] demonstrated a PoC magneto-DNA platform, which simultaneously detected 13 bacterial species within 2 h. First, the total RNA from the bacteria was extracted and amplified by asymmetric RT-PCR. Then, the single-strand DNA from amplification was captured using beads functionalized with capture probes. Next, the nucleotide-modified MNPs were added to form a magnetic sandwich complex. Finally, a micro-NMR (µNMR) system [44] was used to ana-lyze the resultant complex. The micro-NMR technique relies on the decrease in the bulk spin–spin relaxation time of water molecules, when a bioevent (target binding) results in aggregation of MNPs [44]. The downshift of relaxation time can be mea-sured by using an on-board circuit.

Finally, MNPs can also be combined with giant magnetoresistive (GMR) to create PoC systems. The principle here is to measure the local magnetic field changes (as electrical resistance) when MNPs bind to the sensor surface [45]. Gaster et al. [46] developed a battery-powered hand-held platform based on MNP–GMR technology, where human immunodeficiency virus (HIV) p24 proteins were detected at concen-trations down to picograms per milliliter.

8.4.2.2 NEMS/MEMS Biosensors

Nano-/micro-electro-mechanical systems (N/MEMS) are widely considered powerful candidates for clinical diagnostics because they provide easy integration with microfluidics, fast screening (high-density array), high resolution (up to a single molecule), label-free sensing, and a broad range of biorecognition possibilities. These systems can be fabricated on the nano-/microscale using conventional nano- and microfabrication technologies. Their small sizes make them ideal as low mass resonator devices. These devices function by detecting small amounts of mass absorbed on a cantilever, which results in a shift in its resonant frequency; the smaller (lower mass) the device, the higher the resolution. Since these devices can directly measure biomolecules or other species (label-free sensing) absorbed on it, they do not need intermediary labels (such as fluorescent molecules). The biorecognition event can be realized by coating the device with monolayer or multilayer molecules that specifically bind to the analyte of interest. High-density arrays of such MEMS/NEMS cantilevers coated with different layers can be fabricated, which can create high throughput and multiplexing. MEMS/NEMS fabrication is also compatible with microfluidics integration, which can further reduce diagnosis time and cost. Fritz et al. [47], using a cantilever device, detected a single-base mismatch between two 12-mer oligonucleotides. The silicon cantilever was first coated with Au on one side. Thiol-modified oligonucleotides were then bound to gold to form the receptor. The device was then immersed in a hybridization buffer. The bending of the cantilever (down to 0.1 nm) was measured *in situ* by measuring the reflection of an incident laser. To estimate the quantity of an absorbed species, Stoney's formula has been used to convert the measured bending into the amount of molecular coverage [48]. Varshney et al. [49] fabricated a silicon nitride cantilever resonator with multistep surface treatment to create reactive amine groups on the nitride surface. Gluteraldehyde was used to bind amine groups on the cantilever and primary amines on the antibodies. The addition of biotinylated secondary antibodies to streptavidin-conjugated nanoparticles (addition of more mass) decreased the limit of detection to 2 ng/mL. In similar literature, 50 fg/mL prostate-specific antigen (PSA) [50] and 200 pg/mL of PrPc [51] were detected in blood serum.

The second example of NEMS diagnostic biosensor is the application of field effect transistors (FETs), as shown in Figure 8.3. An FET device has a backside gate, source, drain, and a channel between the source and drain. FETs sense conductance changes caused by electrical properties (charge carrier density and mobility) in the channel. Biomolecules binding to an FET channel can change their electrical properties. The selectivity of the sensor is decided by the specificity of the binding agent to

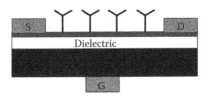

FIGURE 8.3 Schematic of an FET biosensor.

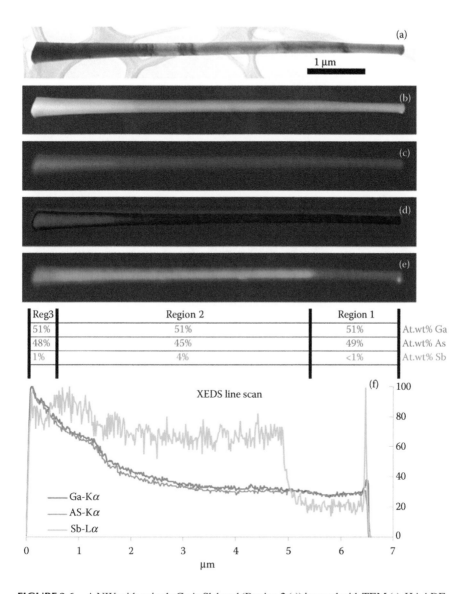

FIGURE 3.6 A NW with a single GaAsSb band (Region 2 (e)) imaged with TEM (a), HAADF–STEM (b), and XEDS–STEM mapping (c) through (e) of Ga, As, and Sb, respectively. An XEDS line scan (f) shows the relative elemental x-ray count distribution along the NW axis. The collective x-ray count increases with NW diameter in the direction of the NW base.

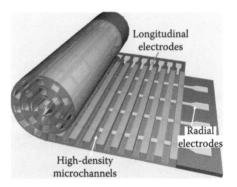

FIGURE 4.1   ReME interface is shown as both rolled 3D structure and partially unrolled 2D microchannels with electrodes.

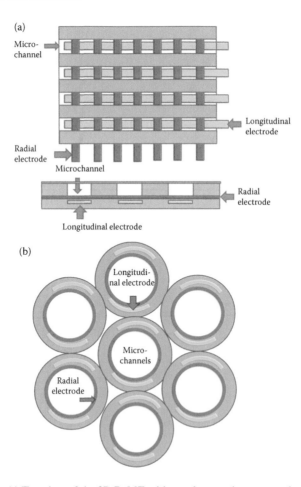

FIGURE 4.2   (a) Top view of the 2D ReME with parylene as the structural material and Au electrodes, and (b) cross-sectional view of 3D ReME microchannels with embedded stimulation and recording electrodes.

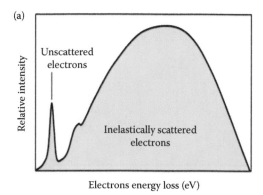

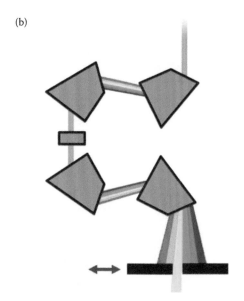

FIGURE 5.8 Energy filtering TEM for contrast tuning. (a) Plot (not real data) of electron energy loss spectrum, showing unscattered and inelastically scattered populations. (b) "Omega filter" schematic, showing prismatic effect on electron beam in line with the optical axis. The aperture (bottom) can be moved to select for a given energy range (i.e., a given vertical slice through the plot shown in (a)).

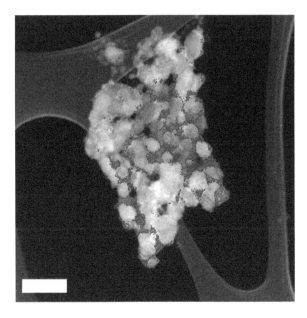

FIGURE 5.9 EELS imaging. Dark-field TEM image of nanoparticles on a lacy carbon grid. Red overlay shows the location of Fe as determined by energy-filtered EELS imaging. Scale bar is 200 nm.

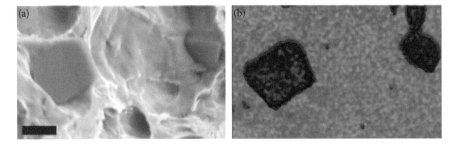

FIGURE 5.14 Energy-dispersive x-ray mapping. (a) SEM image of an alloy material. Scale bar is 2 µm. (b) EDX map of the same field of view, showing the location of Al (green) and Mg (blue).

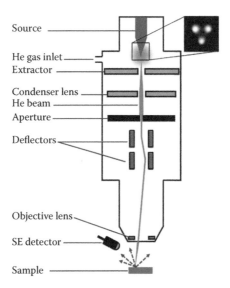

FIGURE 5.15 Helium ion microscope. A simplified cross-sectional illustration of the components in HIM. Top inset: view of three atoms forming the atomically small source tip.

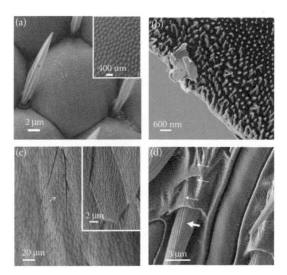

FIGURE 6.3 Examples of insect epicuticle nanostructures. (a) Nanonipples cover the surface (arrow) on the corneal surface of the eye of the fruit fly, *D. melanogaster*. The insert shows greater detail on the corneal nipple array. (b) Wing surface of the dog day cicadae, *Tibieins* ssp, is decorated with tall conical nanonipples (arrow); scale bar 600 nm. (c) Scaled surface (arrow) of the rostral plate of the European hornet, *Vespa crabro*. The insert shows that each has a vermiform, irregular nanopatterned surface. (d) The mouthpart of the European hornet, *V. crabro*, exhibits bristles with nanogrooves (thick arrows) that are surrounded by a epicuticle covered with nanopores (thin arrows).

FIGURE 7.1 This figure illustrates the tunability of quantum dots. Five different nanocrystal solutions are shown excited with the same long-wavelength UV lamp; the size of the nanocrystal determines the color. (Reproduced from Life Technologies. www.invitrogen. com/site/us/en/home/brands/Molecular-Probes/Key-Molecular-Probes-Products/Qdot/ Technology-Overview.html. With permission.)

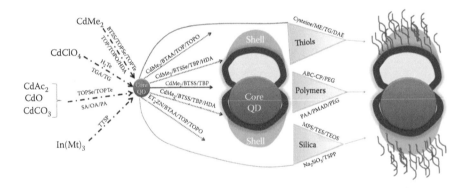

FIGURE 7.2 A schematic illustration of the synthesis of core QDs, coating with materials to form a shell, and overlaying the core/shell QDs with additional layers to facilitate additional conjugation and improve biocompatibility. (SA, stearic acid; OA, oleic acid; TOP, trioctylphosphine; TOPO, trioctylphosphine oxide; BTSS, bis(trimethylsilyl)sulfide; TGA, thioglycolic acid; TG, thioglycerol; TBP, tributyl phosphine; HAD, hexadecyl amine; MPA, methiopropamine; DHLA, dihydrolipoic acid; PAA, polyacrylic acid; PMAD, poly(maleic anhydride alt-1-tetradecane); PEG, polyethylene glycol; MPS, 3-mercaptopropyl trimethoxysilane; TES, tetraethoxysilane; TEOS, tetra ethyl orthosilicate.) (Adapted from Biju, V., Itoh, T., and Ishikawa, M., *Chemical Society Reviews*, 39, 3031, 2010.)

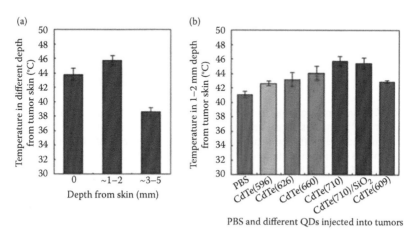

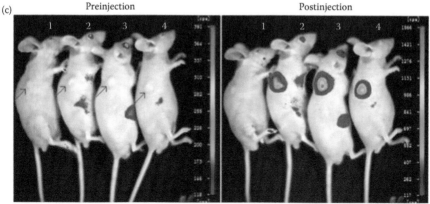

FIGURE 7.3 (a) CdTe(710) QDs was injected on the tumor site and temperature at different depths from the irradiated surface are displayed after a 20 min irradiation. (b) A plot representation after 20 min irradiation of the temperature of the tumors injected with PBS, PBS-dispersed CdTe, CdTe/SiO$_2$ and CdSe QDs. (c) *In vivo* images of tumors (see arrows) pre- and postinjection of (1) CdTe(626), (2) CdTe(660), (3) CdTe(710) and (4) CdTe(710)/SiO$_2$ QDs. (Reproduced from Chu, M. et al., *Biomaterials*, 33, 7071, 2012. With permission.)

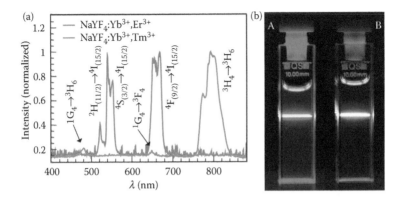

FIGURE 7.4 (a) Emission spectrum of (red) NaYF$_4$:Yb^{3+},Er^{3+} nanocrystals and (green) NaYF$_4$:Yb^{3+},Tm^{3+} nanocrystals excited by a 980 nm laser. (b) A digital image of the upconversion luminescence of the rare-earth co-doped NaYF$_4$ nanocrystals in chloroform after excitation with a 980 nm laser diode (A) Luminescence of NaYF$_4$:Yb^{3+},Er^{3+} nanocrystals; (B) luminescence of NaYF$_4$:Yb^{3+},Tm^{3+} nanocrystals. (Reproduced from Wang, H.Q. and Nann, T., *ACS Nano*, 3, 3804, 2009. With permission.)

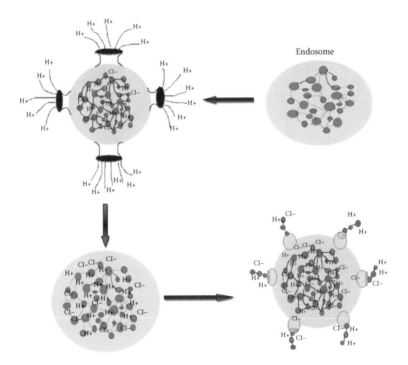

FIGURE 7.5 An illustration of the proton sponge effect that enhances delivery of payloads (polyamines) into the cytosol of the cell trapped inside the endosomes by buffering H$^+$ and subsequent accumulation of Cl$^-$. (Adapted from Biju, V., Itoh, T., and Ishikawa, M., *Chemical Society Reviews*, 39, 3031, 2010.)

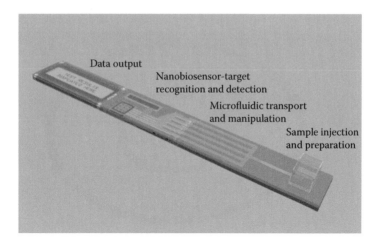

FIGURE 8.1 Schematic of a portable PoC diagnostic system.

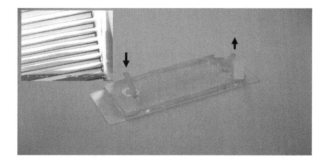

FIGURE 8.2 The image shows a passive microfluidic sample preparation section with inset showing the polydimethylsiloxane (PDMS) microchannels.

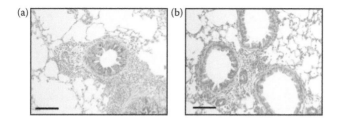

FIGURE 9.2 Novel fullerene derivative inhibits airway inflammation and bronchoconstriction in a murine model of asthma. Lung tissue fixed, sectioned, and stained with PAS. Representative photographs show airway hyperresponsiveness of OVA sensitized mice treated with untreated controls (a) and FD (b).

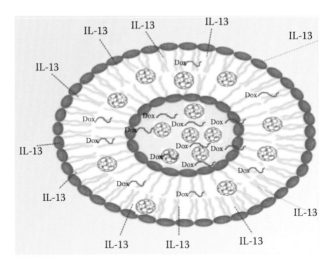

FIGURE 9.7 Representative schematic of glioblastoma targeting theranostic (GTNN). The TMS is incorporated within and/or in the liposome bilayer along with doxorubicin (Dox) and are targeted to the glioma cells by IL-13 (not drawn to scale).

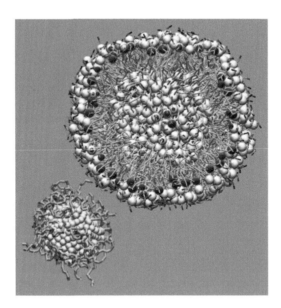

FIGURE 11.3 Initial configuration of copolymer micelle and lipid vesicle used in the CG MD simulations is shown. A cross section of the lipid vesicle is shown to reveal inner components of the vesicle.

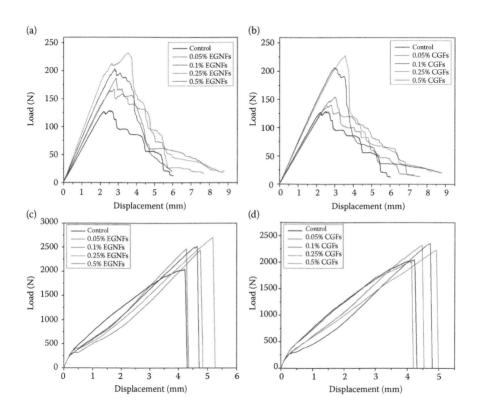

FIGURE 12.9 Typical experimental load–displacement curves recorded from the three-point bending test (a and b) and tension test (c and d) of the multiscale glass fiber-reinforced composite laminates.

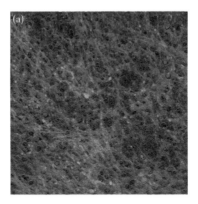

FIGURE 14.5 Fluorescence image of 3T3 cells: (a) healthy cells and (b) cells exposed to gold nanorods at 0.5 nM.

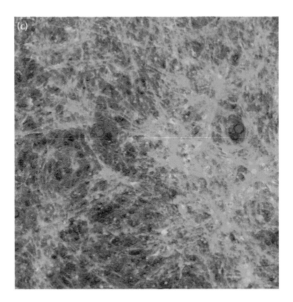

FIGURE 14.7 SEM image of (a) 3T3 cells exposed to silica NPs at 75 μg/mL, (b) SEM image of silica NPs, and (c) fluorescence image of 3T3 cells exposed to silica NPs.

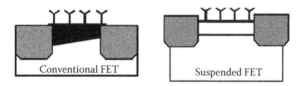

FIGURE 8.4 Comparison between a conventional FET and suspended FET device.

the target. Nano-FET biosensors have used silicon nanowires or nanoribbons, graphene, or a carbon nanotube as the channel material to lower the limit of detection.
Lin et al. [52] developed a polysilicon nanowire FET to detect avian influenza DNA.
Stern et al. [53] developed a microfluidic purification chip with silicon nanoribbon for
label-free detection in whole blood. Zhang and Cui [54] built a suspended graphene
nanoribbon ion-sensitive FET using shrink lithography. Figure 8.4 illustrates the difference between a conventional FET and suspended FET device, as demonstrated in
Ref. [54]. Poly-L-lysine is used to bind the anti-PSA on the suspended ribbon. The
detection limit of PSA was estimated at 0.4 pg/mL. The suspended configuration also
shows larger current at the same conditions compared to a conventional configuration. The suspended FET provides more mechanical stress, which in turn significantly
increases charge carrier mobility. In other examples, nonlinear operation sensors
bifurcation-based and flexure–FET biosensors have been implemented [55,56]. Jain
et al. [56] simulated a flexure–FET nanobiosensor with exponentially high sensitivity.

  In most instances, NEMS/MEMS biosensors are either cantilever-based or suspended (double-clamped) structures such as beams, belts, or nanowires. The fabrication methodology can be top-down semiconductor manufacturing methods,
micromachining, or bottom-up assembly. Figure 8.5 illustrates example process

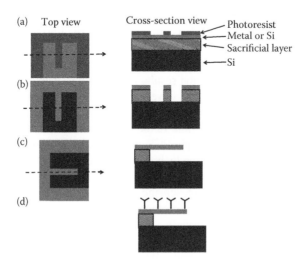

FIGURE 8.5 Examples of top-down bio-N/MEMS fabrication processes. (a) Deposition
and patterning of cantilever, (b) deep trench etching, (c) released cantilever, and (d) functionalization of cantilever.

flows of top-down bio-NEMS/MEMS devices. Lithography and dry-etching methods are used to define NEMS/MEMS patterns. Sacrificial material is used to create suspended structures. Finally, the suspended structure is functionalized with the appropriate receptor (proteins or DNA strands).

On the other hand, when nanowires are used to fabricate the NEMS device, then pure top-down subtractive process is just not enough. There can be two possible process methodologies to integrate nanowires in NEMS device. In the first process, catalyst dots (or electrodes) are first patterned using semiconductor processes. Then, nanowires are synthesized between the catalyst dots [57]. In case of using presynthesized nanowires, the nanowires can be placed between the electrodes by either nanomanipulation [58] or by dielectrophoretic assembly [59]. Finally, the nanowire can be suspended by lithography followed by wet etching. In the second process, dispersion or manipulation methods are used to place nanowires on a wafer, but the location is not precisely controlled. Then, by using e-beam lithography, electrodes can be created to suspend the nanowires [60]. The first process is ideal for large-scale manufacturing, while the second process is simple and straightforward.

Lastly, nanopore-based DNA sequencing is an emerging PoC technique to analyze DNA without PCR. Here, the DNA molecule is analyzed by measuring the current change or photon emission when the molecule passes through a nanosized pore. Branton et al. [61] predicted that this technology would be able to accomplish diploid mammalian genome sequencing in 1 day at a cost of about $1000. In 2012, Oxford Nanopore Technologies developed a USB- (universal serial bus) sized, hand-held nanopore-sequencing device [62].

8.5   CURRENT TRENDS AND FUTURE PERSPECTIVE

The high sensitivity achieved by the current nanobiosensors still faces a number of challenges in clinical samples, such as whole blood. For example, the nanocantilever is an excellent mass sensor but, with a complex sample, the limit of detection is low and high selectivity is hard to achieve. Other technologies that do not need complicated sample preparation steps also suffer significant loss in resolution (low limit of detection) [41]. The loss in resolution results in a failure to detect diseases at their early stages. Currently, the key is to develop reliable on-chip sample preparation technology, which can enable specific binding relations, isothermal PCR, and high sensitive sensing without amplification. In the future, we will evidence more efforts on sample preparation, such as simple PCR processes, novel specific and selective binding technologies, and integration technologies. Innovative integration technologies are required so that all the individual components can be seamlessly assembled into one PoC platform. Furthermore, sensing technologies that can output results in an easy, readable visual form will receive more attention in low-resource settings due to their simplicity and cost-effectiveness.

In summary, clinical diagnostics has a long history of dependence on lab tests, radiological images, biopsy results, and differential diagnosis. The diagnosis process is long and complicated, especially when several diseases result in similar test results. Biomarkers and biosensors have provided an unprecedented opportunity to simplify the diagnostic process. However, most of these current methods still depend on well-trained technicians, complicated procedures, and lab-based equipment. Recent developments in PoC nanodiagnostic systems have helped with these drawbacks. On-chip sample preparation enabled by microfluidics and nanobiosensing using functionalized nanoparticles or fabricated nanodevices, system integration, and multiplexing technologies have all led to novel innovations in clinical diagnostic systems.

REFERENCES

1. Jones, C.M., Diagnostic tests. In: *Evidence Synthesis in Healthcare*. Jones, C.M., Darzi, A., Athanasiou, T., (eds.), Springer-Verlag, London, 2011.
2. Henry, N.L. and Hayes, D.F., Cancer biomarkers, *Mol. Oncol.*, 6, 140, 2012.
3. Lewandrowski, K., Point-of-care testing: An overview and a look to the future (circa 2009, United States), *Clin. Lab. Med.*, 29(3), 421–432, 2009.
4. Graeber, M.B., Biomarkers for Parkinson's disease, *Exp. Neurol.*, 216, 249, 2009.
5. Misek, D.E., Kondo, T., and Duncan, M.W., Proteomics-based disease biomarkers, *Int. J. Proteom.*, 2011, 1, 2011.
6. Craig-Schapiro, R., Fagan, A.M., and Holtzman, D.M., Biomarkers of Alzheimer's disease, *Neurobiol. Dis.*, 35, 128, 2009.
7. Easton, D.F., Ford, D., and Bishop, D.T., Breast and ovarian cancer incidence in BRCA1-mutation carriers. Breast Cancer Linkage Consortium, *Am. J. Hum. Genet.*, 56, 265, 1995.
8. Edwards, S.A., Southern, Northern, Western, and Eastern blots, *Scientia* 2012. http://membercentral.aaas.org/blogs/scientia/southern-northern-western-eastern
9. Lequin, R.M., Enzyme immunoassay (EIA)/enzyme-linked immunosorbent assay (ELISA), *Clin. Chem.*, 51, 2415, 2005.
10. Burlingame, A.L., Boyd, R.K., and Gaskell, S.J., Mass spectrometry, *Anal. Chem.*, 70, 647, 1998.
11. Rovensky, J. and Payer, J., Radioimmunoassay, *Dictionary of Rheumatology*, Springer-Verlag, Germany, p. 182, 2009.
12. McNaught, A.D. and Wilkinson, A., IUPAC, *Compendium of Chemical Terminology*, 2nd ed, Blackwell Scientific Publications, Oxford, 1997.
13. Kissinger, P.T., Biosensors—A perspective, *Biosens. Bioelectron.*, 20, 2512–2516, 2005.
14. Luong, J.H.T., Male, K.B., and Glennon, J.D., Biosensor technology: Technology push versus market pull. *Biotechnol. Adv.*, 26,492–500, 2008.
15. Durner, J., Clinical chemistry: Challenges for analytical chemistry and the nanosciences from medicine, *Angew. Chem. Int. Ed.*, 49, 1026–1051, 2010.
16. Gubala, V., Harris, L.F., Ricco, A.J., Tan, M.X., and Williams, D.E., Point of care diagnostics: Status and future. *Anal. Chem.*, 84(2), 487–515, 2011.
17. Yager, P., Edwards, T., Fu, E., Helton, K., Nelson, K., Tam, M.R., and Weigl, B.H., Microfluidic diagnostic technologies for global public health, *Nature*, 442(7101), 412–418, 2006.
18. Aravamudhan, S., Kumar, A., Mohapatra, S., and Bhansali, S., Sensitive estimation of total cholesterol in blood using Au nanowires based microfluidic platform, *Biosens. Bioelectron.*, 22, 2289–2294, 2007.
19. Whitesides, G.M., The origins and the future of microfluidics, *Nature*, 442, 368, 2006.
20. Manz, A., Graber, N., and Widmer, H.M., Miniaturized total chemical-analysis systems— A novel concept for chemical sensing, *Sens. Actuat. B—Chemical*, 1, 244–248, 1990.
21. Reyes, D.R., Iossifidis, D., Auroux, P.A., and Manz, A., Micro total analysis systems. 1. Introduction, theory, and technology, *Anal. Chem.*, 74, 2623–2636, 2002.
22. Auroux, P.A., Iossifidis, D., Reyes, D.R., and Manz, A., Micro total analysis systems. 2. Analytical standard operations and applications, *Anal. Chem.*, 74, 2637–2645, 2002.

23. Haeberle, S. and Zengerle, R., Microfluidic platforms for lab-on-a-chip applications, *Lab Chip,* 7, 1094–1110, 2007.

24. Mark, D., Haeberle, S., Roth, G. et al. Microfluidic lab-on-a-chip platforms: Requirements, characteristics and applications, *Chem. Soc. Rev.,* 39, 1153–1182, 2010.

25. Le, H.P., Progress and trends in ink-jet printing technology, *J. Imag. Sci. Technol.,* 42, 49–62, 1998.

26. Ahn, C.H., Choi, J.W., Beaucage, G. et al. Disposable smart lab on a chip for point-of-care clinical diagnostics, *Proc. IEEE,* 92, 154–173, 2004.

27. Vo-Dinh, T., Griffin, G., Stokes, D.L., and Wintenberg, A., Multi-functional biochip for medical diagnostics and pathogen detection, *Sens. Actuat. B: Chemical,* 90, 104–111, 2008.

28. Soo Ko, J., Yoon, H.C., Yang, H. et al. A polymer-based microfluidic device for immunosensing biochips, *Lab Chip,* 3, 106–113, 2003.

29. Lemoff, A.V. and Lee, A.P., An AC magnetohydrodynamic micropump, *Sens. Actuat. B: Chemical,* 63, 178–185, 2000.

30. Chou, H.P., Unger, M.A., and Quake, S.R., A microfabricated rotary pump, *Biomed. Microdev.,* 3, 323–330, 2001.

31. Martinez, A.W., Phillips, S.T., Whitesides, G.M., and Carrilho, E., Diagnostics for the developing world: Microfluidic paper-based analytical devices, *Anal. Chem.,* 82, 3–10, 2010.

32. Becker, H. and Locascio, L.E., Polymer microfluidic devices, *Talanta,* 56, 267–287, 2002.

33. Metz, S., Holzer, R., and Renaud, P., Polyimide-based microfluidic devices, *Lab Chip,* 1, 29–34, 2001.

34. Wu, Z., Xanthopoulos, N., Reymond, F. et al. Polymer microchips bonded by O_2-plasma activation, *Electrophoresis,* 23, 782–790, 2002.

35. Sin, M.L., Gao, J., Liao, J.C., and Wong, P.K., System integration—A major step toward lab on a chip, *J. Biol. Eng.,* 5, 6, 2011.

36. Kim, J. and Gale, B.K., Quantitative and qualitative analysis of a microfluidic DNA extraction system using a nanoporous AlOx membrane, *Lab Chip,* 8, 1516, 2008.

37. Krivitsky, V. et al. Si nanowires forest-based on-chip biomolecular filtering, separation and preconcentration devices: Nanowires do it all, *Nano Lett.,* 12(9), 4748, 2012.

38. Kulinsky, L., Noroozi, Z., and Madou, M., Present technology and future trends in point-of-care micro fluidic diagnostics, *Methods Mol. Biol.,* 949, 3, 2013.

39. Elghanian, R. et al. Selective colorimetric detection of polynucleotides based on the distance-dependent optical properties of gold nanoparticles, *Science,* 277, 1078, 1997.

40. Storhoff, J.J. et al. One-pot colorimetric differentiation of polynucleotides with single base imperfections using gold nanoparticle probes, *J. Am. Chem. Soc.,* 120, 1959, 1998.

41. Kalidasan, K., Neo, J.L., and Uttamchandani, M., Direct visual detection of *Salmonella* genomic DNA using gold nanoparticles, *Mol. Bio. Syst.,* 9, 618, 2013.

42. Mancuso, M. et al. Multiplexed colorimetric detection of Kaposi's sarcoma associated herpesvirus and Bartonella DNA using gold and silver nanoparticles, *Nanoscale,* 5, 1678, 2013.

43. Chung, H.J. et al. A magneto-DNA nanoparticle system for rapid detection and phenotyping of bacteria, *Nat. Nanotechnol.,* 8, 369, 2013.

44. Lee, H. et al. Chip–NMR biosensor for detection and molecular analysis of cells, *Nat. Med.,* 14, 869, 2008.

45. Koh, I. and Josephson, J., Magnetic nanoparticle sensors, *Sensors,* 9(10), 8130, 2009.

46. Gaster, R.S., Hall, D.A., and Wang, S.X., nanoLAB: An ultraportable, handheld diagnostic laboratory for global health, *Lab Chip,* 11, 950, 2011.

47. Fritz, J. et al. Translating biomolecular recognition into nanomechanics, *Science,* 288, 316, 2000.

48. Zang, J. and Liu, F., Theory of bending of Si nanocantilevers induced by molecular adsorption: A modified Stoney formula for the calibration of nanomechanochemical sensors, *Nanotechnology*, 18, 405501, 2007.

49. Varshney, M. et al. Prion protein detection using nanomechanical resonator arrays and secondary mass labeling, *Anal. Chem.*, 80, 2141, 2008.

50. Waggoner, P.S., Varshney, M., and Craighead, H.G., Detection of prostate specific antigen with nanomechanical resonators, *Lab Chip*, 9, 3095, 2009.

51. Varshney, M. et al. Prion protein detection in serum using micromechanical resonator arrays, *Talanta*, 80, 593, 2009.

52. Lin, C. et al. Poly-silicon nanowire field-effect transistor for ultrasensitive and label-free detection of pathogenic avian influenza DNA, *Biosens. Bioelectron.*, 24(10), 3019, 2009.

53. Stern, E. et al. Label-free biomarker detection from whole blood, *Nat. Nanotechnol.*, 5, 138, 2010.

54. Zhang, B. and Cui, T., Suspended graphene nanoribbon ion-sensitive field-effect transistors formed by shrink lithography for pH/cancer biomarker sensing, *J. Microelectromech. Syst.*, 22(5), 1140–1146, 2013.

55. Kumar, V. et al. Bifurcation-based mass sensing using piezoelectrically-actuated microcantilevers, *Appl. Phys. Lett.*, 98, 153510, 2011.

56. Jain, A., Nair, P.R., and Alam, M.A., Flexure–FET biosensor to break the fundamental sensitivity limits of nanobiosensors using nonlinear electromechanical coupling, *Proc. Natl. Acad. Sci. U S A*, 109, 9304, 2012.

57. Kong, J. et al. Synthesis of individual single-walled carbon nanotubes on patterned silicon wafers, *Nature*, 395, 878, 1998.

58. Sosnowchik, B., Chang, J., and Lin, L., Pick, break, and placement of 1D nanostructures for direct assembly and integration, *Appl. Phys. Lett.*, 96, 153101, 2010.

59. Pathangi, H., Groeseneken, G., and Witvrouw, A., Dielectrophoretic assembly of suspended single-walled carbon nanotubes, *Microelectron. Eng.*, 98, 218, 2012.

60. Chiu, H. et al. Atomic-scale mass sensing using carbon nanotube resonators, *Nano Lett.*, 8, 4342, 2008.

61. Branton, D. et al. The potential and challenges of nanopore sequencing, *Nat. Biotechnol.*, 26, 1146, 2008.

62. Eisenstein, M., Oxford nanopore announcement sets sequencing sector abuzz, *Nat. Biotechnol.*, 3, 295, 2012.

9 Fullerenes and Their Potential in Nanomedicine

Christopher Kepley and Anthony Dellinger

CONTENTS

9.1 FULLERENES AS A PLATFORM FOR NEW SOLUTIONS IN SEVERAL SCIENTIFIC AREAS

Fullerenes are closed carbon spheres that are being actively pursued globally for a wide range of applications. Empty cage fullerenes (Figure 9.1a and b) have unique electrochemical properties and have a wide range of potentially beneficial biologic properties. Another type of fullerene can have metals enclosed inside them (metallo-fullerenes). Fullerenes have a unique cage structure with delocalized π molecular orbital electrons. This structure confers unusual activity in electron transfer systems due to their low reorganization energy, low lying excited states (singlet and triplet), and extended triplet lifetimes. Further, the spherical configuration of the planar benzene rings imposes an unusual constraint on these π electron orbitals. The fullerene carbon cage is insoluble and thus must be derivatized (simply meaning moieties or side groups must be added to the carbon cage) in order to make them water soluble and compatible in biological systems. The ability of fullerenes to be derivatized with side chains provides opportunities to diversify, manipulate, and harness the

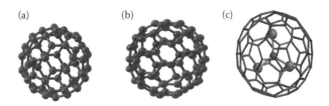

FIGURE 9.1 Representative fullerene structures: (a) empty cage C_{60} fullerene, (b) empty cage C_{70} fullerene, and (c) metallo-fullerene.

electronic properties of the cage for selected applications. Of course, each derivation results in changes in the compounds physical and chemical property including; for example, particle size/length, zeta potential, net charge, molecular weight, purity, surface characteristics, and solubility. Too often, results from studies examining the effects of fullerenes on biological systems tend to be extrapolated into other applications. As demonstrated below, each fullerene derivative (FD) must be assessed separately depending on the biological application. Even extremely similar derivatized fullerenes can have completely opposite results, which stimulates efforts to understand how changing the chemical composition and structural arrangement of fullerenes affects molecular interactions at cellular, tissue, and organ system levels. Consequently, their inherent properties described below combined with their ability to be derivatized with side chains results in almost limitless new chemical structures making them ideal platform molecules for innovative new solutions to basic biological problems. This chapter will focus on the applications of both empty cage and metallo-fullerenes for therapeutic, diagnostic, and theranostic (combination of diagnostic and therapeutic) applications.

9.2 FULLERENES FOR THERAPEUTICS

Fullerenes have been termed "free radical sponges" and described most frequently as antioxidants, although in biological systems they paradoxically can act as both oxidants and antioxidants. The generation of free radicals such as reactive oxygen species (ROS) and reactive nitrogen species (RNS) occurs naturally in cells and their presence at sites of disease pathologies suggests that they contribute to disease progression. The term free radical refers to a molecular species that possess an unpaired electron, which makes them highly reactive. Many of the most common ROS or RNS species that contribute to oxidative or nitrosative stress in biological systems are free radicals, including hydroxyl radical (OH•), superoxide anion (O2−.), and peroxynitrite (ONOO−). Other ROS are not free radicals, including singlet oxygen and hydrogen peroxide (H_2O_2), which are considered ROS because they can generate oxygen radicals like superoxide via the Fenton reaction. These ROS can react with, cross link, and alter the function of many macromolecules and negatively affect a wide variety of biological processes, but are also used in beneficial biological processes such as signaling and cellular defenses. Antioxidants are molecules that can eliminate or neutralize free radical electrons (e.g., vitamins A and C). This has led to a tremendous amount of research on preventing damage using antioxidants theorized to, in turn, counteract disease pathologies. Indeed, the perceived benefits of antioxidants are a widely accepted concept, yet their use has been mainly centered on over-the-counter supplements for general health and antiaging benefits.

The carbon cage (usually C_{60} and C_{70}) of empty cage fullerenes (Figure 9.1) can function as antioxidants based on their ability to absorb electrons and disperse them through the benzene-like rings distributed over its surface. This ability to scavenge free radicals has led to their potential as a new way for treating a wide range of diseases and pathologies. These include multiple sclerosis [1], neurodegenerative [2], HIV infection [3,4], cancer [5], radiation exposure [6], ischemia [7], and general inflammation [8]. Interestingly, mice and rats chronically treated with a water-solubilized carboxylated fullerenes have significantly extended lifespans compared to littermate controls [9,10]. There are almost endless possibilities for adding side groups to the carbon cage to induce functionality.

9.2.1 THE ABILITY OF FULLERENE FDS TO AFFECT MAST CELL-DRIVEN, ALLERGIC INFLAMMATORY DISEASE

Allergic reactions are the result of B cell-produced, specific IgE antibody to common, normally innocuous antigens. In simplistic terms, mast cells (MC), peripheral blood basophils (PBB), natural killer (NK) cells, T cells, and even B cells are responsible for driving the initial, allergen-inducing reaction through the production of IL-4, and other T_H2-specific cytokines that result in IgE sensitization. Reexposure to the allergen triggers an allergic response through the release of inflammatory mediators from MC and PBB. The IgE produced binds to FcεRI on MC and PBB and the release of pre-allergic mediators is induced when two or more IgE molecules are cross-linked with allergen. Indeed, most allergy medications are aimed at neutralizing (antihistamines, H_1-receptor blockers) or preventing ("Omalizumab") MC and PBB FcεRI responses. Recent research has demonstrated that specifically engineered FD are taken up and can stabilize human tissue MC and prevent the release of pro-inflammatory mediators from these cells making them ideal candidates for those disease mediated by MC mediators [11–13]. Thus, it was hypothesized that stabilization of FD *in vitro* could translate to blocking MC-driven diseases *in vivo*. To test this hypothesis, it was examined the ability of FD to block MC-driven disease using mouse models of arthritis and asthma.

9.2.2 MC AND PBB IN ASTHMA

MC are ubiquitously expressed in tissues and are uniquely able to initiate and propagate certain inflammatory responses and offer an interface between innate and adaptive immunity [14]. Mice without MC fail to develop asthma-like pulmonary disease when sensitized with less-aggressive immunization protocols and challenged with aerosolized allergens [15,16]. However, the strongest evidence that MC are critical for human asthma comes from the many therapeutics used to treat the disease in humans. In general, two categories of therapies were developed to control asthma; blocking MC activation before it occurs (stabilizers) and blocking MC activation after it occurs. Cromolyn (Nasalcrom, Intal; Opticrom) is an MC stabilizer used as an oral inhaler and controls episodes of asthma caused by foreign allergens such as pollens by preventing spasm and narrowing of the breathing tubes of the lungs. In addition, the anti-IgE therapy Omalizumab (Xolair) blocks MC activation indirectly by preventing the accumulation of IgE and thus preventing FcεRI-IgE-mediated MC activation to occur. Similarly, therapies targeting the low-affinity IgE receptor-CD23 are also aimed at reducing IgE levels [17]. Syk, a tyrosine kinase that is inextricably linked to lung MC [18] and PBB [19–21]. FcεRI degranulation is also being targeted for new asthma therapies [17]. MC are the only cells in the lung that have pre-stored TNF-α in their granules that can be immediately released upon allergen provocation [22] and blockade of this cytokine is a valid target being investigated [23]. Therapies that block the effects of MC after activation occurs are well known, commonly used, and include antihistamines and H_1-receptor blockers. Thus, the importance of MC in asthma pathogenesis is well established and controlling the amount of mediator release from these cells is a proven drug development strategy.

The role of basophils in asthma is less clear. Its role in allergic disease has largely been viewed as redundant to that of the tissue MC. This line of thought, however, is changing with evidence that has emerged during the last 15 years. These cells are a significant source of the cytokines IL-4 and IL-13 both of which are vital to the pathogenesis of allergic disease including asthma. It has been demonstrated that increased levels of basophils were present in the lungs of deceased asthmatics [24] using a basophil-specific antibody developed for detecting these cell types immunologically [25]. Thus, the infiltration of basophils into allergic lesion sites has sparked greater interest in this once overlooked immune cell, both in adaptive as well as in innate immunity [26].

There is a strong need for novel therapeutics to treat asthmatic disease; indeed up to 55% of patients receiving treatment for asthma have uncontrolled symptomology [27]. In some situations, especially with milder disease, asthma control is achieved with the use of nonspecific anti-inflammatories such as corticosteroids, or by the use of antihistamines or leukotriene inhibitors. However, these therapies are generally less effective in severe asthmatics. As shown below, FD are able to suppress both disease onset and reverse established disease in murine asthma models. The latter is especially important as human asthma treatment always involves established disease.

The utility of the FD role in asthma prevention and reversal can be validated in mice treated with FD throughout an ovalbumin (OVA) challenge. Treated animals have significantly less airway inflammation and bronchoconstriction compared to untreated animals. In fact, total inflammation and bronchoconstriction in FD-treated animals is not only significantly reduced, but is similar to that seen in non-sensitized controls [28]. Thus, symptoms of disease are largely reversed in these animals. Note that these studies used a model previously shown to utilize MC [16]. In the established disease model, as when mice were treated throughout disease development, FD dampened eosinophilia and cytokine levels significantly in the bronchoalveolar lavage (BAL) fluid. Lung sections show massive cellular infiltration in untreated animals, while those receiving FD have minimal cellular infiltration surrounding the airways (Figure 9.2). This led to reduced airway hyper-responsiveness in FD-treated animals.

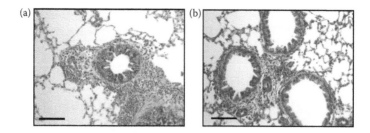

FIGURE 9.2 **(See color insert.)** Novel fullerene derivative inhibits airway inflammation and bronchoconstriction in a murine model of asthma. Lung tissue fixed, sectioned, and stained with PAS. Representative photographs show airway hyperresponsiveness of OVA sensitized mice treated with untreated controls (a) and FD (b).

Thus, FD may be useful in a clinical setting to reverse asthma pathogenesis and limit exacerbation of symptoms.

While previously published *in vitro* studies [11,13] suggested that MC inhibition may be the predominant mechanism of FD inhibition, *in vivo* studies suggest multipotent effects of these unique compounds. FD treatment reduces the levels of BAL Th2 pro-inflammatory cytokines and reduces lung inflammation. While IL-4 stimulates IgE production by B cells, IL-5 both recruits and activates eosinophils at the site of inflammation. FD treatment significantly reduces both IL-4 and IL-5 in the BAL fluid. Additionally, serum IgE levels were significantly reduced following FD treatment.

Importantly, the use of FD revealed a novel mechanism of action and opened up new avenues of research for asthma. Several eicosanoids (EETs) derived from the cytochrome P450 pathway are relatively stable and were measured in BAL fluid samples using mass spectrometry. The EETs are consistently associated with relaxation of the bronchi and other anti-inflammatory actions *in vivo* [29,30]. The 11, 12 EET was consistently upregulated in BAL fluid from FD-treated mice. Further *in vivo* studies demonstrated that the EETs play a major role in dampening the asthma phenotype as inhibitors of the EETs prevented the FD-induced modulation of the OVA-induced asthma model [28]. Certain EETs stabilize human lung MC through the inhibition of FcεRI-mediator release and upregulate the CYP1B1 gene in these cells (Figure 9.3). Thought to be produced by lung epithelial and endothelial cells, the EETs have been shown to relax histamine-precontracted guinea pig and human bronchi [31,32]. Further, they can inhibit the upregulation of VCAM-1, E-selectin, and ICAM-1, thus potentially limiting cellular infiltration of the lung [33]. Consequently, 11, 12 EET upregulation is playing a significant role in dampening airway inflammation and bronchoconstriction in these models. Thus, FD upregulation of EETs is a novel mechanism for controlling asthma and suggests strategies that induce the production of EETs may be a viable therapeutic strategy for treating asthmatics.

In conclusion, the efficacy of FD for the treatment of asthma is illustrated through a mechanism involving the dampening of eosinophilia and cytokine levels, reducing airway hyperresponsiveness, and upregulating EETs. Thus, FD compounds have the potential to become novel therapeutics for the treatment of asthma and pave the way to new research efforts focusing on the role of EETs in human disease.

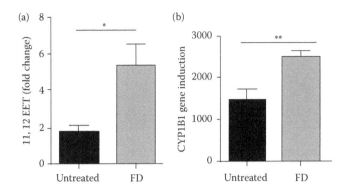

FIGURE 9.3 Novel fullerene derivative upregulate EETs in BAL fluid compared to untreated controls as measured by mass spectroscopy (a) and induce CYP1B1 gene induction as measured by gene microarray analysis (b). * indicates statistical significance.

9.2.3 MCs IN ARTHRITIS

MCs play a critical role in the pathogenesis of synovitis in a murine model of rheumatoid arthritis (RA) [34,35]. The synovium of patients with RA is chronically inflamed and characterized by an expanded population of MC, as in the mouse model. MCs are markedly increased in number and can make up 5% or more of the expanded population of total synovial cells. The number of accumulated MC differs substantially from patient to patient; in general, varying directly with the intensity of joint inflammation [36–38]. MC mediators (histamine and tryptase) are also present at higher concentrations in the synovial fluid of inflamed human joints [39].

MC degranulation has long been associated with arthritis in several animal models, but a critical functional role in the disease was established in the K/BxN mouse model [35,40]. This arthritis model closely mimics human RA via symmetric joint involvement, chronicity, a distal-to-proximal gradient of joint involvement, and histological features, including synovial infiltrates, pannus, and erosions of cartilage and bone [41]. Mice deficient in MC are highly resistant to arthritis, whereas reconstitution with normal MC restores the wild-type phenotype. Furthermore, degranulation of MC in the synovium is the first event observed histologically, occurring within 1–2 hours of administration of K/BxN serum [42]. Recent studies reaffirmed a role for MC mediators in arthritis [34]. Thus, MCs are normal cell populations within the human synovium and have a critical role in the pathogenesis of inflammatory arthritis.

In preliminary studies, it has been demonstrated that FD play a key role in inhibiting human and mouse MC through arthritis-relevant stimulation. The ability of FD to inhibit synovial fibroblast cytokine production and osteoclast development are important prerequisites for predicting *in vivo* efficacy. This is accomplished through the ability of specific fullerene constructs to inhibit IgE-mediated degranulation and cytokine production, but not released in response to other non-IgE-mediated secretagogues. This is done via an interaction between the side chain moieties conjugated to the FD and the electrons [43] on the fullerene cage.

The mitochondrial electron transport is the machinery that orchestrates one of the most fundamental of chemical processes—the generation of cellular energy from oxygen resulting in the fuel that supports all eukaryotic life. However, it is a highly sensitive process and, unbalanced, leads to the generation of free radicals or ROS which have been linked as a mechanism underlying many chronic human diseases, including MC activation and inflammatory arthritis [44]. Fullerenes can be specifically designed to home and accumulate in the internal mitochondrial membrane bilayers. Once incorporated, FD are positioned to neutralize superoxide molecules, reactive lipid radicals, and radicals that have formed on transmembrane proteins at the site where they are generated, thus inhibiting inflammation [8,45]. This in turn is predicted to impact diseases whose pathologies stem from radical injury.

To this end, researchers designed FD that significantly block ROS production [28,46]. While it has been shown previously that human MC degranulation in response to FcεRI and FcεR1 signaling involves ROS [45,46], it is not clear if blocking ROS directly blocks degranulation and cytokine production. However, results suggest that blocking ROS using FD in response to immune complex (IC) (an FcγRIIA-dependent stimuli [47]) parallels inhibition of mediator release. This is in line with previous work suggesting that fullerenes interfere with the generation of mitochondrial-derived ROS [48,49]. It is also been demonstrated that MMP is a critical determinant in human MC FcεR-mediated degranulation.

Arthritic joint tissues demonstrate a striking predilection for uptake of the specifically derivatized fullerenes. Indeed, this strong uptake may provide the basis of their efficacy in ameliorating arthritis. It was also demonstrated that FD inhibited the onset of arthritis in the K/BxN, but not the collagen-induced arthritis (CIA), model of arthritis (not shown). The K/BxN serum transfer model induces a rapid and severe synovitis, similar to human RA, which is dependent on neutrophils, MC, and macrophages. Its MC dependence was revealed from studies in which two strains of mice deficient in this cell type, W/Wv and Sl/Sld, were resistant to disease induction following serum transfer [40]. Susceptibility to disease is restored in the W/Wv strain by MC engraftment. Thus, in the K/BxN model, MC function as a link between the serum transfer induction of auto-antibodies, soluble mediators, and other effector populations. In contrast, MC-deficient mice are still susceptible to CIA [50] as well as anticollagen antibody/lipopolysaccharide (LPS)-induced arthritis [41]. Taken together, results suggest that FD block some component of MC activation leading to TNF-α release. Given that TNF-α blocking agents, such as Etanercept (Amgen Inc., Thousand Oaks, CA), Infliximab (Jassen Biotech, Inc., Horsham, PA), and Adalimumab (AbbVie Inc., North Chicago, IL) are widely used for this disease, compounds such as FD which block production of TNF-α may have therapeutic potential as a candidate for blocking inflammatory arthritis.

Further studies using FD for treatment in RA did not detect any acute liver or kidney toxicity using repeated dosing of concentrations 100× higher than that needed for *in vivo* efficacy (not shown). The *in vivo* imaging studies also demonstrated a lack of uptake in other organs, which portends well for a favorable toxicity profile in clinical development of some FD. Of course, more advanced toxicity studies are needed to assess these FD as is the case with any new therapeutic.

It is demonstrated that the not all FD exhibit the same ability to inhibit inflammatory mediator release from MC and synovial fibroblasts. Two FD were able to significantly block the onset of serum-induced arthritis in a MC-dependent disease model suggesting FD stabilize MC *in vivo* leading to a blunted inflammatory response. More studies are needed to identify those structure–activity relationships that are dependent on the moieties added to the fullerene carbon cage so that precise mechanisms can be found of how this class of compounds inhibits inflammatory disease.

9.2.4   MCs in Multiple Sclerosis and Fullerenes

Experimental allergic encephalomyelitis (EAE) is a rodent model of human multiple sclerosis (MS) characterized by inflammation in the central nervous system (CNS) [51]. Like the human disease, EAE is associated with an early breach of the blood–brain barrier, focal perivascular mononuclear cell infiltrates, and demyelination leading to paralysis of the extremities. While CD4-positive T cells have been implicated, the underlying cause of increased vascular permeability that facilitates the entry of T cells into the CNS is unknown.

MC contribution to the pathogenesis of MS has been hypothesized based on their presence in CNS plaques of MS patients and the correlation between the number, distribution, or MC markers and MS, or EAE pathology [52]. Further evidence for MC involvement in EAE/MS came from studies using MC-deficient mice [53]. The MC-deficient W/W^v mice exhibited significantly reduced disease incidence,

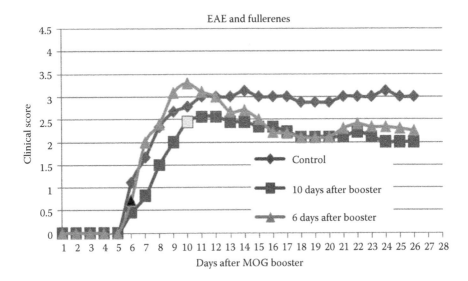

FIGURE 9.4   Fullerene derivative inhibit EAE. To induce disease 300 μg of myelin oligodendrocyte glycoprotein CFA with *Mycobacterium tuberculosis* was injected subcutaneously into the flank on days 0 and 7. Pertussis toxin was given intravenously into the tail vein on days 0 and 2. Disease becomes apparent at approximately day 1. FD treatment inhibited the onset of EAE in mice assessed at day 6 (triangles) and day 10 (squares) compared to non-treated animal (diamonds).

delayed disease onset, and decreased mean clinical scores when compared with their wild-type congenic littermates. No differences were observed in the T and B cell responses between the two groups and reconstitution of the MC population in W/W^v mice restores induction of early and severe disease to wild-type levels. These data provide a new mechanism for immune destruction in EAE and indicate that MC may be sentinels of neurologic inflammation.

Given that MC are important regulators of multiple sclerosis and oxidative stress, through the generation of ROS, is an underlying mechanism that mediates MC signaling and MS pathology, FD were tested for their ability to inhibit EAE. As shown in Figure 9.4, intriguing results in the EAE model that suggest this class of rationally designed compounds may be used as a platform for new areas of therapeutic research.

9.3 FULLERENES FOR DIAGNOSTICS

9.3.1 METALLO-FULLERENES AS NEW CONTRAST AGENTS

Another type of fullerene can have metals enclosed inside them (*metallo-fullerenes*). This class is typified by Trimetaspheres (TMS), which are C_{80}-based with gadolinium (Gd) enclosed within the cage (Figure 9.1, right). One field in which they are being investigated for use is in magnetic resonance imaging (MRI)—a widely used diagnostic procedure that utilizes Gd-based contrast agents (e.g., Magnevist). However, several issues have limited Gd-chelate-based contrast agents (i.e., Magnevist) for use as image-guided interventions with MRI including Gd toxicity [54–57], rapid clearance from the body, poor relaxivity-dependent sensitivity, and inability to be targeted to disease-specific biomarkers. The TMS solve many of the problems associated with current Gd-based contrast agents. First, the toxic Gd (inside cage) is separated from active targeting moieties (outside cage) by an extremely stable carbon shell thus potentially increasing safety. Unlike current chelates adding targeting ligands/moieties to the TMS, it does not affect the ability of Gd to become free of the compound. Second, Gd-fullerenes are more sensitive. Targeted imaging agents require strong signals by which to report the presence of an agent at a particular location. The TMS achieves 25- to 50-fold greater relaxivity compared to traditional Gd contrast agents. Third, the fullerene cage can be targeted to disease biomarkers as targeting moieties can be attached to the cage (including empty cage fullerenes). Finally, the TMS are safe; no toxicity has been observed *in vivo* or *in vitro* [46,58].

The crux of the TMS platform for diagnosis utilizes specific biomarkers conjugated to the fullerene MRI contrast agents and, in theory, could be used for any disease for which specific expression of suitable receptors exist. The two disease processes evaluated, atherosclerosis and glioblastoma, possess such disease-specific biomarkers that can be targeted. Based on working prototypes, TMS-specific fullerenes have been developed to study their cellular uptake by macrophage foam cells (for atherosclerosis) and cancer (glioblastoma cells) *in vitro*. This prototype takes advantage of already identified biomarkers on diseased cells, followed by chemically attaching ligands to target the TMS to these cells (characterizing the molecules), and use *in vitro* assays to determine if they bind to and enter the cells and assess any toxicological effects.

9.3.2 Metallo-Fullerenes MRI Contrast Agents as Diagnostics for Atherosclerotic Plaque

Atherosclerosis is an inflammatory disease representing a major healthcare problem in the United States. Atherosclerosis begins when blood cells (monocytes) "stick" to blood vessel walls as a result of an individual's increased cholesterol consumption. Over time the cells continue to accumulate on vessel walls and engulf the fat droplets until large structures called plaque begin to form. The plaque can rupture and lead to complete blockage of blood flow through the vessels involved—leading to heart attack and stroke. To be able to noninvasively determine the extent of plaque buildup in vessel walls of at-risk patients, using already established diagnostic procedures, would facilitate treatment options that could greatly reduce heart attacks and stroke.

To be able to noninvasively determine the extent of plaque burden in vessel walls of at-risk patients would facilitate treatment options that could greatly reduce myocardial infarction and stroke incidence, which represent the most frequent cause of death in our society. While several molecules have already been identified that are selectively expressed or highly upregulated, on plaque lesions [59–61], preliminary findings in the field focus on CD36 (macrophage scavenger receptor) as the target receptor. The CD36 receptor is involved in foam cell formation, mediating the influx of lipids into the macrophages and possessing a promiscuous ligand binding. It is also highly expressed on lipid-laden macrophages human atherosclerotic aorta [62–65]. Different forms of oxidized phosphatidylcholine (oxPC) and oxidixed LDL (oxLDL) are high-affinity ligands for CD36 [66]. The basic structure

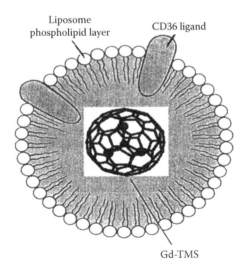

FIGURE 9.5 Basic structure of atherosclerotic-targeting contrast agents. The illustration is not drawn to scale.

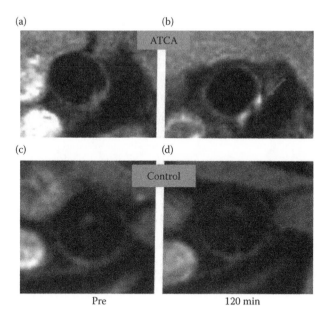

FIGURE 9.6 Atherosclerotic-plaque targeting contrast agents (ATCA) can detect plaque *in vivo*. ApoE -/- with plaque were imaged using MRI. The ATCA (a,b) or non-targeted controls (c,d) were injected i.v. and images of the descending aorta acquired before (a,c) or 120 minutes after (b,d). The arrow indicates the ATCA binding to the plaque in the blood vessel while the controls do not.

of the compounds is the TMS enclosed within liposomes, demonstrated previously [45], with CD36 ligands in the liposomes (Figure 9.5). As seen in Figure 9.6, APOE knockout mice [67] were used to test the atherosclerotic plaque-targeting contrast agent (ATCA), which serves as a working prototype for a new class of diagnostic agents that specifically attach to atherosclerotic plaque so that previously undetected lesions can be diagnosed using MRI. The ATCA clearly reveals plaque accumulation by MRI that is not seen using the same molecule without targeting ligands.

9.4 FULLERENES FOR THERANOSTICS

9.4.1 METALLO-FULLERENES MRI CONTRAST AGENTS AS THERANOSTICS FOR BRAIN CANCER

Glioblastoma multiforme is an aggressive high-grade brain tumor with a poor prognosis. Ideally, a multifunctional nanoscale compound is needed that can simultaneously diagnose tumor progression, but can also specifically target and kill the tumor. Interleukin 13 (IL-13) receptors are selectively expressed on astrocytoma cells in glioma. It has been previously reported that IL-13 receptor-targeted chemotherapies (diphtheria toxin) [68,69] delivered through lipid nano-vesicles (liposomes) were

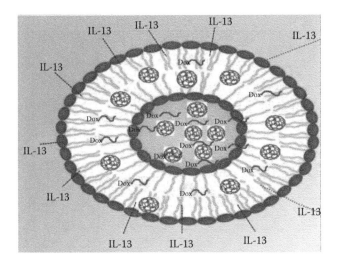

FIGURE 9.7   (**See color insert.**) Representative schematic of glioblastoma targeting theranostic (GTNN). The TMS is incorporated within and/or in the liposome bilayer along with doxorubicin (Dox) and are targeted to the glioma cells by IL-13 (not drawn to scale).

effective in targeting and shrinking glioma tumors in a subcutaneous tumor model. Magnetic resonance imaging using gadolinium (Gd)-based contrast agents has been used for diagnosing glioblastoma. IL-13 receptors are expressed on glioblastoma multiforme (GBM) cells and are not significantly expressed in normal tissue [70]. In fact, many clinical trials are underway using IL-13 as a targeting moiety for various interventions of the disease (see clinicaltrials.gov).

Current platforms are being developed utilizing novel glioblastoma-targeting theranostics (GTTN), which the diagnostic TMS is enclosed within glioblastoma-targeting IL-13-liposomes that are capable of delivering a therapeutic payload of doxorubicin (dox) to the cancerous cells (Figure 9.7). *In vitro* binding efficiencies of GTTN to human glioblastoma cells were verified, followed by *in vivo* evaluation, as seen in Figure 9.8. The glioblastoma theranostics can target and shrink human brain tumors that were transplanted in mice.

9.5   FUTURE DIRECTIONS WITH FULLERENES

Of course, toxicity considerations are implicit when contemplating human use for novel compounds such as fullerenes. To this end there have been a number of studies examining the toxicity using a myriad of fullerene preparations. However, the results of most of these studies are conflicting, inconclusive, and the subject of much debate. One of the most "damaging" fullerene studies (that was subsequently proven to be unfounded) exposed juvenile bass to un-derivatized C_{60}, which is insoluble in water [71]. Unfortunately, the authors did not include a control that provided insights into whether the observed effects were simply due to large aggregated particles/THF contaminants or whether they were specific to the chemical nature of C_{60}. Yet there

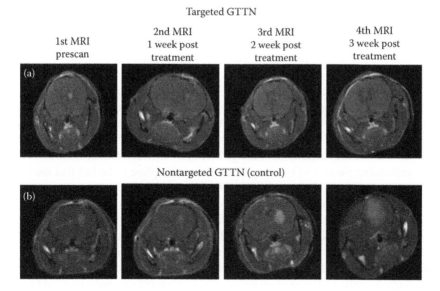

FIGURE 9.8 Glioblastoma targeting nanovesicles with a therapeutic payload attached can shrink brain tumors. Mice with human glioblastoma cells injected into the brain have uncontrolled growth (white mass) as seen in the time lapse in (b). However, when the nanovesicle injected the tumors shrink and disappear (a) as determined with MRI.

was widespread publicity that concluded fullerenes, as a class, could be toxic [72]. A more recent publication by the same group (without publicity) demonstrated that the originally observed "toxicity" was due to impurities in the sample [73]. Even more recently these original studies were "formally" debunked by a follow-up publication that stated the original Oberdorster studies where "compromised by experimental artifacts" [74]. On the opposite end of the spectrum, studies in mice demonstrate that similar C_{60} preparations significantly increase the lifespan of mice [9]. Thus, it is difficult for researchers, and the general public, to determine if fullerenes are dangerous nanostructures that should be banned or a potential novel platform for developing new fountain-of-youth medicines. How can a class of compounds be simultaneously toxic and lifespan extenders?

Contributing to the confusion is that few studies examining toxicity use well-characterized and highly purified material that are not mixtures of many isomers, aggregate sizes, and impurities. Most studies in which fullerenes are deemed to be toxic use starting material with little or no characterization (DLS, zeta potential, FTIR, etc.). In addition, no studies have compared differences in cage sizes (C_{60} vs. C_{70} vs. C_{80}); the latter two are much more likely to become U.S. Food and Drug Administration (FDA) approved products due to their size being conducive to fewer numbers of isomers when adding side chain moieties. The FDA requires that every new chemical entity (NCE) must be evaluated separately; extrapolating toxicity (or nontoxicity) by categorizing compound mixtures and making generalizations about classes of compounds (as is the case with fullerenes) with many different isomers is not acceptable to the FDA. Anyone with a high school-level understanding of basic chemistry is aware that even extremely similar molecules can have different biologic activities. There are many examples where two isomers that are very similar have completely different biological behavior. For example, the tragedy of the drug tha-lidomide, where one isoform was an effective sedative and the mirror image isoform was teratogenic resulting in fetal defects, changed the face of drug testing. This applies to the studies with fullerenes where even extremely small changes in the core fullerene structure (through the addition of side-chain moieties) can result in the FD having completely different biological properties. This has been demonstrated repeatedly by several laboratories, which highlights the difficulty in interpreting data gathered from extrapolating findings between even very similar compounds.

Complicating the task of bringing FD to the marketplace is the fact that the FDA does not have specific guidelines for products containing nanoscale materials. A report issued by the FDA Nanotechnology Task Force (July 2007) recommends guidance by various centers within the FDA for industries working with nanomaterials. Unlike "standard" drug products, it is increasingly evident that at least in the area of charac-terization of nanomaterials used in drug products, different standards apply. Applying small molecule principles and methodologies to nanomaterials cannot be extrapolated in biological settings. The study stressed that biodistribution analysis should be at the core of any evaluation of products containing nanomaterials. These biodistribu-tion studies, as recommended by the FDA, provide valuable information on where the nanoparticles are traveling and possibly accumulating, therefore, subjecting those sites to an increased likelihood of toxicological effects. It was also stressed in the 2007 report that most studies (using nanomaterials) are limited in that they are short-term, and might leave long-term effects unevaluated, especially because the long-term toxicity and effects for most nanoscale materials remain unknown. Furthermore, appropriate endpoints for *in vitro* assays are difficult to determine, as single cell types are often not sufficient for evaluation on the function or health of organs or tissues that are made up of multiple cell types, given that numerous types of tissues are exposed to in the body. The major recommendation was that nanoscale material be characterized with respect to size (surface area and size distribution), chemical composition (such as purity and crystallinity), surface structure (surface reactivity, surface groups, coat-ings, etc.), solubility, shape, and aggregation. The protocols developed at the National Cancer Institute's, Nanotechnology Characterization Laboratories was recommended as being very useful in helping to characterize nanoscale materials and to develop standards and standardized methods for measuring nanoscale materials.

The potential for using fullerene-based medicines is substantial, but concerns of toxicity have slowed the initial enthusiasm that surrounded their discovery. Only those studies using well characterized, single species, "lead candidate" fullerene formula-tions can provide meaningful information regarding potential toxicological effects. Such studies are needed as the state of research today with fullerenes is shaped by studies such as these that address the observation "that extrapolation across similar nanoparticles will be dependent upon surface chemistry and concentration which may affect the degree of agglomeration and thus biological effects" [75]. Thus, more thorough studies will serve as a building block in developing a database that links surface functionalization chemistry of fullerene compounds to biological function.

REFERENCES

1. Basso AS, Frenkel D, Quintana FJ, Costa-Pinto FA, Petrovic-Stojkovic S, Puckett L et al. Reversal of axonal loss and disability in a mouse model of progressive multiple sclerosis. *J Clin Invest*. 2008;118(4):1532–1543.

2. Dugan LL, Turetsky DM, Du C, Lobner D, Wheeler M, Almli CR, et al. Carboxyfullerenes as neuroprotective agents. *Proc Natl Acad Sci USA*. 1997;19;94(17):9434–9439.

3. Bosi S, Da RT, Spalluto G, Balzarini J, Prato M. Synthesis and anti-HIV properties of new water-soluble bis-functionalized[60]fullerene derivatives. *Bioorg Med Chem Lett*. 2003;13(24):4437–4440.

4. Marchesan S, Da RT, Spalluto G, Balzarini J, Prato M. Anti-HIV properties of cationic fullerene derivatives. *Bioorg Med Chem Lett*. 2005;15(15):3615–3618.

5. Berger CS, Marks JW, Bolskar RD, Rosenblum MG, Wilson LJ. Cell internalization studies of gadofullerene-(ZME-018) immunoconjugates into A375 m melanoma cells. *Transl Oncol*. 2011;4(6):350–354. Epub 2011/12/23.

6. Daroczi B, Kari G, McAleer MF, Wolf JC, Rodeck U, Dicker AP. *In vivo* radioprotection by the fullerene nanoparticle DF-1 as assessed in a zebrafish model. *Clin Cancer Res*. 2006;12(23):7086–7091.

7. Lai YL, Murugan P, Hwang KC. Fullerene derivative attenuates ischemia-reperfusion-induced lung injury. *Life Sci*. 2003;72(11):1271–1278.

8. Dellinger A, Zhou Z, Lenk R, MacFarland D, Kepley CL. Fullerene nanomaterials inhibit phorbol myristate acetate-induced inflammation. *Exp Dermatol*. 2009;18(12):1079–1081. Epub 2009/06/27.

9. Quick KL, Ali SS, Arch R, Xiong C, Wozniak D, Dugan LL. A carboxyfullerene SOD mimetic improves cognition and extends the lifespan of mice. *Neurobiol Aging*. 2008;29(1):117–128.

10. Baati T, Bourasset F, Gharbi N, Njim L, Abderrabba M, Kerkeni A et al. The prolongation of the lifespan of rats by repeated oral administration of [60]fullerene. *Biomaterials*. 2012;33(19):4936–4946. Epub 2012/04/14.

11. Ryan JJ, Bateman HR, Stover A, Gomez G, Norton SK, Zhao W et al. Fullerene nanomaterials inhibit the allergic response. *J Immunol*. 2007;179(1):665–672. Epub 2007/06/21.

12. Dellinger A, Zhou Z, Norton SK, Lenk R, Conrad D, Kepley CL. Uptake and distribution of fullerenes in human mast cells. *Nanomedicine*. 2010;6(4):575–582. Epub 2010/02/09.

13. Norton SK, Dellinger A, Zhou Z, Lenk R, Macfarland D, Vonakis B, et al. A new class of human mast cell and peripheral blood basophil stabilizers that differentially control allergic mediator release. *Clin Transl Sci*. 2010;3(4):158–169. Epub 2010/08/20.

14. Galli SJ, Tsai M, Piliponsky AM. The development of allergic inflammation. *Nature*. 2008;454(7203):445–454.

15. Kobayashi T, Miura T, Haba T, Sato M, Serizawa I, Nagai H et al. An essential role of mast cells in the development of airway hyperresponsiveness in a murine asthma model [In Process Citation]. *J Immunol*. 2000;164(7):3855–3861.

16. Williams CM, Galli SJ. Mast cells can amplify airway reactivity and features of chronic inflammation in an asthma model in mice. *J Exp Med*. 2000;192(3):455–462.

17. Adcock IM, Caramori G, Chung KF. New targets for drug development in asthma. *Lancet*. 2008;372(9643):1073–1087.

18. Kepley CL, Cohen N. Evidence for human mast cell nonreleaser phenotype. *J Allergy Clin Immunol*. 2003;112(2):457–459. Epub 2003/08/05.

19. Kepley CL, Youseff L, Andrews RP, Wilson BS, Oliver JM. Syk deficiency in non-releaser basophils. *J Allergy Clin Immunol*. 1999;104(2):279–284.

20. Kepley CL, Youseff L, Andrews RP, Oliver JM. Turning off IgE-mediated signaling natures way; reversible Syk deficiency in non-releaser basophils. *Allergy Clin Immunol Int.* 2001;13:11–17.
21. Kepley CL, Youssef L, Andrews RP, Wilson BS, Oliver JM. Multiple defects in Fc epsilon RI signaling in Syk-deficient nonreleaser basophils and IL-3-induced recovery of Syk expression and secretion. *J Immunol.* 2000;165(10):5913–5920. Epub 2000/11/09.
22. Gordon JR, Galli SJ. Mast cells as a source of both preformed and immunologically inducible TNF-alpha/cachectin. *Nature.* 1990;346:274–276.
23. Brightling C, Berry M, Amrani Y. Targeting TNF-alpha: A novel therapeutic approach for asthma. *J Allergy Clin Immunol.* 2008;121(1):5–10.
24. Kepley CL, McFeeley PJ, Oliver JM, Lipscomb MF. Immunohistochemical detection of human basophils in postmortem cases of fatal asthma. *Am J Respir Crit Care Med.* 2001;164(6):1053–1058. Epub 2001/10/06.
25. Kepley CL, Craig SS, Schwartz LB. Identification and partial characterization of a unique marker for human basophils. *J Immunol.* 1995;154(12):6548–6555. Epub 1995/06/15.
26. Schroeder JT. Basophils beyond effector cells of allergic inflammation. *Adv Immunol.* 2009;101:123–161.
27. Jackson DJ, Sykes A, Mallia P, Johnston SL. Asthma exacerbations: Origin, effect, and prevention. *J Allergy Clin Immunol.* 2011;128(6):1165–1174. Epub 2011/12/03.
28. Norton SK, Wijesinghe DS, Dellinger A, Sturgill J, Zhou Z, Barbour S et al. Epoxyeicosatrienoic acids are involved in the C(70) fullerene derivative-induced control of allergic asthma. *J Allergy Clin Immunol.* 2012;130(3):761–769.e2.
29. Sudhahar V, Shaw S, Imig JD. Epoxyeicosatrienoic acid analogs and vascular function. *Curr Med Chem.* 2010;17(12):1181–1190.
30. Pfister SL, Gauthier KM, Campbell WB. Vascular pharmacology of epoxyeicosatrienoic acids. *Adv Pharmacol.* 2010;60:27–59. Epub 2010/11/18.
31. Morin C, Rousseau E. Effects of 5-oxo-ETE and 14,15-EET on reactivity and Ca^{2+} sensitivity in guinea pig bronchi. *Prostaglandins Other Lipid Mediators.* 2007;82(1–4):30–41. Epub 2006/12/14.
32. Morin C, Sirois M, Echave V, Gomes MM, Rousseau E. EET displays anti-inflammatory effects in TNF-alpha stimulated human bronchi: Putative role of CPI-17. *Am J Resp Cell Mol Biol.* 2008;38(2):192–201. Epub 2007/09/18.
33. Node K, Huo Y, Ruan X, Yang B, Spiecker M, Ley K et al. Anti-inflammatory properties of cytochrome P450 epoxygenase-derived eicosanoids. *Science.* 1999;285(5431):1276–1279. Epub 1999/08/24.
34. Nigrovic PA, Lee DM. Synovial mast cells: Role in acute and chronic arthritis. *Immunol Rev.* 2007;217:19–37.
35. Woolley DE. The mast cell in inflammatory arthritis. *N Engl J Med.* 2003;348(17):1709–1711.
36. Irani AA, Golzar N, DeBlois G, Gruber B, Schwartz LB. Distribution of mast cell subsets in rheumatoid arthritis and osteoarthritis synovia. *Arthritis Rheum.* 1987;30:66.
37. Bridges AJ, Malone DG, Jicinsky J, Chen M, Ory P, Engber W et al. Human synovial mast cell involvement in rheumatoid arthritis and osteoarthritis: Relationship to disease type, clinical activity, and antirheumatic therapy. *Arthritis Rheum.* 1991;34:1116–1124.
38. Gotis-Graham I, Smith MD, Parker A, McNeil HP. Synovial mast cell responses during clinical improvement in early rheumatoid arthritis. *Ann Rheum Dis.* 1998;57(11):664–671.
39. Lavery JP, Lisse JR. Preliminary study of the tryptase levels in the synovial fluid of patients with inflammatory arthritis. *Ann Allergy.* 1994;72:425–427.
40. Lee DM, Friend DS, Gurish MF, Benoist C, Mathis D, Brenner MB. Mast cells: A cellular link between autoantibodies and inflammatory arthritis. *Science.* 2002;297(5587):1689–1692.

41. Dinser R. Animal models for arthritis. *Best Pract Res Clin Rheumatol.* 2008;22(2):253–267.
42. Nigrovic PA, Lee DM. Mast cells in autoantibody responses and arthritis. *Novartis Found Symp.* 2005;271:200–209.
43. Phillips DC, Dias HK, Kitas GD, Griffiths HR. Aberrant reactive oxygen and nitrogen species generation in rheumatoid arthritis (RA): Causes and consequences for immune function, cell survival, and therapeutic intervention. *Antioxid Redox Signal.* 2010;12(6):743–785.
44. Winyard PG, Ryan B, Eggleton P, Nissim A, Taylor E, Lo Faro ML, Burkholz T et al. Measurement and meaning of markers of reactive species of oxygen, nitrogen and sulfur in healthy human subjects and patients with inflammatory joint disease. *Biochem Soc Trans.* 2011;39(5):1226–1232.
45. Zhou Z, Lenk RP, Dellinger A, Wilson SR, Sadler R, Kepley CL. Liposomal formulation of amphiphilic fullerene antioxidants. *Bioconjug Chem.* 2010;21(9):1656–1661. Epub 2010/09/16.
46. Ehrich M, Van Tassell R, Li Y, Zhou Z, Kepley CL. Fullerene antioxidants decrease organophosphate-induced acetylcholinesterase inhibition in vitro. *Toxicol In Vitro.* 2011;25(1):301–307. Epub 2010/10/05.
47. Zhao W, Gomez G, Macey M, Kepley CL, Schwartz LB. *in vitro* desensitization of human skin mast cells. *J Clin Immunol.* 2012;32(1):150–160.
48. Chirico F, Fumelli C, Marconi A, Tinari A, Straface E, Malorni W et al. Carboxyfullerenes localize within mitochondria and prevent the UVB-induced intrinsic apoptotic pathway. *Exp Dermatol.* 2007;16(5):429–436.
49. Fumelli C, Marconi A, Salvioli S, Straface E, Malorni W, Offidani AM et al. Carboxyfullerenes protect human keratinocytes from ultraviolet-B-induced apoptosis. *J Invest Dermatol.* 2000;115(5):835–841.
50. Pitman N, Asquith DL, Murphy G, Liew FY, McInnes IB. Collagen-induced arthritis is not impaired in mast cell-deficient mice. *Annals Rheum Dis.* 2011;70(6):1170–1171. Epub 2010/12/07.
51. Brown MA, Tanzola MB, Robbie-Ryan M. Mechanisms underlying mast cell influence on EAE disease course. *Mol Immunol.* 2002;38(16–18):1373–1378.
52. Gregory GD, Bickford A, Robbie-Ryan M, Tanzola M, Brown MA. MASTering the immune response: Mast cells in autoimmunity. *Novartis Found Symp.* 2005;271:215–225; discussion 225–231:215–225.
53. Secor VH, Secor WE, Gutekunst CA, Brown MA. Mast cells are essential for early onset and severe disease in a murine model of multiple sclerosis. *J Exp Med.* 2000;191(5):813–822.
54. Chewning RH, Murphy KJ. Gadolinium-based contrast media and the development of nephrogenic systemic fibrosis in patients with renal insufficiency. *J Vasc Interv Radiol.* 2007;18(3):331–333.
55. Runge VM. Gadolinium and nephrogenic systemic fibrosis. *AJR Am J Roentgenol.* 2009;192(4):W195–W196.
56. Ledneva E, Karie S, Launay-Vacher V, Janus N, Deray G. Renal safety of gadolinium-based contrast media in patients with chronic renal insufficiency. *Radiology.* 2009;250(3):618–628.
57. Cowper SE. Nephrogenic systemic fibrosis: A review and exploration of the role of gadolinium. *Adv Dermatol.* 2007;23:131–154.
58. MacFarland DK, Walker KL, Lenk RP, Wilson SR, Kumar K, Kepley CL et al. Hydrochalarones: A novel endohedral metallofullerene platform for enhancing magnetic resonance imaging contrast. *J Med Chem.* 2008;51(13):3681–3683. Epub 2008/06/19.

59. Ouimet T, Lancelot E, Hyafil F, Rienzo M, Deux F, Lemaitre M et al. Molecular and cellular targets of the MRI contrast agent p947 for atherosclerosis imaging. *Mol Pharmaceutics*. 2012;9(4):850–861. Epub 2012/02/23.
60. Uno K, Nicholls SJ. Biomarkers of inflammation and oxidative stress in atherosclerosis. *Biomark Med*. 2010;4(3):361–373.
61. Sadeghi MM, Glover DK, Lanza GM, Fayad ZA, Johnson LL. Imaging atherosclerosis and vulnerable plaque. *J Nucl Med*. 2010;51 Suppl 1:51S–65S.
62. Nergiz-Unal R, Rademakers T, Cosemans JM, Heemskerk JW. CD36 as a multiple-ligand signaling receptor in atherothrombosis. *Cardiovasc Hematol Agents Med Chem*. 2011;9(1):42–55.
63. Collot-Teixeira S, Martin J, Rmott-Roe C, Poston R, McGregor JL. CD36 and macrophages in atherosclerosis. *Cardiovasc Res*. 2007;75(3):468–477.
64. Ge Y, Elghetany MT. CD36: A multiligand molecule. *Lab Hematol*. 2005;11(1):31–37.
65. Silverstein RL, Febbraio M. CD36, a scavenger receptor involved in immunity, metabolism, angiogenesis, and behavior. *Sci Signal*. 2009;2(72):re3.
66. Podrez EA, Poliakov E, Shen Z, Zhang R, Deng Y, Sun M et al. Identification of a novel family of oxidized phospholipids that serve as ligands for the macrophage scavenger receptor CD36. *J Biol Chem*. 2002;277(41):38503–38516.
67. Kolovou G, Anagnostopoulou K, Mikhailidis DP, Cokkinos DV. Apolipoprotein E knockout models. *Curr Pharm Des*. 2008;14(4):338–351.
68. Li C, Hall WA, Jin N, Todhunter DA, Panoskaltsis-Mortari A, Vallera DA. Targeting glioblastoma multiforme with an IL-13/diphtheria toxin fusion protein *in vitro* and *in vivo* in nude mice. *Protein Eng*. 2002;15(5):419–427.
69. Todhunter DA, Hall WA, Rustamzadeh E, Shu Y, Doumbia SO, Vallera DA. A bispecific immunotoxin (DTAT13) targeting human IL-13 receptor (IL-13R) and urokinase-type plasminogen activator receptor (uPAR) in a mouse xenograft model. *Protein Eng Des Sel*. 2004;17(2):157–164.
70. Madhankumar AB, Slagle-Webb B, Mintz A, Sheehan JM, Connor JR. Interleukin-13 receptor-targeted nanovesicles are a potential therapy for glioblastoma multiforme. *Mol Cancer Ther*. 2006;5(12):3162–3169.
71. Oberdorster E. Manufactured nanomaterials (fullerenes, C60) induce oxidative stress in the brain of juvenile largemouth bass. *Environ Health Perspect*. 2004;112(10):1058–1062.
72. Barnaby JF. Study raises concerns about carbon particles. *New York Times*. 2004.
73. Zhu S, Oberdorster E, Haasch ML. Toxicity of an engineered nanoparticle (fullerene, C60) in two aquatic species, Daphnia and fathead minnow. *Mar Environ Res*. 2006;62 Suppl:S5–S9.
74. Henry TB, Petersen EJ, Compton RN. Aqueous fullerene aggregates (nC60) generate minimal reactive oxygen species and are of low toxicity in fish: A revision of previous reports. Current opinion in biotechnology. 2011;22(4):533–537. Epub 2011/07/02.
75. Saathoff JG, Inman AO, Xia XR, Riviere JE, Monteiro-Riviere NA. *In vitro* toxicity assessment of three hydroxylated fullerenes in human skin cells. *Toxicol in Vitro*. 2011;25(8):2105–2112. Epub 2011/10/04.

Section IV

Nanomodeling

10 Modeling at Nano Scale

Material Chemistry Level Modeling in Processing and Mechanics of Engineered Materials

Ram V. Mohan and Ajit D. Kelkar

CONTENTS

10.1 INTRODUCTION

The properties and behavior of materials are strongly dependent upon their fundamental building blocks and their characteristics starting from their molecular and sub-molecular structures. This dependency is further strengthened in engineered/

nanoengineered multi-scale, heterogeneous material systems (e.g., polymer composites, cementitious materials) where the mechanical properties, behavior, and damage propagation depends upon the interdependency, and material morphology transcending across the nano, micro, to macro length scales. The properties and behavior of the multi-scale complex and heterogeneous materials with material phases at varying length scales including nanomaterial constituents is strongly influenced by material interactions during the processing, as well as damages and defects in the associated constituent nanomaterials starting from their material chemistry structure levels. These processing and material-induced variations subsequently influence their engineering scale properties, strength, and failure behavior associated at various length scales of the multi-scale composite material system. A bottom-up modeling approach starting from the material chemistry at the nano length scale is effective in providing an insight into the material level interactions that exist due to variations in the processing methods, their influence on the associated material properties, functions well as an effective methodology to provide predictive engineering properties, behavior and the influence of these material properties, and behavior due to the variations in the material chemistry.

With an emphasis on the material chemistry level, this chapter focuses on the atomistic/molecular nano scale level modeling and their applications in the processing and mechanics of engineered material systems. The material chemistry level nano length scale modeling in the processing and mechanics of three engineered material systems are discussed. They are

1. Processing of alumina nanoparticulate hybrid composites
2. Epoxy–carbon nanotube (CNT) polymer nanocomposite
3. Cementitious materials

The first two polymer composite material systems discussed in this chapter clearly elucidate the applicability of low length-scale molecular dynamics (MD) modeling at the material chemistry level of these material systems. This provides insight into the molecular level interactions that are influenced by functionalization during the processing, and molecular defects that impact the engineering scale mechanical properties of nanocomposites. These approaches are also effective in the mechanics of several other engineered material systems as the atoms and bonds and their interaction captured at their associated length scale defines the material behavior and characteristics at their engineering scales. The application of the material chemistry level modeling to obtain the mechanical properties and stress–strain behavior from our recent work on heterogeneous cementitious material is also presented and discussed. Such material modeling based on material chemistry enables us to understand the expected predictive properties and behavior of the materials due to changes in material chemistry, nanomaterial additives, and so on, for tailored material development and in coupling material science aspects into the engineering design analysis. The theoretical background of the MD method and details of the molecular chemistry structures of the material constituents in the three material systems; different material interactions as related to the variations in the processing methods, associated potential energy, and modeling approaches employed are discussed.

10.2   COMPUTATIONAL MODELING AND NANO-LENGTH SCALE

Computational modeling is playing an important role in science and engineering of materials. Computational models by definition are mathematical representation of a phenomenon for a system. Computational models along with experimental investigations and analytical theory are the three corners required to understand processing–structure–property triangle of material systems. Due to the rapid growth of computational materials science (CMS) in the field of materials research [1] and a need to understand the scale size of the associated material chemistry, researchers and engineers are now challenged to investigate beyond the macro- and micro-scale levels. Computational simulations at the molecular/nano scale level enable us to study the behavior of materials at their molecular, material chemistry level. This is critical not only to predict but also to understand and improve the mechanical properties through appropriate material additives and molecular structural changes. MD modeling is an effective methodology for such analysis at the molecular, nano scale level and is one of the effective computational methodologies that can be employed at nano-length scale. In most materials modeling applications of mechanical behavior and interactions, MD modeling methods provide an effective means to include the energies of the molecular interactions associated with the material chemistry in a computationally efficient manner; neglecting the electronic degree of freedom that are included with quantum mechanics-based methods, but at a higher computational cost. As a computational method that is applicable in several material systems including bio systems based on the fundamental building block of atoms and bonds, the following chapter on bio-nano interfaces also employs the same computational methodology, and a further coarse-grained formulation to simulate larger bio-molecular material systems. A brief background of the MD methodology is presented next for completeness.

10.2.1   MD MODELING APPROACH

MD modeling is a computer simulation of atoms and molecules interacting with each other. MD methodology was formulated in the late 1950s and early 1960s by Alder and Wainwright [2]. Currently, MD is applied in materials science and biomolecules, and allows the study and investigation of the structure and behavior of interacting atoms in any molecular system. The methodology is built upon transient dynamic analysis of atoms represented by a system of particles based on classical Newtonian mechanics. The transient dynamic atomistic configurations under varying thermodynamic state conditions is coupled with statistical mechanics to obtain predicted thermo-physical properties, as well as the molecular level behavior of material systems based on their atomistic positions and velocities. The associated Newton's equation of motion can be represented by

$$F_i = m_i a_i = m_i \frac{dv_i}{dt} = m_i \frac{d^2 r_i}{dt^2} \qquad (10.1)$$

where (m_i, v_i, r_i) are atom mass, velocities, and position, respectively; F is associated force, defined as the gradient of the potential energy representing the bonded and nonbonded energy associated with the material's molecular configuration.

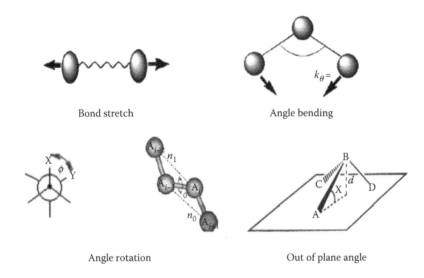

Bond stretch

Angle bending

Angle rotation

Out of plane angle

FIGURE 10.1   Potential energy atom bond interaction energy components.

The total potential energy for the material system varies for various molecular types due to the different associated molecular parameters and atoms. Several forms of the potential energy function tailored for specific material systems that gives good conformation to the physical properties built upon first principle calculations have been formulated and are used to define the total potential energy. These include mathematical terms corresponding to major molecular interactions, which are bonded terms, nonbonded terms, and cross terms. Bond terms include bond stretching, angle bending, angle rotation, and out-of-plane angle terms, associated with configurations of the molecular material system [3]. Figure 10.1 illustrates the bonded atom interaction components associated with bond stretching, angle bending, torsion, and out-of-plane rotation. Nonbonded energies include van der Waals and electrostatic energy as shown in Figure 10.2.

In addition, cross terms are important for predicting the vibration frequencies and structural variation, and include the combination of internal coordinates such as bond–bond, bond–angle, and bond–torsion.

Molecular dynamics: MD is a combination of molecular modeling, computer simulation, and statistical mechanics. MD modeling analysis consists of three steps: energy minimization, dynamic analysis, and post analysis for the calculation of the mechanical properties.

van der Waals interactions

Electrostatics

FIGURE 10.2   Potential energy nonbond components.

Energy minimization is finding the stable structure configurations, which corresponds to the lowest energy for the molecular system. Some of the methods that can be used in the static minimization are: steepest descents, conjugate gradient, Newton–Raphson, and simplex method. Steepest descents and conjugate gradient are gradient methods, which depend on the direction of the first derivative of the potential energy function, and indicate where the minimum lies.

Dynamic analysis: The dynamics of atom motion based on the initial atom velocities that are dictated by thermodynamic state temperature and pressure conditions are determined by solving the coupled Newton's equation of motion for all the atoms. The time-dependent atom configuration at each time step forms the trajectory, with molecular configurations at each time step forming a possible phase-state configuration of the molecular system, following the principles of statistical mechanics. Time integration algorithms are used to integrate the equation of motion. Some of the time integration methods are Euler method, second-order Runge–Kutta method (RK2) or sometimes called modified Euler method, fourth-order Runge–Kutta method (RK4), Verlet algorithm, velocity-Verlet algorithm, and predictor–corrector methods. The modeling analysis applications presented in this chapter employed velocity-Verlet algorithm. The time dynamic trajectory consisting of several snapshots of atomic positions and velocities of the material molecular structure are used for the predictive mechanical property determination in conjunction with Ergodic hypothesis linking the time average from classical mechanics to the phase averages following statistical mechanics. In the first of the two molecular modeling applications relevant to polymer nanocomposites discussed in this chapter, the interface energies of different material phases at the material chemistry level are based on their associated potential energy, while the mechanical properties are evaluated based on their dynamics time trajectories and associated properties following Ergodic hypothesis. The details the three material chemistry level modeling applications identified earlier are discussed and presented next.

10.3 MATERIAL CHEMISTRY LEVEL MODELING OF MATERIAL INTERFACES IN ALUMINA NANOPARTICULATE HYBRID COMPOSITES

Hybrid epoxy fiber composites with alumina nanoparticles are formed via nano-material integration through either fiber or resin modification [4,5]. To improve the compatibility of inorganic alumina nanoparticulates with epoxy polymer, alumina was functionalized using silane (tris-2-methoxyethoxy-vinyl-silane). The engineering scale mode-I fracture toughness values were found to be significantly higher with functionalized alumina in our experimental investigations [6]. Engineering macro-scale material behavior depends upon the material interactions at the material chemistry level. In this case, the functionalization of alumina with silane changes the material interface configuration to silane–alumina from the original epoxy–alumina, altering the molecular interfacial energy across the associated material chemistry level configurations. Molecular mechanics modeling for the stable interaction energy of different material interfaces in the hybrid epoxy fiber composite that exist due to functionalization during the processing are studied and correlated. These are built upon the recent advances and interest in the understanding of the influence of material chemistry on the larger engineering scale behavior of materials [5,7–11].

The primary and secondary constituent material chemistry structures involved are

1. Epoxy resin based on epoxy end chain with bisphenol-A inner organic group
2. Curing agent butan-diamine, an amorphous organic compound
3. Alumina (aluminum oxide, Al_2O_3), an inorganic compound
4. Tris-2-methoxyethoxy-vinyl-silane, with an organo-functional group linked to vinyl silane

The cured molecular models of epoxy consisting of EPON™ 9554 and curing agent butan-diamine (BDA) were developed using Accelrys Material Studio (Accelrys MS) development tool [12]. The molecular models are configured for various degrees of networking of the cross-linked epoxy structure. MD models of the silane agent tris-2-methoxyethoxy vinyl silane (T2MEVS) and unbonded alumina structure were also developed using Accelrys MS. A parameterized COMPASS (Condensed-phase Optimized Molecular Potentials for Atomistic Simulation Studies) force-field term [13] was employed in the modeling analysis. MD analysis employed a canonical (NVT, constant number of atoms, constant volume, and constant temperature) thermodynamic state ensemble.

10.3.1 MOLECULAR MODELING OF MATERIAL CONSTITUENTS

10.3.1.1 EPON™ 9554, BDA, and Cured Epoxy

EPON™ 9554 is an epoxy end chain (di-epoxy) with a bisphenol-A inner organic group. The epoxy structure is an amorphous polymer with high propensity for cross-linking polymer chains to form a thermoset polymer compound. The di-epoxy ends provide the potential for a polymerization cure by the cross-linking reaction. Figure 10.3 shows the molecular/chemical structure of the epoxy and BDA. The molecular weight of epoxy is 340.419 amu.

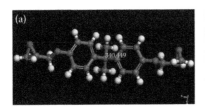

FIGURE 10.3   Material chemistry molecular structure: (a) epoxy resin and (b) BDA.

FIGURE 10.4 (a) Rhomboid, (b) triangle, and (c) rectangle molecular configurations of cured epoxy structure.

The curing compound used in the processing is butan-diamine (BDA). It is a non-aromatic polyamine and an amorphous organic compound. It is very volatile, readily breaking down to give off toxic ammonia gas. The end di-amine chain serves as a cross-linker in a complex polymeric structure due to its ability to give off its hydrogen to the epoxy molecule to form a hydroxyl and bond to the carbon atom in the epoxy ends. The molecular weight of butan-diamine is 88.154 amu.

MD models of the cured epoxy were built using the configurations for the epoxy resin chemically bonded with BDA curing agent. Using a stoichiometric ratio of 1:1 to concur with the mixing weight ratio employed during processing, molecular unit consisting of 4, 6, and 8 molecules of epoxy and curing agents were considered along with rhomboid, triangular, and rectangular structures of cured epoxy structure with 4, 6, and 8 molecule cells. These epoxy cell structures are shown in Figure 10.4.

10.3.1.2 Atomistic Analysis of Material Constituent Interactions

The interaction energy analysis between the material constituents employed layered molecular cell configurations. The molecular cells used are interfaced material chemistry structures placed in a layered vacuum slab. The two molecular cells of any two primary material constituents are interfaced with each other to form a layered cell. The layered cells used in this study are: epoxy–alumina representing the interface material interaction between the epoxy and nonfunctionalized alumina; silane–alumina, representing the material interaction between functionalized alumina and epoxy resin; epoxy–silane and epoxy–epoxy layered configurations were studied as control modeling analysis. Figure 10.5 presents the four different configurations of layered molecular interactions for the different material interaction interfaces.

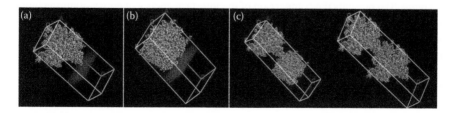

FIGURE 10.5 Molecular layer configurations: (a) epoxy–alumina (nonfunctionalized interaction); (b) silane–alumina (functionalized interaction); (c) epoxy–silane and epoxy–epoxy.

10.3.1.3 Surface Interaction Energy of Constituent Chemistry Level Material Interactions

Total potential energies are obtained after canonical molecular mechanics for the periodic cells of the layered molecular models shown in Figure 10.5. The potential energy of the four material interaction configurations shown in Figure 10.5 is used to compute the interface energy of the interactions. The interface energy between two material constituents A and B is computed as follows for a layered cell that contains these two molecular material components. The total energy of the layered cell is the sum of the energy of each component and the energy at the interface. The surface energy at the interface of a layered molecular cell can thus be defined by

$$E_{A-B} = E_T - (E_A + E_B) \tag{10.2}$$

Subscripts A and B denote two molecular material constituents in a layered cell, E_T is the total energy of the entire layered cell, while E_A and E_B are the energies of the individual materials, and E_{A-B} defines the energy at their interface. The interaction energy at the interface is determined by the surface interaction energy per unit area given by

$$\gamma = \frac{E_{A-B}}{NS} \tag{10.3}$$

In the above equation, N is the number of moles and S is the interfacing lattice area in squared angstroms. All MD analysis were performed using the Discover module in Accelrys [12,14].

In the analysis of organic–inorganic interactions, the energy contribution of alumina is negative due to the neglecting of the ionic bond interaction in the nonbonded alumina molecular models. All the interactions involving alumina neglected ionic bond interaction consistently. This results in a net negative energy of the system. This net repulsion is indicative of the energy between the alumina with either epoxy or silane T2MEVS that does not include the ionic bond interaction within alumina. Both alumina-layered molecular material configurations employed consistent noninclusion of ionic bond interactions between alumina molecules. Complete details of the molecular modeling are presented in reference [15].

Figure 10.6 presents the comparison of the interface energy of epoxy–alumina (nonfunctionalized interaction) and silane–alumina (functionalized interaction) interfaces for 10 cured epoxy molecular units. The energy level comparisons show silane to have improved adhesion to alumina with a higher energy level than a direct EPON–alumina.

The interaction of epoxy and silane agent T2MEVS were analyzed with a canonical ensemble in a molecular vacuum slab employing these material chemistry structures with equivalent cell sizes. The equivalences used are epoxy interacting with epoxy as a control, and epoxy interacting with an equivalent silane cell. Table 10.1 presents the computed interface energy for the epoxy–epoxy and epoxy–silane configuration using a 10-cell epoxy configuration for the three cured epoxy molecular configurations (Figure 10.4). There was a very little difference in interface binding

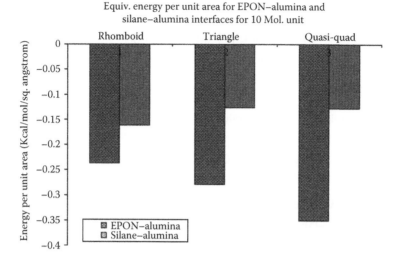

FIGURE 10.6 Comparison of epoxy–alumina and silane–alumina interfaces energies.

energy between epoxy–epoxy and epoxy–silane material chemistry configurations indicating that these two material constituents are equally compatible.

It can be inferred from Figure 10.6, that silane–alumina interface showed better adhesion compared to epoxy–alumina. This improved adhesion of silane–alumina could potentially be attributed to the increase in fracture toughness energy release rate values with functionalization that were obtained for the mode-I fracture behavior of the alumina nanoparticulate epoxy glass hybrid composite laminates in the experimental studies [6]. Table 10.2 presents the mode-I fracture toughness values and their comparison between functionalized and nonfunctionalized alumina nanoparticulates obtained from engineering scale material characterization based on double cantilever beam (DCB) testing using modified beam theory. It can be inferred from the interaction energy of the material constituents obtained through molecular modeling that functionalization employed during the processing improves the chemical affinity between the constituent materials involved. This can be a potential contributor to the improved mode-I fracture toughness values with functionalization, observed in the macroscopic, engineering scale experimental studies.

TABLE 10.1

Interface Binding Energy for Epoxy–Epoxy and Epoxy–Silane

Epoxy Structure	Rhomboid	Triangle	Quasi-Quad
Epoxy–epoxy energy (KCal/Mol/$Å^2$)	−0.00744	−0.00737	−0.00838
Epoxy–silane energy (KCal/Mol/$Å^2$)	−0.00835	−0.00895	−0.01103

TABLE 10.2

Mode-I Fracture Toughness at Crack Initiation in (J/m²)

	Resin Modification with Nonfunctionalized Alumina Nanoparticulate	Resin Modification with Silane Functionalized Alumina Nanoparticulate	Fiber Modification with Nonfunctionalized Alumina Nanoparticulate	Fiber Modification with Silane Functionalized Alumina Nanoparticulate
$G_{IC\,MBT}$	368.70 ± 4.73	450.60 ± 10.30	338.58 ± 14.47	422.64 ± 16.42

Source: Adapted from Akinyede, O. et al., *Journal of Composite Materials*, 2010. 43(7): 769–781.

10.4 MECHANICAL PROPERTY PREDICTIONS OF EPOXY–CARBON NANOTUBE COMPOSITES

Computational techniques such as MD simulations based on material chemistry are an alternative to traditional experimental and theoretical methods for estimating mechanical properties of the epoxy–carbon nanotube (CNT) composite systems. However, differences have been observed between results obtained from engineering scale experiments and those obtained from material chemistry models via MD simulations, with the experimental values being lower. The effect of carbon vacancy defects in the single-walled carbon nanotube (SWCNT) on the elastic Young's modulus of the EPON 862-DETDA-SWCNT composite evaluated through MD simulations is presented in this section.

CNTs have gained significant attention over the years because of their superior chemical, mechanical, and thermo-physical properties. Inclusion of CNTs in polymer matrices have shown significant improvement when comparing properties with the parent polymers; however, defects in these CNTs have also been observed to have detrimental effects on the mechanical properties of the composites [16–18]. Material chemistry level modeling as demonstrated in this chapter helps to understand how the variations in material chemistry influence engineering scale properties of material systems. The elastic modulus evaluated in the material chemistry level models following MD analysis is based on the associated molecular-level potential energy that depends on the dynamic atomic positions.

Mechanical properties: Elastic constants can be determined based on the associated potential energy of the molecular system, and is given by

$$C_{ij} = \frac{1}{V} \frac{\partial^2 U}{\partial \varepsilon_i \partial \varepsilon_j} \tag{10.4}$$

where ε_i, ε_j are lattice strain components, U is potential energy, and V is the MD molecular cell volume. The above equation can be used to obtain the elastic stiffness matrix (C) and the elastic compliance matrix (S) (the inverse of C). The elastic modulus (E) can be computed from elastic compliance matrix (S).

$$E_x = S_{11}^{-1}, \quad E_y = S_{22}^{-1}, \quad E_z = S_{33}^{-1} \tag{10.5}$$

10.4.1 Modeling Methodology

All molecular models were created and analyzed in Materials Studio/Discover by Accelrys [12]. The potential energy of the molecular structures was characterized by the COMPASS force field, with the nonbond energies characterized by the van der Walls and Coulomb's interactions. Figure 10.7 shows the molecular structures of EPON 862 and DETDA.

The recommended weight ratio of EPON 862 to DETDA for a fully cured composite during processing is 100:26.4. Because EPON 862 has a molecular weight of 312, and DETDA has a molecular weight of 178, the molecular ratio of the fully cured composite was approximated to 2 molecules of EPON 862 to 1 molecule of DETDA. The fully cured composite chemistry level molecular structure was therefore constructed with 8 molecules of EPON 862 and 4 molecules of DETDA.

Three nanocomposite models with SWCNT weight percentages between 7% and 12% were investigated. The three molecular models of the defective SWCNT (DSWCNT) cured epoxy system used had the following configurations and weight percentages:

1. 3 unit cells of DSWCNT and 2 fully cured epoxy matrix corresponding to the CNT weight percentage of 11.28–11.58%.
2. 4 unit cells of DSWCNT and 3 fully cured epoxy matrix corresponding to the CNT weight percentage of 10.34–10.49%.
3. 4 unit cells of DSWCNT and 4 fully cured epoxy matrix corresponding to the CNT weight percentage of 7.95–8.08%.

The molecular material systems were minimized to obtain the lowest energy configuration. A cascade of the steepest descent minimization method and the Fletcher–Reeves method were used for the minimization. The minimized molecular material configurations were then equilibrated with the NVT ensemble for 100 ps at 298 K. A sample simulation cell of cross-linked epoxy and CNT is shown in Figure 10.8.

Simulated annealing was used to mimic the curing cycle of EPON 862-DETDA-SWCNT composite and to ensure that the final configuration had the lowest energy possible. A characteristic property of simulated annealing is lowering the temperature slowly in stages to allow thermal equilibrium to be attained at each stage. The cell was heated to 498 K, and the temperature was dropped to 298 K in steps of 10 K using the NPT thermodynamic state ensemble with a specified pressure of 0.0001 GPa (1 atm). MD analyses were performed at each temperature for a total

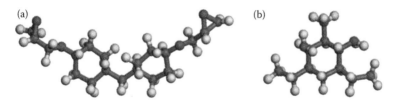

FIGURE 10.7 (a) EPON 862 and (b) DETDA material chemistry structures.

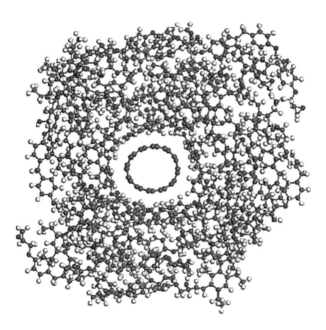

FIGURE 10.8 Simulation cell showing the CNT embedded in the fully cured epoxy matrix.

time duration of 200 pico seconds (ps) with a time step of 1 femto second (fs). The final structure of each temperature step was used as the starting structure of the next step. At 298 K, an analysis of the elastic properties was performed using 10 trajectories from the dynamic analysis for the estimation.

10.4.1.1 Defect Types

The effect of 2- and 4-carbon vacancy defects in the SWCNT on the mechanical properties of the nanocomposite are investigated. The removal of two adjacent vertical carbon atoms on one side of the nanotube resulted in 2 defects. Because of the short length of the nanotube, 4 defects were formulated by removing two adjacent vertical carbon atoms on opposite sides of the nanotube. Figure 10.9 shows an image of the DSWCNT with 2 adjacent carbon atoms removed.

The rule of mixtures density for the nanocomposite was computed using a SWCNT density of 1.9 g/cm³ and the epoxy resin density of 1.2 g/cm³. To obtain the Young's modulus at the exact densities for each of the epoxy–CNT molecular structures, simulated annealing runs were conducted at three different lattice configurations of

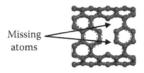

Missing atoms

FIGURE 10.9 SWCNT with two missing adjacent carbon atoms.

the molecular cell for each epoxy–CNT molecular structure, and interpolated to the exact density to obtain the corresponding elastic Young's modulus.

10.4.1.2 Predictive Elastic Modulus of Epoxy: CNT Nano Composite

Table 10.3 presents the predicted mechanical properties of the epoxy–CNT nanocomposites obtained from the MD simulations, and can be interpreted as follows. For the 11.87% SWCNT composite, introduction of 2.78% of defects into the SWCNT resulted in 17.9% overall reduction in the elastic modulus, while introduction of 5.56% of defects into the CNT resulted in 29.6% reduction in the elastic modulus. For the 10.67% SWCNT composite, introduction of 2.08% of defects into the SWCNT resulted in 15.7% overall reduction in the elastic modulus of the nanocomposite, while introduction of 4.17% of defects into the CNT resulted in 25.1% reduction in the elastic modulus. For the 8.24% SWCNT composite, introduction of 2.08% of defects into the SWCNT resulted in 13.6% overall reduction in the elastic modulus of the composite, while introduction of 4.17% of defects into the CNT resulted in 21.9% reduction.

Atomistic material chemistry level modeling for the predictive properties of epoxy–CNT nanocomposites clearly indicated that 2-carbon vacancy defects in the SWCNT resulted in a reduction in Young's modulus between 13% and 18% when compared with the defect-free CNT–epoxy nanocomposite molecular models, while four vacancy defects resulted in a reduction of Young's modulus value between 21% and 30% when compared with the defect-free CNT–epoxy nanocomposite molecular models. There is a great potential for the damage, and introduction of such defects during the experimental processing of CNT–epoxy nanocomposites. This coupled with the multiple orientations of the CNT distribution in the macro, engineering scale specimens potentially contribute to the disparity seen in the predictive molecular modeling properties and the experimental macro properties of these nanocomposites.

Clearly, material chemistry level modeling based on MD provides an effective means for predictive material modeling and understanding the effect of material chemistry structure changes, as well as nano additives on the mechanical properties. Such material modeling is also effective in obtaining not only the material properties but also their deformation characteristics in different material systems. Material chemistry modeling to obtain the material stress–strain behavior in cementitious material systems from our recent work is presented next.

TABLE 10.3
Reduction in Young's Modulus with the Introduction of Defects

CNT Weight% (No Defect)	Young's Modulus (GPa)	
	% Reduction with 2 Defects in Composite	% Reduction with 4 Defects in Composite
11.87	17.9 (2.78% defects in CNT)	29.6 (5.56% defects in CNT)
10.69	15.7 (2.08% defects in CNT)	25.1 (4.17% defects in CNT)
8.24	13.6 (2.08% defects in CNT)	21.9 (4.17% defects in CNT)

10.5   MATERIAL CHEMISTRY LEVEL MODELING IN MECHANICS OF CEMENTITIOUS MATERIALS

Concrete, one of the most used materials in the world is a mixture of cement, water, fine and coarse aggregate, and is an excellent example of highly heterogeneous composite material system. Cement paste consisting of a dry cement powder and water mix starting configuration, by itself is a complex, multi-scale composite material system that evolves during the hydration process. Though commonly used over the years, cementitious materials are one of the complex material systems in terms of material morphology and structure; more so than fiber-reinforced composite materials due to the presence of multiple and highly heterogeneous material phases with no specific repeatability. In addition, cementitious materials undergo chemical and morphological changes gaining strength during the transient hydration process. Hydration in cement is a very complex process creating complex microstructures and the associated molecular structures that vary. A fundamental understanding can be gained through nano to continuum multi-scale level modeling for the behavior and properties of both hydrated and unhydrated cement material chemistry level constituents at the atomistic/nano length scale in order to further explore their role and the manifested effects at larger engineering length scales.

Portland cement in powder form consists of four different major constituents: tricalcium silicate (C_3S), di-calcium silicate (C_2S), tri-calcium aluminate (C_3A), and tetra calcium aluminoferrite (C_4AF) [19]. Different mixture percentages of these constituents produce different types of Portland cement, which is the most commonly used cement [20]. Cement and water mixture (cement paste) is the binder for the concrete. Hydration of cement is the chemical reaction between cement compounds and water, which cause the hardening of cement forming the heterogeneous composite material. The hydration process is highly complex and produces complicated products that control the strength. The most important hydrated cement product is calcium silicate hydrate (C–S–H). Owing to the complexity of C–S–H, molecular structure of C–S–H has not been resolved yet. Other naturally occurring minerals Jennite [21] and Tobermorite14 [22] molecular structures are the closest representation of C–S–H crystal that are accepted in the field. In the above unhydrated and hydrated constituents, C refers to CaO, A refers to Al_2O_3, S refers to SiO_2, H refers to H_2O, and F refers to Fe_2O_3 notation commonly used in cementitious material community.

While the prior section of this chapter discussed the application of MD modeling based on material chemistry structures for the mechanical property evaluation, the application of the MD-based material modeling to obtain the stress–strain shear deformation and shear strength of C–S–H Jennite molecular structure is presented next.

The determination of the shear strength and shear stress–strain deformation behavior is obtained by using a series of canonical material chemistry molecular structure ensembles. The number of particles, temperature, and volume are kept constant during the MD analysis, and a series of deformations are applied to the material chemistry molecular systems. The material chemistry molecular system represented by C–S–H Jennite is deformed due to the shear force, and the shear stress is obtained as the time-average value from the corresponding potential energy following Ergodic hypothesis after the dynamic analysis. This material chemistry level material modeling process employed for the cementitious material chemistry structure C–S–H Jennite is schematically illustrated in Figure 10.10. The application of the shear deformation results in obtaining the shear stress–strain deformation behavior of the material (C–S–H Jennite in the present application) solely based on its material chemistry structure. This provides an effective method to understand and predict the effect of material chemistry and additives on this deformation behavior aiding the material design of tailored material systems.

In this application for cementitious materials, MD modeling analysis was implemented using the Discovery module of Accelrys Material Studio®; details of this MD implementation are presented in reference [23]. The molecular structure of C–S–H Jennite used for this study was the same one published in reference [21]. The material chemistry computational model consisted of one unit cell of C–S–H Jennite (68 atoms) modeled using periodic boundary conditions in all directions. Figure 10.11 presents the shear stress–shear strain curves obtained from the material chemistry models of cementitious material C–S–H Jennite. The average value of the shear strength from these molecular modeling analyses was 3.75 ± 0.95 GPa, which is comparable to the value reported in reference [13] for cement paste C–S–H gel. Additionally, the average shear modulus obtained from the shear stress–strain material deformation model was 29.6 ± 5.3 GPa, which is also in good agreement with values obtained using other MD approaches [24,25]. Nano scale modeling based on material molecular structures are effective in understanding the mechanics of materials and provide an effective predictive means to predict and understand the effect of material chemistry and additive changes on the mechanical properties for their tailored design and associated material deformation models.

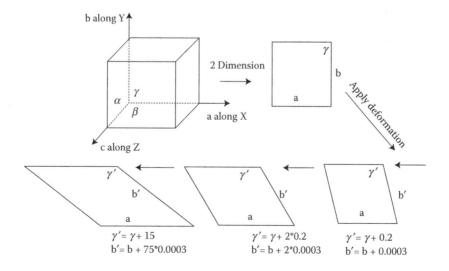

FIGURE 10.10 Schematics of shear deformation.

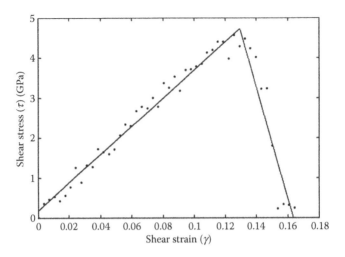

FIGURE 10.11   Shear stress–strain deformation for C–S–H Jennite obtained from material molecular chemistry structure.

10.6   CONCLUDING REMARKS

Computational material chemistry level modeling at the nano length scale of materials is effective to provide insight into the influence on their engineering scale properties, and behavior due to variations in the material chemistry and nanomaterial additions in the development of nanoengineered and high-performance material systems. The application of the material chemistry level modeling and its effectiveness to provide further insight of: (1) processing effects on the engineering scale properties; (2) material molecular defects on their mechanical properties; and (3) a predictive material modeling methodology for the deformation behavior was discussed in this chapter. The material chemistry level modeling of material interfaces in the first nanoengineered material system of hybrid composites (including functionalized and nonfunctional nanomaterial additives to traditional woven fabric thermoset composites) demonstrated the applicability of the material chemistry level modeling to understand the potential influence of functionalization on the engineering scale properties. The effectiveness of material chemistry level modeling at the material's nano scale to understand the influence of molecular structure changes due to the presence of molecular defects in SWCNT on the elastic modulus of epoxy–CNT nanocomposites was demonstrated. The potential of material chemistry modeling to predict the deformation (stress–strain) behavior of a hydrated cementitious material constituent, C–S–H Jennite, solely based on its molecular structure was discussed. Clearly, predictive modeling at the nanoscale material chemistry level based on MD modeling provides an effective methodology to understand the effect of material chemistry; changes and additives aiding the design of tailored, high-performance material systems. Furthermore, the methodology and discussions presented can be extended and effectively applied to understand the processing and mechanic effects in not only the material systems discussed but also in several other material systems.

ACKNOWLEDGMENTS

The authors acknowledge the graduate student research contributions from Dr. O. Akinyede, Mr. E. Fefey, and Dr. A. Mohamed. Financial support in parts from the Office of Naval Research (Award No. N00014-09-1-0842), the Army Research Office (Award No. W911NF-11-2-0043), and the Air Force Research Laboratory through Clarkson Aerospace are acknowledged.

REFERENCES

1. Roberto Gomperts, E.R. and M. Mehta, Enabling technologies for innovative new materials. *American Laboratory*, 2005. 37(22): 12–14.
2. Alder, B.J. and T.E. Wainwright, Studies in molecular dynamics. I. General method. *Journal of Chemical Physics*, 1959. 31(2): 459.
3. Allen, M.P. and D.J. Tildesley, *Computer Simulation of Liquids*. 1989, New York, USA: Clarendo Press.
4. Akinyede, O. et al., Processing and thermo-physical characterization of alumina particulate reinforced 3-phase hybrid composite material system. *Journal of Advanced Materials*, 2010. 42(3): 5–19.
5. Akinyede, O., R. Mohan, and A. Kelkar, Static and fatigue behavior of epoxy/fiber glass composites hybridized with alumina nanoparticles. *Journal of Composite Materials*, 2009. 43(7): 769–781.
6. Akinyede, O. et al., Static and fatigue behavior of epoxy/fiber glass composites hybridized with alumina nanoparticles. *Journal of Composite Materials*, 2010. 43(7): 769–781.
7. Gou, J. et al., Computational analysis of effect of single walled nanotube rope on molecular interactions and load transfer of nanocomposites. *Composites Part B: Engineering*, 2005. 35: 524–533.
8. Wong, M. et al., Physical interactions at carbon nanotube–polymer interfaces. *The Journal of Physical Chemistry B*, 2003. 106(12): 3046–3048.
9. Frankland, S.J.V. et al., Molecular simulations of the influence of chemical cross-links on the shear strength of carbon nanotube–polymer interfaces. *The Journal of Physical Chemistry B*, 2002. 106(12): 3046–3048.
10. Frankland, S.J.V. and V. Harik, Analysis of carbon nanotube pull out from polymer matrix. *Surface Science Letters*, 2003. 525: 303–308.
11. Pan, E., R. Zhu, and A.K. Roy, Molecular dynamics study of the stress–strain behavior of carbon nanotube reinforced EPON 862 composite. *Material Science and Engineering A*, 2007. 44: 51–57.
12. Accelrys, *Material Studio Online Help Manual*, Accelrys Software Inc., Release 5.5, San Diego, 2011.
13. Sun, H., *COMPASS: An ab Initio Force-Field Optimized for Condensed-Phase Applications Overview with Details on Alkane and Benzene Compounds*. Molecular Simulations Inc., 9685 Scranton Road, San Diego, California, 1998.
14. Hasnip, P., *A Beginner's Guide to Materials Studio and DFT Calculations with Castep*. 2007, pp. E3-1–E3-25, http://www.tcm/phy.cam.uk/castep/CASTEP_talks_07/Tuesday_exercises.pdf.
15. Akinyede, O. et al., Molecular dynamics simulations and analysis of material interactions in alumina particulate hybrid composites, in *AIAA Structural Dynamics and Materials Conference*, AIAA, Editor 2008, Chicago, IL: AIAA.
16. Mielke, S.L. et al., The role of vacancy defects and holes in the fracture of carbon nanotube. *Chemical Physical Letters*, 2004. 290: 413–420.

17. Li, Z. et al., First principles study for transport properties of defective carbon nanotubes with hydrogen absorption. *The European Physical Journal B*, 2009. 89: 375–382.

18. Mielke, L.S. et al., Mechanics of defects in carbon nanotubes: Atomistic and multi-scale simulaitons. *Physical Review B*, 2005. 71: 115403.

19. C01, A.I.C., Standard Specification for Portland Cement. ASTM C150, April 15, 2012.

20. Construction, C.E.-M.f.C., Cementitious materials for concrete. *ACI Education Bulletin*, 2001, E3-1–E3-25 (Pages 1–25 in Bulletin E3-01). Supersedes E3–E83.

21. Bonaccorsi, E., S. Merlino, and H.F.W. Taylor, The crystal structure of jennite, $Ca_9Si_6O_{18}(OH)_6 \cdot 8H_2O$. *Cement and Concrete Research*, 2004. 34(9): 1481–1488.

22. Bonaccorsi, E., S. Merlino, and A.R. Kampf, The crystal structure of tobermorite 14 A (Plombierite), a C-S-H phase. *Journal of the American Ceramic Society*, 2005. 88(3): 505–512.

23. Mohamed, A., *Computational Material Modeling for Mechanical Properties Prediction and a Methodology for Mie Gruneisen Equation of State Characterization via Molecular/Nano Scale Cementitous Material Constituents, in Computational Science and Engineering*. 2013, North Carolina A&T State University: Greensboro.

24. Manzano, H., J.S. Dolado, and A. Ayuela, Elastic properties of the main species present in Portland cement pastes. *Acta Materialia*, 2009. 57(5): 1666–1674.

25. Wu, W. et al., Computation of elastic properties of Portland cement using molecular dynamics. *Journal of Nanomechanics and Micromechanics*, 2011. 1(2): 84–90.

11 Computational Modeling of Nano-Bio Interfaces

Goundla Srinivas, Ram V. Mohan, and Ajit D. Kelkar

CONTENTS

11.1 INTRODUCTION

Many intriguing biological phenomena including protein folding, protein interaction with lipids, and other biomolecules occur on nanoscale [1]. Rapid growth in nanotechnology combined with biological applications revolutionized disease detection and treatment methodologies. For example, functional nanoparticles found to be effective in tumor detection and their treatment procedure as well [2]. Nano-bio interface involves a fruitful combination of nanotechnology and biomaterials [3]. Probing nano-bio interface using advanced experimental technologies has been described in previous chapters. In this chapter, a detailed description of computational efforts with a focus on understanding nano-bio interface from a molecular viewpoint is provided. Examples include model biological cell interaction with nanomaterials and other biomolecules using multi-scale molecular dynamics (MD) simulation studies. This chapter also presents a systematic comparison between experimental and computational studies, highlighting the need and usefulness of computational tools in providing a comprehensive picture of nano-bio interface.

Despite rapid progress in experimental advances, much of the biological phenomena is still considered to be under the twilight zone. Thanks to advances in experimental techniques, a broader picture is emerging. Nevertheless, intriguing underlying details are not yet fully understood. Large-scale computational modeling, in combination with experimental advances, provides an alternative measure of understanding such problems from a molecular viewpoint. Computational modeling can be advantageously combined with experimental techniques to obtain detailed atomic and/or molecular level picture of complex nano-bio interface. Advantages of computational modeling are that on one hand, it can reproduce experimental results (and thus validating the model itself) and on the other, it can provide critical insights into the observed experimental phenomena, which otherwise would have been near impossible to probe using the experimental techniques alone [4].

As mentioned above, many important biological phenomena such as nucliopore formation, membrane fusion, lipid raft formation, and so on, occur on hundreds of nanometer to micrometer length scales. To put things into prospective, a single protein spans a length of few tens of nanometers while typical biological cell sizes range between 1 and 100 μm. The length scales grow rapidly when the system under investigation involves other biological components such as enzymes and lipids in addition to proteins and solvent. Typically, such systems involve millions of atoms making it computationally difficult to simulate over relevant time scales. Hence, the time and length scales present a major hurdle for the computational modeling of systems belonging to the nano-bio interface.

Computational modeling and simulation play an integral role in scientific and technological advances. In particular, its application in biomedical research and nanotechnology revolutionized drug discovery. A fundamental assumption for computational modeling and simulation is that insight into system behavior can be developed or improved from a model that adequately represents a selected subset of the system's attributes. Depending on properties and applicability, simulation techniques can be broadly categorized into three classes: (1) quantum simulations, (2) classical simulations, and (3) mesoscopic simulations. Each method has its merits and limitations as they have been designed to efficiently handle different spatial and temporal scales. For example, for studying chemical bond formation and bond dissociation between atoms and molecules involving electronic interactions, a simulation method based on quantum mechanics is an appropriate choice. However, due to complex interactions and computationally intense numerical calculations, this method can only be applicable for systems with relatively smaller length and time scales. Nevertheless, many interesting phenomena relevant to biological and material sciences involves time and length scales over micrometers and microseconds and beyond. Such problems cannot be directly dealt with methods based on quantum mechanics. On the other hand, classical simulations based on Newtonian mechanics, are highly suitable to deal with such large time and length scales appropriately. For this very reason, classical MD is the most commonly used simulation method. The classical MD approach is routinely used to explore tens of nanosecond times and nanometer length scales. However, such efficiency over quantum simulations comes with a penalty of relatively lower accuracy. Naturally, more efficient methods such as mesoscopic simulations that explore time and length scales beyond classical/atomistic simulation capabilities are relatively less accurate. Bridging these disparate spatial and temporal scales is essential to explore complex biomedical engineering problems in detail.

11.2 BRIDGING EXPERIMENTS AND SIMULATIONS THROUGH HIGH-PERFORMANCE COMPUTING

Experimental work on complex biomedical and material systems spans a broad range of temporal and spatial scales, from femtosecond dynamics and atomistic detail to real-time macroscopic phenomena. Simulation methods in which each atom is explicitly represented are well established, but have difficulty addressing many cooperative effects of experimental and theoretical interest. There is simply too large a gap between the timescale and spatial scale that govern typical intramolecular events and those that are relevant for collective motions. Often, the timescale gap between simulation and experiment is about six orders of magnitude [5]. Existing simulation techniques for specific timescales and spatial scales are illustrated schematically in Figure 11.1. These techniques make use of a variety of approaches to reduce the level of detail in the representation of the system under study as the time and length scales grow. Bridging these disparate scales is possible with multi-scale modeling [6] in which the various levels of treatment are coupled and fed back into one another in an iterative fashion. In particular, reduced models that retain close connections to the underlying atomistic representation have been promising. Nano-bio interface simulations involve timescales of hundreds of nanoseconds to milliseconds and spatial

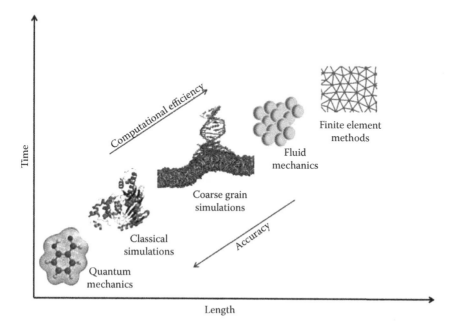

FIGURE 11.1 Various simulation techniques are depicted along with the representative systems that can be explored efficiently. Note the decrease in computational accuracy as the efficiency increase.

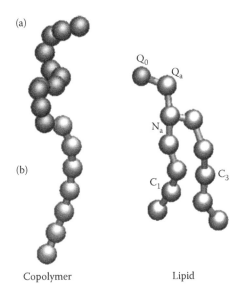

FIGURE 11.2 CG structures for copolymer and lipid molecules. Hydrophilic and hydrophobic blocks are shown as (a) and (b) respectively. For the lipid molecule polar head group is composed by Q_0, Q_a, and N_a while C_1 and C_3 represent hydrocarbon tails.

scales of microns. Hence, in this chapter we focus on the development strategies and applications of coarse grain (CG) simulation method that has ready access to events on these scales; CG models are gaining widespread usage in the material [7] and biophysical communities [8] as explained in the following (Figure 11.2).

11.3 CHALLENGES FOR COMPUTATIONAL MODELING

There are many phenomena that lie within the mesoscopic spatio-temporal scale that might be explored with CG methods. Examples of such phenomena from biology include protein–protein, protein–lipid, and membrane–membrane interactions. From a materials perspective, the optimal design of nanosyringes that penetrate membranes are of interest, as well as the design and properties of artificial polymer-based membranes which can act as carriers and controlled release vehicles for drug delivery. From a computational viewpoint there exists lot of commonality between material and biological systems. In both the cases, underlying systems are still made of atoms and molecules that are connected through chemical bonds. Broadly speaking, basic atomic and molecular structure and function determines large-scale macroscopic properties in a given system. Note that the computational description of either system involves a similar set of physical principles and mathematical equations. Moreover, both systems span similar spatial and length scales. Hence, the development of CG models for both systems involves a similar procedure.

A detailed atomic and molecular level understanding of material and biological processes occurring on a mesoscopic scale has been the focus of many large-scale computational modeling studies. Experimental studies over the years have shown that vesicles similar to biological vesicles can be synthesized from organic super amphiphiles such as block copolymers [9]; diblock copolymers in particular have architecture similar to that of natural lipids [10]. Many biological membrane processes such as protein integration, fusion, DNA encapsulation, and compatibility can be reliably mimicked by synthetic polymer vesicles. The hydrophilic/hydrophobic ratio can be selected with ease. In addition, block copolymers have the intrinsic ability to self-organize into membranes and offer fundamental insight into natural design principles for biomembranes. While current computational resources allow the study of such large pre-assembled systems, it becomes impractical studying dynamical phenomena such as polymer-assisted drug delivery, which occurs typically on hundreds of nanoseconds to microsecond timescale. Existing simulation studies of such large systems have been mostly carried out using arbitrary potentials. The CG approach described in this chapter was proven to be effective and reliable in providing a quantitative comparison with experimental results. In other words, CG simulations are capable of providing microscopic insight into the corresponding experimental systems. In the following, we describe the efficient polymeric micellar nanocarriers for drug delivery, which were explored using CG MD simulations.

11.4 POLYMER MICELLAR NANOCARRIERS FOR DRUG DELIVERY

Micellar drug carriers also known as nanocarriers, offer a promising approach for formulating and achieving improved delivery of drug molecules in comparison to conventional methods. An ideal drug delivery system should be composed of biocompatible and biodegradable materials, encapsulate a wide range of drugs and drug classes, maintain particle size in biological media, have the ability to attach cell-specific targeting groups, and release the therapeutic at the site of disease. Polymer micelles have received much attention over the past couple of decades as drug delivery vehicles due to similarity and compatibility with lipid structures [11]. However, conventional micelles do not have long-term stability in complex biological environments such as plasma. Recent studies showed that block copolymer micelles encapsulate several different hydrophobic drugs [12]. They can be engineered to be stable at low concentrations even in complex biological fluids, and to release cargo in response to low pH environments, such as in the tumor microenvironment or in tumor cell endosomes. In the following, we describe the computational efforts to study such polymer micelle-assisted drug delivery using the CG MD simulations.

11.4.1 SIMULATION OF POLYMER MICELLAR NANOCARRIER FOR DRUG TRANSPORT

Recently, the CG simulation approach was successfully applied to study polymer micelle-assisted drug transportation across lipid vesicle membrane. The lipid vesicle, containing 877 DPPC (1,2-dipalmitoyl-*sn*-glycero-3-phosphocholine) lipids, was solvated with 190,000 water molecules. The total system, including polymer micelle, contained more than 300,000 atoms. Simulating such large systems over hundreds of nanosecond timescales with complete atomistic details is near impossible. Hence, the previously described CG simulation methodology was considered as an alternative. The fore field parameter set for lipids and water was adapted from the "MARTINI" force field as the basis [13], which is a previously developed CG force field with an emphasis on biological molecules.

11.4.2 CG SIMULATION APPROACH

As mentioned above, various approaches have been utilized in developing efficient alternative simulation models, albeit with lower resolution. For example, dissipative particle dynamics (DPD) [14–17], discontinuous molecular dynamics (DMD) [18], and Brownian dynamics simulations [19,20] have been used to study self-assembly of biomolecules and nanomaterials. In particular, a simplified model for amphiphiles and CG models for lipid molecules proved to be highly successful in studying self-assembled biological membranes [21–25]. Marrink et al. demonstrated the efficiency of CG models by studying membrane fusion and vesicle fusion [26,27]. All these studies utilize the basic CG approach built on pioneering works of Smit and coworkers [28,29]. Along similar lines, Klein et al. have developed a CG model for diblock super amphiphiles [23–25]. The CG polymer model was successful in reproducing the spontaneous self-assembly of diblock copolymers with a lipid-like hydrophobic/hydrophilic ratio into membrane bilayer structures. In this chapter we consider the same CG model for the copolymers [30].

The polymer micelle-assisted drug incorporation and transportation can be conveniently simulated in two separate steps: (1) simulate a lipid vesicle–polymer micelle system in water, and (2) simulation after loading drug-like molecules inside the polymer vesicle. The first step provides an opportunity to understand how polymer micelle interacts with lipid vesicle, while the second step includes details on drug incorporation, transportation, and delivery mechanism. After the equilibration, a pre-assembled polymer micelle was added to the lipid vesicle–water system. While constructing such a system, minimum distance between the polymer micelle and lipid vesicle need to be ensured to exceed the cut-off distance for nonbonded interactions. Either the conjugated-gradient method or steepest descent method can be used for minimization. It is important to equilibrate the systems at least for a nanosecond after the minimization. In this example, all the simulations were carried out using GROMACS [31] software in NPT ensemble with 10 fs timestep at 300 K temperature and 1 atm pressure.

11.4.3 CG SIMULATION EFFICIENCY

By adapting the CG approach, simulations gain at least three orders of computational efficiency. Such efficiency comes from a combination of factors. First, the size of the system reduces by at least 10-fold by using CG representation compared to atomistic structure. This size reduction increases computational efficiency around 10 times. Second, due to wider potential wells, CG simulations allow the use of much larger timesteps (10–40 fs) compared to 1 or 2 fs used in atomistic simulations. Together, these factors make CG simulations highly efficient compared to their atomistic counter parts.

11.5   SIMULATION RESULTS AND DISCUSSION

Here, we discuss the results from the CG simulation study described above. We begin with the results of binary copolymer micelle interaction with lipid vesicle in water. At the beginning of the simulation study, polymer micelle and lipid vesicle were separated by 5 nm (surface-to-surface distance) as shown in Figure 11.3. During the course of simulation, the polymer micelle moved into lipid vesicle proximity. When the micelle moves within the interaction range of lipid vesicle, favorable interaction between lipid head groups and hydrophilic polymers initiates the association. Due to this favorable interaction, hydrophilic polymer blocks start penetrating into the lipid head region of the vesicle. As the penetration progresses, a narrow path into the vesicle was formed. This widens the path, and as a result the polymer micelle penetrates into lipid vesicle. Snapshots of micelle–vesicle interaction are shown in Figure 11.4. Around 110 ns micelle was found to penetrate inside the vesicle. During penetration process, lipid vesicle showed significant changes in its morphology. Initial spherical shape of vesicle transformed to near ellipsoidal. This morphological change was found to be driven by favorable lipid head and hydrophilic polymer interaction. At 150 ns, half of the micelle was found to penetrate lipid vesicles. The final snapshot at 210 ns shows most of the micelle penetrated inside the vesicle. During the penetration process, we observed few individual polymers detaching themselves from the micelle and distributing into the lipid vesicle.

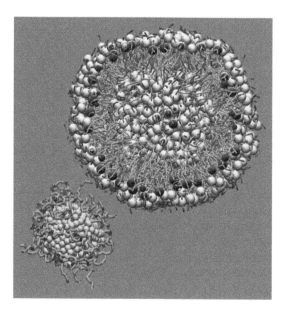

FIGURE 11.3   (**See color insert.**) Initial configuration of copolymer micelle and lipid vesicle used in the CG MD simulations is shown. A cross section of the lipid vesicle is shown to reveal inner components of the vesicle.

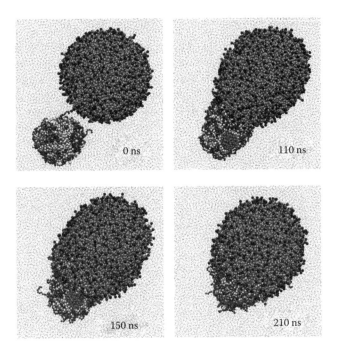

FIGURE 11.4 Simulation snapshots of polymer micelle interaction with lipid vesicle. Initially separated polymer micelle (0 ns) fully adsorbs into the lipid vesicle at the end of the simulation (210 ns).

11.5.1 INCORPORATING DRUG-LIKE MOLECULES INSIDE LIPID VESICLE

In the next step, we describe the results from the study of incorporating hydrophilic molecules into the polymeric micelle. Unconnected CG beads were used as model representative hydrophilic drug-like components. From an experimental viewpoint, incorporating hydrophobic drugs such as doxorubicine can be relatively easy. However, hydrophilic drug incorporation is not straightforward, since polymer core is generally made of hydrophobic blocks. For this purpose, we focus on the study of hydrophilic drug components. The incorporation of hydrophilic molecules inside the polymer micelle was done by replacing three selected copolymers within the hydrophilic patch of the micelle with unconnected CG beads. The interaction energy parameters for the model drug components were mostly hydrophilic to mimic the nature of hydrophilic drugs. The interaction parameters for such hydrophilic CG beads with the rest of the CG system were obtained using mixing rules. After incorporating hydrophilic contents, the polymeric micelle was equilibrated for 1 ns in water. The equilibrated polymer micelle contained 33 hydrophilic components. At the beginning of the CGMD analysis, the equilibrated polymer micelle was placed at a distance greater than 1.2 nm from the surface of the lipid vesicle. The simulation conditions including timestep size, temperature, pressure, and ensemble were chosen to be the same as before. During the course of the CGMD analysis, the polymer micelle interacted with the lipid vesicle, as before. During the micelle interaction

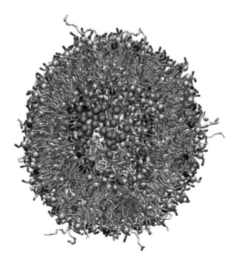

FIGURE 11.5 A cross-sectional view of a CG simulation snapshot showing the incorporation of drug-like molecules (darker spherical beads) inside the lipid vesicles.

with the lipid vesicle, simulations provide the opportunity to monitor relative movements of hydrophilic components. The initial molecular contact was facilitated by the favorable interaction between copolymer of the micelle with the lipid head groups of the vesicle. As a result, the polymer micelle progressively penetrated inside the lipid head group region of the vesicle. During the course of the interaction, the hydrophobic polymers of the micelle got exposed as the micelle opened up. At this stage, the hydrophobic polymer blocks penetrated toward the nonpolar lipid tail region due to favorable interactions. This further exposed the polymer micelle core, thereby releasing the inner contents of the polymer micelle. As described before, the polymer micelle was loaded with hydrophilic components (depicted as darker beads in Figure 11.5), which quickly diffuse into the polar lipid head group region. However, the lipid vesicle head group region is relatively thinner (~1 nm) compared to the hydrophobic tail region (3–4 nm). Hydrophilic contents do not favor the hydrophobic lipid tail region and move out of that region in less than a few nanoseconds. Some of the hydrophilic components found to reverse the course and move out of the vesicle (into the bulk water), while the remaining entered the inner core region of the vesicle, where the confined water exists. The simulations show that approximately 50% of initial hydrophilic contents successfully gets transported across the lipid bilayer and delivered into the inner core of lipid vesicle. CGMD simulations thus demonstrate the incorporation of hydrophilic components in the polymer micelle and their efficient transportation across the lipid bilayer to deliver inside the lipid vesicle.

11.6 NEED FOR THE MULTISCALE SIMULATION METHODS

As discussed before, atomistic simulations and CG simulations proved to be highly successful in exploring biomedical and material phenomena with suitable time and length scales [32–35]. However, many processes that take place at the nano-bio

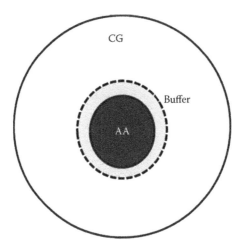

FIGURE 11.6 A schematic representation of a hybrid AA/CG simulation approach with a defined buffer zone for the smooth interchange is shown.

interface cannot be fully studied using either one of the methods. For example, a large-scale protein or nanotubular motion occurs on time and length scales that can be efficiently studied using CG simulations, while the highly important atomistic level interactions within these molecules can be best explored by classical MD simulations. Hence, there is a need and opportunity for combining such disparate simulation methods using a hybrid multi-scale methodology. In principle, such methods should be able to combine advantages of atomistic and CG simulations without sacrificing efficiency or losing the resolution. This is similar to the development of the hybrid quantum mechanics/molecular mechanics (QM/MM) method that successfully combines atomistic and quantum mechanical simulations. In QM/MM simulations, a relatively small, highly specific region is defined where QM simulations will be carried out while the rest of the system will be treated using classical simulations. In addition, a buffer zone between QM and MM is created for smooth transition of QM and MM treatments as molecules and atoms moves in and out of either region. A hybrid AA/CG method may utilize similar principles. However, unlike QM/MM methods, the AA/CG hybrid method involves structural changes as well. In other words, in the buffer region between AA and CG regions molecules will have part atomistic and part CG structure. Because of this, the buffer region needs to be relatively thicker compared to the case of QM/MM. A pictorial representation of a possible hybrid AA/CG method is shown in Figure 11.6.

Recently, Voth et al. [36] proposed a multiscale CG method and applied it to 2- and 4-site hexane models based on underlying atomistic data. Nevertheless, as the authors note, the method suffers from sampling issues. Moreover, its applicability for larger systems has not been tested, which is crucial to explore systems at nano-bio interface. Hence, we would like to further our present CG methodology to include hybrid AA/CG in an efficient fashion to explore nano-bio interface on relevant time and length scales.

11.7 CONCLUSION

In this chapter we have described polymer micelle-assisted drug incorporation, transport, and delivery by using large-scale CG MD simulations. Extensive CG MD simulations over 250 ns fully facilitate the study of micelle–vesicle interaction. Simulations revealed polymer micelle absorbance mechanisms in the lipid vesicle, which was difficult to probe using conventional experimental studies.

The results from the above-described CG MD simulations reveal the capability of CG simulation in capturing the underlying molecular level picture of nano-bio interface. In this case, CG simulations provided a detailed mechanism for polymer micelle-assisted hydrophilic content transportation across the lipid membrane and delivery inside the lipid vesicle. For example, CGMD analysis showed that the hydrophilic polymers interact with the lipid head groups of the vesicle to initiate the micelle–vesicle fusion process. Consequently, the hydrophobic inner core of the lipid vesicle gets exposed to the hydrophobic micelle core. During this step, the hydrophilic contents get transported across the lipid membrane via the pathways provided by the hydrophobic polymers. It is near impossible to obtain such molecular level details using the conventional experimental methods. This chapter provides another example to strengthen the argument that the simulations provide a greater opportunity to advantageously combine with experimental results in obtaining a detailed picture of nano-bio interface.

It is important to note that due to complex and interdisciplinary nature of the nano-bio interface phenomena, neither atomistic nor CG simulations will be explored entirely. In order to explore such problems, we propose to develop a hybrid atomistic/CG simulation approach based on the principles of quantum-mechanics/molecular-mechanics method. Such an approach may provide novel and efficient tools to study complex nano-bio interface phenomena on desired spatial and temporal scales.

ACKNOWLEDGMENTS

This work was supported in part by the U.S. Army Research Office under award/contract no. W911NF-11-1-0168. We acknowledge the use of high-performance computational facilities at the North Carolina State A&T University during the course of this study.

REFERENCES

1. Berg J M, Tymoczko J L, and Stryer L. 2002. *Biochemistry*. 5th edition. New York: W H Freeman; Available from: http://www.ncbi.nlm.nih.gov/books/NBK21154/.
2. Dhar S, Liu Z, Thomale J, Dai H, and Lippard SJ. 2008. Targeted single-wall carbon nanotube-mediated Pt(IV) prodrug delivery using folate as a homing device. *J. Am. Chem. Soc.* 130(34), 11467–11476.
3. Duan X, Gao R, Xie P, Cohen-Karni T, Qing Q, Choe S-H, Tian B, Jiang X, and Lieber C M. 2012. Intracellular recording of action potentials by an extracellular nanoscale field-effect transistor. *Nat. Nanotechnol.* 7, 174–179.

4. Allen M P, and Tildesley D J. 1989. *Computer Simulation of Liquids*. New York: Clarendon Press.

5. Li P-C and Makarov D E. 2003. Theoretical studies of the mechanical unfolding of the muscle protein titin: Bridging the time-scale gap between simulation and experiment. *J. Chem. Phys.* 119, 9260.

6. Rafii-Tabar H and Chirazi A. 2002. *Phys. Rep.* 365, 145.

7. Tries V, Paul W, Baschnagel J, and Binder K. 1997. Modeling polyethylene with the bond fluctuation model. *J. Chem. Phys.* 106, 738.

8. Stevens M J, Hoh J H, and Woolf T B. 2003. Insights into the molecular mechanism of membrane fusion from simulation: Evidence for the association of splayed tails. *Phys. Rev. Lett.* 91, 188102.

9. Cornelissen J J L M, Fischer M, Sommerdijk N A J M, and Nolte R J M. 1998. *Science* 280, 1427.

10. Zhang L and Eisenberg A. 1995. *Science* 268, 727.

11. Rios-Doria J, Carie A, Costich T, Burke B, Skaff H, Panicucci R, and Sill K. 2012. *Journal of Drug Delivery*, 2012(2012), Article ID 951741.

12. Kataoka K, Matsumoto T, Yokoyama M, Okano T, Sakurai Y, Fukushima S, Okamoto K, and Kwon G S. 2000. Doxorubicin-loaded poly(ethylene glycol)–poly(β-benzyl-l-aspartate) copolymer micelles: Their pharmaceutical characteristics and biological significance. *J. Controlled Release* 64, 143–153.

13. Marrink S J, Risselada H J, Yefimov S, Tieleman D P, and de Vries A H. 2007. The MARTINI force field: Coarse grained model for biomolecular simulations. *J. Phys. Chem. B* 111, 7812.

14. Fraaije J G E M, and Sevink G J A. 2003. Model for pattern formation in polymer surfactant nanodroplets. *Macromolecules* 36, 7891.

15. Groot R D, Madden T J, and Tidesley D J. 1999. On the role of hydrodynamic interactions in block copolymer microphase separation. *J. Chem. Phys.* 110, 9739.

16. Leibler L. 1980. Theory of microphase separation in block copolymers. *Macromolecules* 13, 1602.

17. Shillcock J C and Lipowsky R. 2002. Equilibrium structure and lateral stress distribution of amphiphilic bilayers from dissipative particle dynamics simulations. *J. Chem. Phys.* 117, 5048.

18. Schultz A J, Hall C K, and Genzer J. 2002. Computer simulation of copolymer phase behavior. *J. Chem. Phys.* 117, 10329.

19. Pastor R, Venable R M, Karplus M, and Szabo A. 1988. A simulation based model of NMR T_1 relaxation in lipid bilayer vesicles. *J. Chem. Phys.* 89, 1128.

20. Nagochi H. 2002. Fusion and toroidal formation of vesicle by mechanical forces: A Brownian dynamics simulation. *J. Chem. Phys.* 117, 8130; Nagochi H, and Takasu M. 2001. Fusion pathways of vesicle: A Brownian dynamics simulation. *J. Chem. Phys.* 115, 9547.

21. Goetz R, Gompper G, and Lipowsky R. 1999. Mobility and elasticity of self-assembled membranes. *Phys. Rev. Lett.* 82, 221.

22. Goetz R and Lipowsky R. 1998. Computer simulations of bilayer membranes: Self-assembly and interfacial tension. *J. Chem. Phys.* 108, 7397.

23. Shelley J C, Shelley M Y, Reeder R C, Bandyopadhyay S, and Klein M L. 2001. A coarse grain model for phospholipid simulations. *J. Phys. Chem. B* 105, 4464.

24. Nielsen S, and Klein M L. 2002. In *Bridging Time Scales: Molecular Simulations for the Next Decade*; Nielaba, P., Mareschali, M., Ciccotti, G., Eds.; Berlin, Germany: Springer-Verlag, pp. 25–63.

25. Nielsen S, Lopez C F, Srinivas G, Klein M L. 2003. *J. Chem. Phys.* 119, 7043; Nielsen S O, Lopez C F, Srinivas G, and Klein M L. 2004. *J. Phys. Condens. Matter* 16, 481.

26. De Vries A H, Mark A E, and Marrink S J. 2004. Molecular dynamics simulation of the spontaneous formation of a small DPPC vesicle in water in atomistic detail. *J. Am. Chem. Soc.* 126, 4488.

27. Tieleman D P, Leontiadou H, Mark A. E, and Marrink S J. 2003. Simulation of pore formation in lipid bilayers by mechanical stree and electric fields. *J. Am. Chem. Soc.* 125, 6382; Marrink S J, Lindahl E, Edholm E, and Mark A E. 2001. *J. Am. Chem. Soc.* 123, 8638; Marrink S. J, and Mark A E. 2003. *J. Am. Chem. Soc* 125, 11144.

28. Smit B, Hilbers P A J, Esselink K, Rupert L A M, Van Os N M, and Schlijper A G. 1991. Structure of water/oil interface in the presence of micelles: A computer simulation study. *J. Phys. Chem.* 95, 6361; Kranenburg M, Venturoli M, and Smit B. 2003. *J. Phys. Chem. B* 107, 11491.

29. Venturoli M and Smit B. 1999. Simulating the self-assembly of model membranes. *Phys. Chem. Comm.* 10.

30. Srinivas G, Mohan R V, and Kelkar A D. 2013. Polymer micelle assisted transport and delivery of model hydrophilic components inside a biological lipid vesicle: A coarse-grain simulation study. *J. Phys. Chem. B,* 117, 12095–12104.

31. Berendsen H J C, Van der Spoel D, and van Drunen R. 1995. GROMACS: A message-passing parallel molecular dynamics implementation. *Comput. Phys. Commun.* 91, 43–56.

32. Müller-Plathe F. 2002. Coarse-graining in polymer simulation: From the atomistic to the mesoscopic scale and back. *Chem. Phys. Chem.* 3, 754.

33. Lyubartsev A P and Laaksonen A. 1995. Calculation of effective interaction potentials from radial distribution functions: A reverse Monte Carlo approach. *Phys. Rev. E* 52, 3730.

34. Srinivas G and Pitera J. 2008. Soft patchy nanoparticles from solution-phase self-assembly of binary diblock copolymers. *Nano Lett.* 8, 611.

35. Srinivas G, Shelley J C, Nielsen S, Discher D E, and Klein M L. 2004. Simulation of diblock copolymer self-assembly using a coarse-grain model. *J. Phys. Chem. B,* 108, 8153–8160.

36. Das A, Lu L, Andersen H C, and Voth G A. 2012. The multiscale coarse-graining method. X. Improved algorithms for constructing coarse-grained potentials for molecular systems. *J. Chem. Phys.* 136, 194115.

Section V

Nanolithography and
Nanofabrication

12 Multiscale Glass Fiber-Reinforced Composite Developed from Epoxy Resin Containing Electrospun Glass Nanofibers

Lifeng Zhang and Hao Fong

CONTENTS

12.1   INTRODUCTION

Fiber-reinforced polymer composites are generally fabricated through impregnation of fiber fillers (with high strength and modulus) into polymeric resin matrices. The integration of filler and matrix phases results in excellent mechanical properties that cannot be achieved from either components alone. Polymer composite laminates are assemblies of layers of fibrous composite. These layers are joined together to provide the required engineering properties, including in-plane stiffness, bending stiffness, strength, and coefficient of thermal expansion. The high ratio of strength/modulus to weight makes the fiber-reinforced polymer composites and laminates usable in a wide range of fields such as aircrafts, automobiles, sports utilities, and satellites [1].

The conventional fibers for making reinforced polymer composites include carbon fibers, glass fibers, and polymer fibers (e.g., Kevlar fibers) with diameters typically in the range of a few to tens of micrometers. With the development of nanomaterials in recent years, innovative nanofibers have attracted growing interests in making polymer composites due to their large specific surface areas that can lead to substantial improvement of interfacial bonding strength between fillers and matrices. To date, the majority of the reported research efforts on nanofiber-reinforced polymer composites have been focused on vapor-grown carbon nanotubes and/or nanofibers [2–6].

The materials-processing technique of electrospinning provides a viable approach for convenient fabrication of polymer, ceramic, and carbon fibers (commonly known as "electrospun nanofibers") with diameters ranging from nanometers to micrometers [7–10]. Electrospinning is a technique that utilizes electric force to drive the spinning process and produce fibers. Unlike conventional spinning techniques such as solution spinning and melt spinning that are capable of producing fibers with diameters in micrometer range (~10–200 μm), electrospinning is capable of producing fibers with diameters in submicrometer and nanometer range (~50–1000 nm). Unlike nanotubes, nanowires, and nanorods, most of which are made of synthetic, bottom-up methods and usually require further expensive purifications, electrospun nanofibers are made through a top-down nanomanufacturing process and are therefore inexpensive, continuous, and relatively easy to be aligned, assembled, and processed into applications. In the recent decade, the technique of "electrospinning" and its unique product of "nanofibers" have been actively researched throughout the world.

Nevertheless, to the best of our knowledge, only limited research endeavors have been devoted to the development of polymer composites with electrospun nanofibers [10–14]. The main reason is that in many cases, electrospun polymer and carbon nanofibers possess lower mechanical properties than those of their conventional counterparts such as vapor-grown carbon nanotubes/nanofibers, whereas electrospun ceramic nanofibers have been primarily developed for electronic and/or catalytic applications [15]. It is noteworthy that electrospun glass nanofibers (EGNFs) possess higher mechanical strength and modulus than polymer nanofibers and they can be used for the fabrication of nanofiber-reinforced polymer composites [16,17].

In our previous research, silica (SiO_2) fibers with diameters of ~500 nm (i.e., glass nanofibers) were readily prepared by electrospinning a spin dope consisting of tetraethyl orthosilicate (TEOS, the alkoxide precursor for making SiO_2) and polyvinylpyrrolidone (PVP, the carrying polymer) in *N,N*-dimethyl formamide and/ or dimethyl sulfoxide (DMF/DMSO, the solvent) followed by pyrolysis at 800°C [18,19]. These electrospun SiO_2 nanofibers (EGNFs) are morphologically uniform and structurally amorphous, and they can retain their fiber morphology when they are subjected to vigorous ultrasonication; therefore, EGNFs are nanoscaled glass fibers. When EGNFs were used to partially replace (up to a mass fraction of 7.5%) the conventional dental glass filler (i.e., the dental glass powder with particle sizes ranging from tens of nanometers to a few microns), the flexural strength, elastic modulus, and work of fracture (WOF) of the resulting dental composites were considerably improved [20].

In this study, EGNFs with an average diameter of 400–500 nm were incorporated at very low mass fractions (up to 1%) into epoxy resin for reinforcement and/ or toughening purposes, and this innovative EGNFs/epoxy resin was further used for the development of high-performance polymer composite laminates containing conventional glass fiber (CGF) fabrics. For comparison, short glass fibers that were chopped from a commercially available glass wool (i.e., [CGFs]) were also incorporated into the epoxy resin to make composites. Mechanical properties including tensile properties, flexural properties, impact adsorption energy, and interlaminar shear strength of the prepared composites were evaluated and scanning electron microscopy (SEM) was employed to examine the micro- and nanoscaled morphologies as well as the fracture surfaces to investigate the corresponding failure mechanisms.

12.2 PREPARATION AND EVALUATION OF EPOXY COMPOSITE RESINS CONTAINING ELECTROSPUN GLASS NANOFIBERS (EGNFs)

In this part, EGNFs with diameters of 400–500 nm were incorporated into epoxy resin for reinforcement and/or toughening purposes, and the motivation was for making innovative epoxy composite resins containing EGNFs, which could be further used for the development of high-performance polymer composites. Two silane-coupling agents with respective end groups of epoxy and amine, that is, 3-glycidoxyl-trimethoxysilane (GPTMS) and 3-aminopropyl triethoxysilane (APTES), were selected for surface treatment of EGNFs. The surface treatment of EGNFs with GPTMS or APTES would improve the interfacial bonding strength between the fibers and the matrix, and also facilitate the uniform dispersion of EGNFs in the resin matrix. The effects of incorporation of EGNFs and the different silanization treatments on the mechanical properties of the resulting epoxy composite resins were investigated, and the results were compared to those acquired from the epoxy composite resins containing CGFs with diameters of ~10 μm.

12.2.1 EXPERIMENTAL

EGNFs were prepared using the spin dope consisting of 13% TEOS and 13% PVP in a mixture solvent of *N,N*-DMF and/or DMSO with volume ratio of 2:1 followed by pyrolysis at 800°C. Two silane-coupling agents, GPTMS, and APTES, with respective end groups of epoxy and amine were selected for surface treatment of EGNFs. All chemicals that are mentioned were purchased from the Sigma-Aldrich Co. (St. Louis, MO) and used without further purification. All concentration/proportion used in this chapter is mass fraction.

Prior to silanization, the prepared EGNFs were first dispersed in water with the mass fraction of 5%; the suspension was then subjected to vigorous ultrasonication with a 100 W ultrasonic probe purchased from the Branson Ultrasonics (Danbury, CT) for three 5-min time periods. The lengths of EGNFs after ultrasonication were a few to tens of micrometers. The CGFs were cut into short fibers with lengths of 1–2 mm from a commercial glass wool. The silanization was conducted by immersing either EGNFs or CGFs into 15% silane solution in ethanol, and the suspension was then heated to 50°C followed by being stirred for 1 h at 125 rpm using a Heidolph RZR 50 Heavy Duty Stirrer. The physically adsorbed silane molecules on fiber surfaces were removed through sonication in ethanol for 10 min followed by being thoroughly rinsed with ethanol. The GPTMS-treated fibers were desiccated under vacuum (~27.9 kPa) at room temperature, while the APTES-treated fibers were desiccated in an oven at 110°C for 15 min.

EGNFs or CGFs (with or without silanization treatment) were first added into the SC-15A epoxy resin (Applied Poleramic, Benicia, CA) at the mass fractions of 0.5% and 1%. The mixtures were then mechanically stirred at 125 rpm for 12 h at 60°C followed by being sonicated for 30 min to uniformly disperse the glass fibers. Subsequently, the corresponding SC-15B hardener (Applied Poleramic, Benicia, CA) was added into each mixture, and the mass ratio of the epoxy resin versus the hardener was set at 100/30. After degassing, each mixture was poured into an aluminum mold followed by being cured at 60°C for 2 h initially and was post-cured at 110°C for 5 h to obtain a composite panel with the length, width, and thickness being 100 and 3 mm, respectively. Finally, specimens with dimensions of $64 \times 12.7 \times 3$ mm^3 were cut from the composite panels for Izod impact test, while the dog-bone-shaped specimens for tension test were machined according to ASTM D 1708.

A Zeiss Supra 40 VP field-emission SEM was employed to examine the morphologies of fibers as well as the fracture surfaces of composite resins. Prior to SEM examinations, specimens were sputter coated with gold to avoid charge accumulations. Fourier transform infrared (FT-IR) spectra of EGNFs before and after silanization treatments with both GPTMS and APTES were acquired from the Bruker Tensor-27 FT-IR spectrometer equipped with a liquid nitrogen cooled mercury–cadmium–telluride (MCT) detector. The samples were prepared by pressing the fibers with potassium bromide (KBr), and the FT-IR spectra were acquired by scanning the samples (64 scans) from 600 to 4000 cm^{-1} with a resolution of 4 cm^{-1}. Measurements of mechanical properties were conducted at room temperature. The Izod impact test was carried out using a Tinius Olsen impact tester (Impact 104) according to ASTM D 256. The standard tension test was performed according to ASTM D 1708 at a strain rate of 1 mm/min using a computer-controlled universal mechanical testing machine (QTEST™/10, MTS Systems, USA). Five specimens for each sample were tested and the mean values and standard deviations were calculated.

12.2.2 Results and Discussion

12.2.2.1 Silanization

Both EGNFs and CGFs had the cylindrical shape with smooth surface, and their diameters were ~400 nm and ~10 μm, respectively, as shown in Figure 12.1(a1 and b1); therefore, the EGNFs were approximately 25 times thinner than the CGFs. Two silane-coupling agents, one with epoxy end group (GPTMS) and the other with amine end group (APTES), were used for surface treatment of both types of glass fibers. No appreciable difference on fiber morphology was identified for both types of glass fibers after silane treatments as shown in Figure 12.1(a2, a3, b2, and b3). The reactions between silane molecules and silanol (Si–OH) groups on the surface of fibers are illustrated in Figure 12.2, as evidenced by FT-IR spectra (Figure 12.3).

The FT-IR spectrum of EGNFs (Figure 12.3) prior to silane treatments exhibited a broad peak centered at 3440 cm^{-1} that was attributed to Si–OH groups. The two strong peaks centered at 1060 and 1170 cm^{-1} were attributed to the vibration

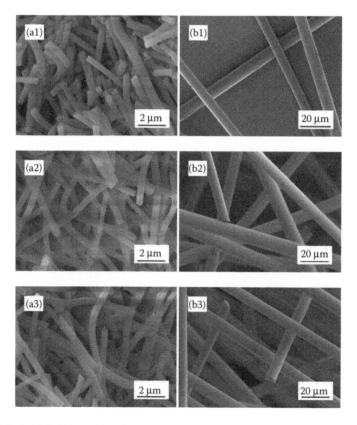

FIGURE 12.1 SEM images showing representative morphologies of EGNFs after ultrasonication (a) and CGFs after being cut from a commercial glass wool (b). (a1), (a2), and (a3) are untreated, GPTMS-treated, and APTES-treated EGNFs, respectively; (b1), (b2), and (b3) are untreated, GPTMS-treated, and APTES-treated CGFs, respectively. (From Chen, Q. et al. *J Appl Polym Sci* 2012, 124, 444 [21]. With permission.)

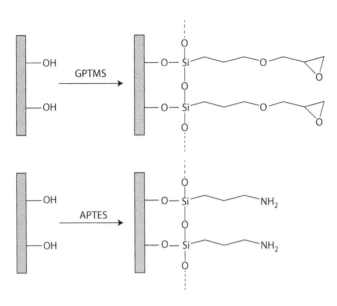

FIGURE 12.2 Schematic diagrams showing the reaction between silane-coupling agents and silanol (Si–OH) groups on the surface of glass fibers (EGNFs and CGFs).

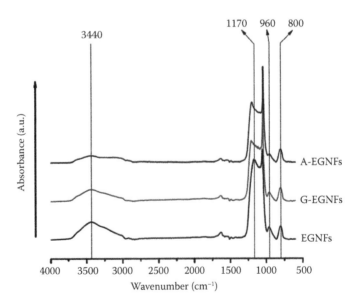

FIGURE 12.3 FT-IR spectra of EGNFs, GPTMS-treated EGNFs, and APTES-treated EGNFs. (From Chen, Q. et al. *J Appl Polym Sci* 2012, 124, 444. With permission.)

splitting of Si–O–Si asymmetric stretching. The other two peaks centered at 960 and 800 cm^{-1} were assigned to the Si–OH stretching and Si–O–Si bending [22,23]. Since the silanization reaction only occurred to some of Si–OH groups on the fiber surface, the epoxy and amino groups from the silane-coupling agents of GPTMS and APTES were overwhelmed and/or overlapped with Si–O–Si and Si–OH absorptions in FT-IR spectra. Hence, instead of the variations of epoxy and amino groups, the variations of Si–O–Si and Si–OH peaks before and after silanization were used as evidence for the silanization reaction. The peak centered at 800 cm^{-1} was selected as the internal reference peak because it did not change appreciably before and after silanization treatments; the reason being this peak was related to Si–O–Si bending vibration, which would not vary substantially since the silanization only occurred on the fiber surface while the SiO$_2$ in the bulk of EGNFs did not change. Through comparison of FT-IR spectra of original EGNFs, GPTMS-treated EGNFs (G-EGNFs), and APTES-treated EGNFs (A-EGNFs), it was evident that the Si–OH absorption between 3000 and 3750 cm^{-1}, as well as the Si–OH absorption centered at 960 cm^{-1}, were distinguishably weakened, and the Si–O–Si asymmetric vibration centered at 1170 cm^{-1} was shifted to higher wave numbers after the silanization treatment. Both indicated that the reactions between silanol groups on EGNFs and silane-coupling agents of GPTMS and APTES did occur as illustrated in Figure 12.2. The CGFs used in this study were a commercial product and gave a complex FT-IR spectrum, probably due to additives in the CGFs; nonetheless, it was reasonable to expect that the silanization reactions occurring to EGNFs could also occur to CGFs.

12.2.2.2 Mechanical Properties

Impact and tension tests on the epoxy composite resins containing low mass fractions (0.5% and 1%) of EGNFs or CGFs with or without silanization treatment were performed, and the acquired results are shown in Figure 12.4. The control sample was the neat epoxy resin that was prepared with the same processing conditions. G-EGNFs and G-CGFs represent the GPTMS-treated fibers, while A-EGNFs and A-CGFs represent APTES-treated fibers.

12.2.2.2.1 Strength

The incorporation of EGNFs or CGFs into the epoxy resin improved the tensile strength of the resulting composite resins, whereas EGNFs outperformed CGFs as the reinforcement agent (Figure 12.4a). The tensile strength of the neat epoxy resin was 41.1 MPa; the incorporation of 0.5% EGNFs or CGFs resulted in 6.3% or 3.4% increase on tensile strength of the respective composite resins. With increasing the mass fraction of EGNFs or CGFs to 1.0%, the composite resins demonstrated 12% or 4.1% increase on tensile strength. G-EGNFs improved the tensile strength by 25% and 31% at the mass fractions of 0.5% and 1.0%, respectively, while G-CGFs resulted in merely a 9.2% increase on tensile strength at the same mass fractions. The APTES-treated fibers showed the largest improvements on tensile strength. The tensile strengths of the resulting epoxy composite resins increased to 54.2 and 57.6 MPa (an increase of 32% and 40%) with the incorporation of 0.5% and 1.0% A-EGNFs, respectively, while the tensile strengths merely increased by 8.5% and 18.5% with the incorporation of the same mass fractions of A-CGFs.

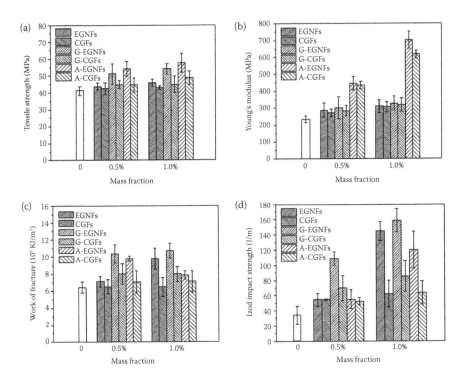

FIGURE 12.4 Tensile strength (a), Young's modulus (b), toughness (c), and impact strength (d) of the neat epoxy resin (control sample) and the composite resins containing EGNFs or CGFs. G- represents the GPTMS-treated fibers while A- represents APTES-treated fibers. (From Chen, Q. et al. *J Appl Polym Sci* 2012, 124, 444. With permission.)

12.2.2.2.2 Stiffness

Young's modulus was used as the measure of stiffness for the epoxy composite resins. EGNFs outperformed CGFs slightly on Young's modulus (Figure 12.4b). The neat epoxy resin exhibited Young's modulus of 234 MPa, while Young's moduli of epoxy composite resins increased with the amount of glass fibers. The incorporation of EGNFs resulted in 23% and 33% increases of modulus for the composite resins at mass fractions of 0.5% and 1.0%, respectively, while CGFs showed 16% and 31% increases correspondingly. G-EGNFs led to 29% and 40% improvements of modulus for the composite resins at mass fractions of 0.5% and 1.0%; for comparison, G-CGFs led to 22% and 37% improvements. The epoxy composite resins containing 1% A-EGNFs had the largest Young's modulus of 704 MPa, an increase by 201%, while Young's modulus of the composite resin with 0.5% A-EGNFs increased by 91%. For comparison, Young's moduli of the composite resins with 0.5% and 1.0% A-CGFs were increased by 86% and 165%, respectively.

12.2.2.2.3 Toughness

In general, the epoxy resin has relatively low toughness; the improvement on toughness for epoxy resin has continuingly been a goal of research efforts. The toughness of the prepared epoxy composite resins was characterized by measuring the area under the stress–strain curve (i.e., WOF) as well as the Izod impact strength.

Compared to the composite resins with CGFs, the composite resins with EGNFs exhibited higher WOF (Figure 12.4c). The WOF value of the neat epoxy resin was 6.4×10^3 kJ/m^3. Incorporation of EGNFs at the mass fractions of 0.5% and 1.0% resulted in 11% and 52% increases of WOF for the respective composite resins. As a comparison, CGFs merely showed ~1.5% increase of WOF at these incorporation levels. The incorporation of 1% G-EGNFs further increased the value of WOF to 1.1×10^4 kJ/m^3 (an increase by 67%), whereas the composite resin containing 1% G-CGFs showed merely 24% increase of WOF. Intriguingly, albeit the incorporation of APTES-treated fibers into the composite resins led to higher WOF than the incorporation of untreated glass fibers, the improvements on WOF were not as much as the incorporation of GPTMS-treated fibers.

The Izod impact strength of the neat epoxy resin acquired in this study was 34.3 J/m. As shown in Figure 12.4d, the incorporation of EGNFs or CGFs into epoxy resin improved the impact strength of the resulting composite resins, and EGNFs outperformed CGFs. With the incorporation of untreated EGNFs, the impact strength increased to 54.3 J/m (an improvement of 58%) at the EGNFs mass fraction of 0.5%, while the impact strength increased to 145 J/m (an improvement of 322%) at EGNFs mass fraction of 1.0%. Correspondingly, the epoxy composite resins showed 58% and 81% increases of impact strength with 0.5% and 1.0% CGFs, respectively. Incorporation of GPTMS-treated fibers further improved the impact strength of the composite resin. The value of impact strength for the composite resin containing 0.5% G-EGNFs increased to 108 J/m, and the value further increased to 159 J/m with the increase of G-EGNFs to 1.0%; this represented 363% increase when compared with that of the neat epoxy resin. For comparison, the impact strengths of the composite resins containing 0.5% and 1.0% G-CGFs were increased by 103% and 149%, respectively. The composite resins containing 0.5% APTES-treated fibers showed similar values of impact strength as those containing the same amount of untreated fibers. At a higher mass fraction of 1.0%, however, A-EGNFs led to 250% increase of impact strength, while A-CGFs resulted in merely 85% increase of impact strength.

12.2.2.3 Effects on Reinforcement and/or Toughening

Through incorporation of EGNFs into epoxy resin, the resulting composite resins achieved simultaneous improvements on strength and toughness, and EGNFs outperformed their conventional counterpart of CGFs in both reinforcement and toughening. This was attributed to the high specific surface area of EGNFs. The average diameter of EGNFs was ~1/25 of that of CGFs; hence, the specific surface area of EGNFs was ~25 times larger than that of CGFs. The larger specific surface area would result in more fiber–epoxy interfacial interactions to facilitate the efficient transfer of stresses. Incorporation of higher amount of fibers from 0.5% to 1% would also result in more interfacial interactions, and in most cases, showed more effective reinforcement and/or toughening.

The surface silanization treatments of fibers improved the interfacial bonding strength between the fiber filler and the resin matrix, and led to higher mechanical properties of the composite resins. Intriguingly, two types of silanized EGNFs exhibited different effects on strength/stiffness and toughness. The G-EGNFs showed a higher degree of toughening effect, while A-EGNFs showed a higher degree of reinforcement effect. This was probably attributed to the different interfacial strengths that resulted from the processing methods. During the preparation of composite resins in this study, the epoxy functional groups on G-EGNFs did not react with epoxy molecules in the matrix, and the chemical bonds between the filler of G-EGNF fibers and the matrix of epoxy resin were not formed until the curing process through linking with hardener molecules. Whereas, the amine functional groups on A-EGNFs would react with epoxy molecules in the matrix and form chemical bonds prior to the curing process, resulting in stronger fiber–matrix interfacial bonding strength than that of G-EGNFs.

12.2.2.3.1 Reinforcement Effect

According to composite theory, the reinforcement of glass fibers for epoxy composites is attributed to the substantially higher strength and modulus of glass fibers than those of epoxy resin. However, the reinforcement potential of glass fibers can only be achieved if an effective load transfer from the epoxy matrix to the glass fibers is available. In regard to the mechanical properties of glass fiber-reinforced epoxy composite resins, the fiber–matrix interfacial bonding strength is essential, and the strong bonding strength would result in high strength and modulus.

In general, the silane-treated EGNFs exhibited higher capability on reinforcement of composite resins than the untreated EGNFs; this is due to the improvement on fiber–matrix interfacial bonding strength. The stronger interface between A-EGNFs and epoxy resin led to more efficient load transfer from the epoxy matrix to the fiber filler; therefore, the higher tensile strength and modulus were observed in the epoxy composite resins containing A-EGNFs other than G-EGNFs.

12.2.2.3.2 Toughening Effect

Fracture surfaces of composite resins can provide valuable information about fracture mechanisms and the influence of fiber surface treatment on fracture behavior. The fracture surfaces of impact test samples were examined by SEM (Figure 12.5). The relatively smooth surface with oriented fracture lines initiated from sites of crack growth was observed on the fracture surface of the neat epoxy resin. The lack of plastic deformation and the smooth fracture surface were in agreement with the typical materials having low toughness. The epoxy composite resins containing EGNFs had rough features on their fracture surfaces, and jagged, short, and multiplane fracture lines were observed on the fracture surfaces of the composite resin with 1.0% EGNFs; this indicated that the crack fronts were deflected and kinked during growth. Thus, the main function of EGNFs in epoxy composite resins was probably to deflect the propagating cracks and force the crack growth to deviate from the existing fracture plane. Additional energy was then necessitated to continually drive crack growths, because the creation of additional fracture surface area would consume more energy. Increasing the mass fraction of EGNFs from 0.5% to 1.0% resulted in a higher degree of deflecting effect because of more EGNFs in the composite resin, which further improved the toughness. It is noteworthy that, due to the

FIGURE 12.5 SEM images showing the representative fracture surfaces of the neat epoxy resin (a), the composite resins with 1% EGNFs (b), 1% G-EGNFs (c), and 1% A-EGNFs (d). (From Chen, Q. et al. *J Appl Polym Sci* 2012, 124, 444. With permission.)

relatively weak interface between untreated EGNFs and the epoxy, a relatively small amount of energy was required to debond EGNFs from the epoxy matrix, leading to moderate improvement on toughness.

The epoxy composite resins containing silanized EGNFs had much improved fiber–matrix interface, and the required debonding energy was much higher; hence, the toughness of composite resins with silanized EGNFs was substantially higher than that of composite resins with untreated EGNFs. Additionally, A-EGNFs formed stronger bonding with the epoxy matrix, and thus, interfacial debonding was more difficult to occur when compared with G-EGNFs. Consequently, G-EGNFs showed the largest improvement on toughness of the epoxy composite resins. The strong interfacial interaction between A-EGNFs and the epoxy resin might enable more crack bridging in the fracture process instead of crack deflecting. This could be responsible for the fact that the fracture surface of epoxy composite resin containing A-EGNFs showed less deflecting lines (Figure 12.5d).

12.3 MULTISCALE GLASS FIBER-REINFORCED COMPOSITE LAMINATES DEVELOPED FROM EPOXY COMPOSITE RESIN CONTAINING EGNFs

In general, fiber-reinforced composite laminates exhibit excellent in-plane properties, whereas the resin matrices dominate out-of-plane properties (e.g., interlaminar shear strength and delamination toughness), which are substantially lower than in-plane properties [24]. To improve the out-of-plane properties of fiber-reinforced composite laminates, nanoscaled materials have been introduced into matrix resins for the development of multiscale composites [25–27]. Numerous research efforts have indicated that the properties of these composites, in which nanoscaled materials are dispersed as the second phase in matrices, are significantly higher [28–30]. Several types of nanoscaled materials including graphite nanofibers, carbon nanotubes/nanofibers, exfoliated graphite nanoplatelets, activated carbon, organoclay, and SiO_2 nanoparticles have been studied to reinforce the matrix-rich interlaminar regions due to their high mechanical properties and large surface-to-mass ratios [31–35].

In this part, EGNFs were studied to reinforce and/or toughen epoxy composite laminates that are made with conventional glass microfiber fabrics. Figure 12.6 is a schematic representation of the multiscale glass fiber-reinforced composites that have been developed and evaluated in this study. The hypothesis was that, by dispersing a small amount of EGNFs into epoxy resin, mechanical properties (particularly out-of-plane mechanical properties) of the resulting multiscale composite laminates would be significantly improved. To test the hypothesis, vacuum-assisted resin transfer molding (VARTM, a commonly used composite-manufacturing technique) was adopted to fabricate the composites, and the reinforcement and/or toughening effects of EGNFs on mechanical properties (including interlaminar shear strength, flexural properties, impact adsorption energy, and tensile properties) of the multiscale epoxy composite laminates were studied. For comparison, short fibers chopped from CGFs were also incorporated into the epoxy resin for making counterpart composite laminates. SEM was employed to examine the micro- and nanoscaled morphologies as well as the fracture surfaces to study the failure mechanisms.

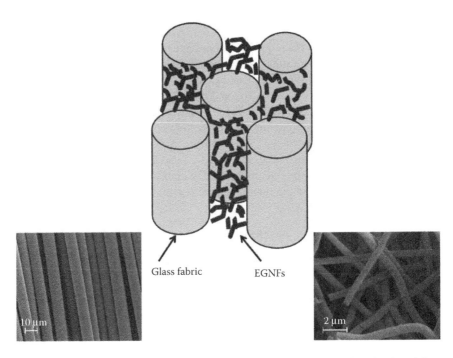

Glass fabric   EGNFs

FIGURE 12.6 Schematic representation of multiscale epoxy composites developed from conventional glass fabric and epoxy resins containing EGNFs.

12.3.1 EXPERIMENTAL

EGNFs with four mass fractions (i.e., 0.05%, 0.1%, 0.25%, and 0.5%) were dispersed into SC-15A epoxy resin by stirring at 125 rpm for 12 h followed by being sonicated for 30 min to achieve uniform dispersion of nanofibers in the resin. Subsequently, SC-15B hardener was added with mass ratio of the hardener versus epoxy resin being set at 30/100, and the mixtures were then hand mixed for 5 min. After degassing for 20 min under vacuum (~27 mm Hg), the prepared epoxy resins containing EGNFs were infused into a vacuum bag containing six plies of conventional woven glass fabrics to fabricate multiscale epoxy composite laminates using the VARTM technique (Figure 12.7). For comparison, chopped CGFs were also dispersed in the epoxy resin using the same procedure to prepare the counterpart epoxy composite laminates.

It is noteworthy that with even a small amount of glass fibers (including both EGNFs and chopped CGFs) incorporated into the epoxy resin, a considerable increase of viscosity was observed; to improve the fluidity, the multiscale glass fiber/epoxy system was kept at 50°C and a vacuum of 27 mm Hg was applied during the initial curing at room temperature for 24 h. The obtained composites were further cured in an oven at 110°C for 5 h prior to the following characterization and evaluation. The control composite laminate that is made of six layers of glass fabrics and epoxy resin without EGNFs and/or CGFs was also fabricated and evaluated.

A Zeiss Supra 40 VP field-emission SEM was employed to examine morphologies of the fibers as well as fracture surfaces of the composites. Prior to SEM examinations, the specimens were sputter coated with gold to avoid charge accumulations.

Mechanical properties of the fabricated composites were tested at room temperature. The impact specimens (64 mm in length, 12.7 mm in width, and 1.6 mm in thickness), the flexural specimens (50.8 mm in length, 12.7 mm in width, and 1.6 mm in thickness), and the short-beam specimens (8 mm in length, 4 mm in width, and 1.6 mm in thickness) were cut from the prepared composite panels by water jet. The impact tests were performed on a Tinius Olsen impact tester (Impact 104) according

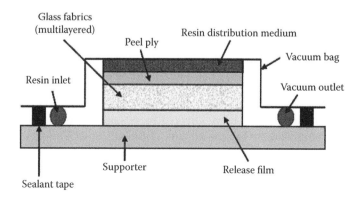

FIGURE 12.7 Schematic of VARTM.

to the ASTM D256. The specimens for flexural and short-beam tests were evaluated in accordance with ASTM D790 and ASTM D2344, respectively. The three-point flexural test with the span distance of 25.4 mm was conducted to fracture specimens at the strain rate of 0.01 mm/min on a QTEST/10 mechanical testing machine purchased from the MTS Systems Co. (Eden Prairie, MN). The short-beam test was carried out at the span-to-thickness ratio of 4 and the cross-head speed of 1 mm/min until the specimens failed. Five specimens of each composite were evaluated, and the mean values and the associated standard deviations of the mechanical properties were calculated.

12.3.2 Results and Discussion

12.3.2.1 Interlaminar Shear Strength

The short-beam shear test was carried out to measure the interlaminar shear strength of the fabricated composite laminates. Interlaminar shear strength is to describe the laminate's resistance against the failure under shear stress; herein, the interlaminar shear strength is calculated according to the following formula:

$$\tau_s = 0.75 \frac{P_m}{b \times h}$$

where τ_s is the short-beam interlaminar shear strength in MPa, P_m is the maximum load that is recorded in the test in N, b is the specimen width in mm, and h is the specimen thickness in mm.

As shown in Figure 12.8a, the interlaminar shear strength of the epoxy composite laminates increased with the increase of EGNFs or CGFs amount up to 0.25%. The value for the control sample (i.e., the epoxy composite with only CGFs) was (17.3 ± 1.1) MPa; for the epoxy composite laminates containing 0.25% EGNFs or CGFs, the respective values were (23.4 ± 1.2) and (22.4 ± 0.4) MPa. Thus, the interlaminar shear strength was improved by 35.3% and 29.5%, respectively. Nonetheless, when EGNFs or CGFs increased to 0.5%, the interlaminar shear strength decreased; this was probably due to the agglomeration of EGNFs or CGFs at higher concentration. It is well known that the agglomerates would act as mechanical weak points (structural defects) in the composites.

12.3.2.2 Flexural Properties

Figure 12.9a and b shows the typical load–displacement curves experimentally acquired from the multiscale glass fiber-reinforced composite laminates with and without EGNFs or CGFs. It was evident that the incorporation of EGNFs or CGFs substantially increased the flexural rigidity (stiffness) and failure load, which would reach the maximum values when the amount of EGNFs or CGFs was 0.25%. The flexural strength and WOF of the composite laminates containing varied mass fractions of EGNFs or CGFs were measured, and the results are shown in Figure 12.8b and c. The values of flexural strength and WOF were substantially increased by the incorporation of small mass fractions (up to 0.25%) of EGNFs or CGFs; if the incorporation amount was even higher (e.g., 0.5%), the values would decrease.

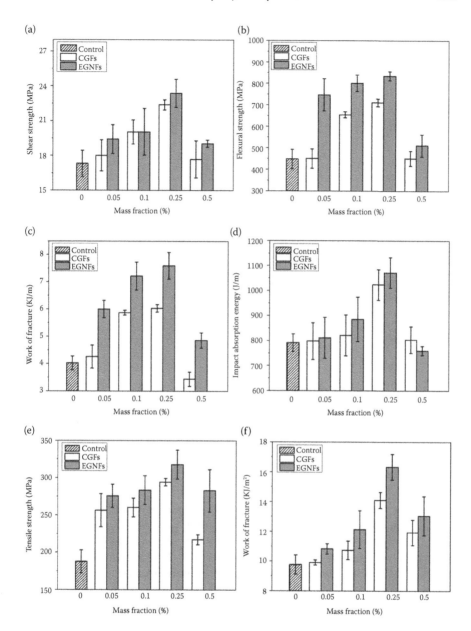

FIGURE 12.8 Mechanical properties of the multiscale glass fiber-reinforced composite laminates acquired from the short-beam shear test, three-point bending test, Izod impact test, and tensile test: (a) shear strength; (b) flexural strength; (c) flexural WOF; (d) impact absorption energy; (e) tensile strength; and (f) tensile WOF.

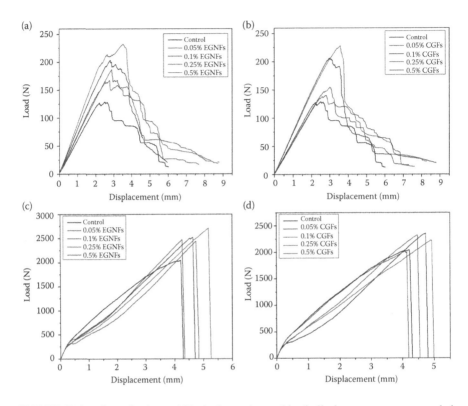

FIGURE 12.9   (**See color insert.**) Typical experimental load–displacement curves recorded from the three-point bending test (a and b) and tension test (c and d) of the multiscale glass fiber-reinforced composite laminates.

The flexural strength and WOF for the control sample were (448.4 ± 45.1) MPa and (4.0 ± 0.3) kJ/m², respectively. For the composite laminates with 0.25% EGNFs, the measured values of flexural strength and WOF were (835.0 ± 20.5) MPa and (7.6 ± 0.5) kJ/m², representing the improvements of 86.2% and 90% compared to the values acquired from the control sample, respectively. For the composite laminates with 0.25% CGFs, the flexural strength and WOF were increased to (711.5 ± 17.6) MPa and (6.0 ± 0.1) kJ/m². Thus, the flexural strength was improved by 58.7%, and the WOF was improved by 50%, respectively. Further increase of the incorporation amount of EGNFs or CGFs did not result in higher mechanical properties. For the composite laminates with 0.5% EGNFs, the flexural strength and WOF were (511.2 ± 50.6) MPa and (4.9 ± 0.3) kJ/m², respectively, while for the composite laminates with 0.5% CGFs, the respective values were (451.2 ± 34.5) MPa and (3.4 ± 0.3) kJ/m². These results indicated that the integration of both nano- and microscaled glass fibers in epoxy resin could substantially improve flexural properties of the resulting composite laminates when the incorporation amount of EGNFs or CGFs was low (e.g., 0.25 wt.%), whereas EGNFs outperformed CGFs. This is primarily due to the following two reasons: (1) the mechanical strength of glass fibers is inversely proportional to the square root of their diameters assuming that the fibers possess the same density and distribution of structural defects [1]; thus, the EGNFs might be stronger than CGFs; and (2) the smaller the fiber diameters are, the larger the total surface areas will be, which would result in the improvement on interfacial bonding strength.

12.3.2.3   Three-Point Bending Fracture Surface and the Reinforcement Mechanism

It is well known that shear stress is typically transferred from layer to layer through resin matrix during interlaminar shear failure and three-point flexural failure of composite laminates. Thus, the main failure mechanism is related to interfacial bonding between the resin and fibers, while the deformation/fracture of the resin matrix may also contribute to the failure [24].

To understand the reinforcement mechanism of EGNFs, the fracture surfaces of three-point bending specimens were examined by SEM. Representative fracture surfaces of the multiscale glass fiber-reinforced composite laminates with 0.25% EGNFs or CGFs as well as the control sample are shown in Figure 12.10. For the control sample, the matrix completely detached from the surface of glass fabrics due to weak adhesion, and the failure surfaces of fibers were smooth without the remnants of the resin (Figure 12.10a). In comparison, the specimens with EGNFs or CGFs could be distinguished from significantly different interfacial microstructure and the deformation of the matrix, as shown in Figure 12.10b–d. These SEM micrographs showed that EGNFs or CGFs were surrounded by and adhered to the resin, indicating that the interfacial bonding between the glass fabrics and the epoxy matrix could be improved by the integration of nano- or microscale glass fibers in the epoxy resin. The results suggested that the presence of EGNFs or CGFs could deflect the microcracks and, thus, the resistance to crack growth was increased. Additionally, EGNFs or CGFs could also be broken and/or detached from the epoxy resin when the load was applied; this would dissipate the strain energy, preventing the failure of the composites and leading to a higher value of WOF. Nonetheless, as shown in Figure 12.10d, the EGNFs appeared to form agglomerates in the composite laminates with 0.5% EGNFs. This was probably the reason why such a composite had lower mechanical strength. For the same mass fraction of EGNFs or CGFs in the epoxy resin, smaller fiber diameter of EGNFs resulted in a larger total surface area that led to better improvement on interfacial bonding strength and, thus, higher mechanical properties (including strength, modulus, and WOF) of the composite laminates.

12.3.2.4   Impact Property

The Izod impact test of notched specimens was conducted to examine fracture behaviors of the fabricated composite laminates by measuring the energy absorption to break the specimens at high strain rates. Upon impact, the total energy can be divided into two components including (1) the elastic energy stored in the specimen in the form of mechanical vibrations, part of which is reflected back to the impactor; and (2) the dissipated energy associated with plastic deformation of the resin,

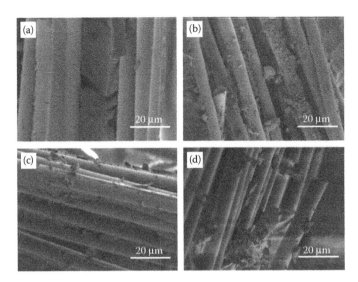

FIGURE 12.10 SEM images showing the representative three-point bending fracture surfaces: (a) The composite laminate of glass fabric/epoxy without EGNFs or CGFs (control sample), (b) the composite laminate with 0.25% CGFs, (c) the composite laminate with 0.25% EGNFs, and (d) the composite laminate with 0.5% EGNFs. (From Chen, Q. et al. *Composites: Part B* 2012, 43, 309 [36]. With permission.)

generation of cracks in the matrix, as well as detachment of fibers from the resin matrix in the interfacial regions [37].

As shown in Figure 12.8d, the values of impact absorption energy increased with the incorporation of EGNFs or CGFs up to 0.25%; the values then decreased with further increase of the incorporation amount. The impact absorption energy of the control sample was (791.9 ± 35.0) J/m. For the composite laminate containing 0.25% EGNFs or CGFs, the respective impact absorption energies were increased to (1072.0 ± 61.0) and (1023.7 ± 61.2) J/m; thus, the impact absorption energies were improved by 35.4% and 29.3%, respectively, and it appeared that EGNFs slightly outperformed CGFs. Nonetheless, the impact absorption energies of the composite laminate with 0.5% EGNFs or CGFs were decreased to (759.9 ± 18.0) and (803.8 ± 53.2) J/m, respectively; and this was also attributed to the formation of agglomerates of EGNFs or CGFs and apparently, it is easier for EGNFs to form agglomerates due to their higher specific surface area.

12.3.2.5 Impact Fracture Surfaces and the Toughening Mechanism

In general, impact energy absorption may result in delamination of composites, breakage and/or pullout of fibers, and deformation of resin matrices. Upon impact, if the energy is lower than the critical value, no impact failure will occur, while the energy will only lead to elastic deformation of the resin matrix. As the incident impact energy increases, delamination starts to occur and/or propagate until the maximum delaminated area is reached. When the impact energy is further increased, localized failures such as breakage and/or pullout of fibers as well as complete delamination will occur.

As shown in Figure 12.11a1, the interface between neat epoxy resin and glass fabrics had the lowest impact resistance, as evidenced by smooth fiber surfaces upon debonding failure. As shown in Figure 12.11a2, the fracture surface in the interlaminar regions was smooth with oriented fracture lines due to extension of crazings

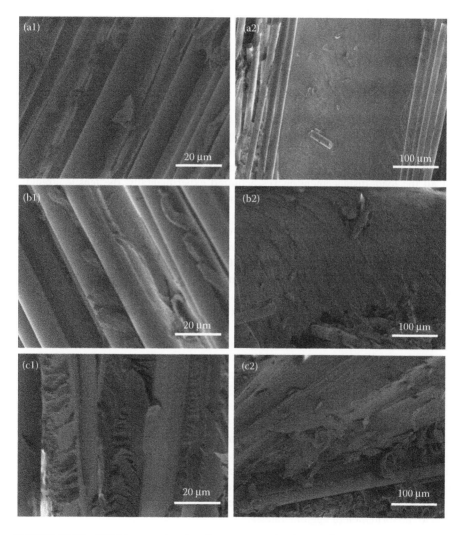

FIGURE 12.11 SEM images showing the representative impact fracture surfaces: (a) The composite laminate of epoxy/glass fabric without EGNFs or CGFs (control sample), (b) the composite laminate with 0.25% CGFs, and (c) the composite laminate with 0.25% EGNFs. a1, b1, and c1 show the regions within composite laminae, while a2, b2, and c2 show the interlaminar regions. (From Chen, Q. et al. *Composites: Part B* 2012, 43, 309. With permission.)

initiated from the locations of stress concentration. In contrast, Figure 12.11b1 and c1 exhibited dimpled/scalloped fracture features of CGFs or EGNFs containing composite laminates. This could explain the formation of tougher interface between epoxy matrix and glass fabrics due to the presence of CGFs or EGNFs. Meanwhile, the fracture surface of the composite with EGNFs was rougher than that of the composite with CGFs, indicating more impact absorption energy. These results suggested that the presence of EGNFs or CGFs could increase the matrix deformation and crack length considerably during the impact fracture, making the delamination more difficult to occur (i.e., more energy would be required for impact fracture). When the cracks finally broke away from the fibers, kinked fracture surfaces were created, suggesting more strain-energy dissipation during cracking, which led to the increase of impact strength/resistance.

12.3.2.6   Tensile Properties

The typical load–displacement curves of tensile tests for the composites with and without EGNFs and CGFs are shown in Figure 12.9c and d, respectively. Similar to the ones shown in Figure 12.9a and b, the multiscale glass fiber-reinforced composite laminate with 0.25% EGNFs had the highest stiffness and failure load. On the basis of all the load–displacement curves acquired experimentally, the tensile strength and WOF of the composite laminates containing varied amounts of EGNFs or CGFs were acquired, as shown in Figure 12.8e and f. Each datum in the plots provided the mean value of five measurements with the error bar representing standard deviation.

As shown in Figure 12.8e and f, the values of tensile strength and WOF were significantly increased when EGNFs or CGFs were employed. The tensile strength and WOF for the glass fabric/epoxy control sample were (187.8 ± 15.4) MPa and (9.8 ± 0.6) kJ/m^2, respectively. For the composite laminate with 0.25% EGNFs, the respective tensile strength and WOF were (318.1 ± 19.2) MPa and (16.3 ± 0.9) kJ/m^2; thus, the tensile strength and WOF were improved by 69.4% and 66.3%, respectively, compared to the control sample. As for the composite laminate with 0.25% CGFs, the tensile strength and WOF were (294.3 ± 5.0) MPa and (14.1 ± 0.5) kJ/m^2, respectively; thus, the tensile strength was improved by 36.2%, and the WOF was increased by 43.9%. It is also evident that EGNFs outperformed CGFs. As explained previously, this might be attributed to the following two reasons: (1) ENGFs might possess higher mechanical strength than CGFs, since the mechanical strength of glass fibers would be inversely proportional to the square root of the fiber diameter; and (2) the smaller the fiber diameters are, the larger the total surface areas will be and this would result in the improvement on interfacial bonding strength. Similar to other tests, the mechanical properties of composite laminates with 0.5% EGNFs or CGFs were lower than those of composite laminates with 0.25% EGNFs or CGFs due to the formation of agglomerates.

12.4   CONCLUSION

EGNFs with diameters of 400–500 nm were prepared and evaluated for composite purpose. The incorporation of EGNFs into epoxy resin resulted in simultaneous improvements on both strength and toughness of the resulting composite resins at small mass fractions of up to 1%. Compared to the neat epoxy resin, the incorporation of 1% EGNFs increased the tensile strength by 12%, Young's modulus by 33%, WOF by 52%, and impact strength by 322%. The silanization treatment of EGNFs improved the fiber–matrix interfacial bonding strength, and resulted in even higher mechanical properties of the corresponding composite resins. The salinized EGNFs with epoxy end groups (G-EGNFs) demonstrated a higher degree of toughening effect. Compared to the neat epoxy resin, the incorporation of 1% G-EGNFs increased the tensile strength by 31%, Young's modulus by 40%, WOF by 67%, and impact strength by 363%. The silanized EGNFs with amine end groups (A-EGNFs) demonstrated a higher degree of reinforcement effect. Compared to the neat epoxy resin, the incorporation of 1% A-EGNFs increased the tensile strength by 40%, Young's modulus by 201%, WOF by 22%, and impact strength by 250%. Additionally, EGNFs substantially outperformed CGFs with diameters of ~10 μm in all cases. Further study revealed that the epoxy resins containing EGNFs could be applied in conventional glass fabric/epoxy composite laminates to generate multiscale glass fiber-reinforced composite laminate. 0.25% EGNFs in epoxy resin could substantially improve the mechanical properties of the resulting composite laminates including the increase of interlaminar shear strength by 35.3%, flexural strength by 86.2%, impact strength by 35.4%, and tensile strength by 69.4%. The comparison showed that EGNFs outperformed CGFs once again on all property improvements. It is evident that EGNFs have the potential to be utilized as innovative reinforcement and/or toughing agent in polymer composite materials. The continuing research on EGNFs for composite purpose will pave the road toward next-generation fiber-reinforced high-performance nanocomposites.

REFERENCES

1. Mallick, P. K. *Fiber Reinforced Composites: Materials, Manufacturing, and Design*; Marcel Dekker: New York, 1993, pp. 1–11.
2. Hammel, E., Tang, X., Trampert, M., Schmitt, T., Mauthner, K., Eder, A., Potschke, P. Carbon nanofibers for composite applications. *Carbon* 2004, 42, 1153–1158.
3. Tibbetts, G. G., Lake, M. L., Strong, K. L., Rice, B. P. A review of the fabrication and properties of vapor-grown carbon nanofiber/polymer composites. *Compos Sci Technol* 2007, 67, 1709–1718.
4. Jiang, H.-X., Ni, Q.-Q., Natsuki, T. Tensile properties and reinforcement mechanisms of natural rubber/vapor-grown carbon nanofiber composite. *Polym Compos* 2010, 31, 1099–1104.
5. Chou, T. W., Gao, L., Thostenson, E. K., Zhang, Z., Byun, J. H. An assessment of the science and technology of carbon nanotube-based fibers and composites. *Compos Sci Technol* 2010, 70, 1–19.
6. Paiva, M. C., Novais, R. M., Araujo, R. F., Pederson, K. K., Proenca, M. F., Silva, C. J. R., Costa, C. M., Lanceros-Mendez, S. Organic functionalization of carbon nanofibers for composite applications. *Polym Compos* 2010, 31, 369–376.
7. Dzenis, Y. Spinning continuous fibers for nanotechnology. *Science* 2004, 304, 1917–1919.
8. Greiner, A., Wendorff, J. H. Electrospinning: A fascinating method for the preparation of ultrathin fibres. *Angew Chem Int Ed* 2007, 46, 5670–5703.
9. Fong, H. In *Polymeric Nanostructures and Their Applications*; Nalwa, H. S., ed.; American Scientific Publishers: Los Angeles, 2005; Vol. 2, pp. 451–474.

10. Huang, Z. M., Zhang, Y. Z., Kotaki, M., Ramakrishna, S. A review on polymer nanofibers by electrospinning and their applications in nanocomposites. *Compos Sci Technol* 2003, 63, 2223–2253.

11. Fong, H. Electrospun nylon 6 nanofiber reinforced Bis-GMA/TEGDMA dental restorative composite resins. *Polymer* 2004, 45, 2427–2432.

12. Lin, S., Cai, Q., Ji, J., Sui, G., Yu, Y., Yang, X., Ma, Q., Wei, Y., Deng, X. Electrospun nanofiber reinforced and toughened composites through in situ nano-interface formation. *Compos Sci Technol* 2008, 68, 3322–3329.

13. Romo-Uribe, A., Arizmendi, L., Romero-Guzman, M. E., Sepulveda-Guzman, S., Cruz-Silva, R. Electrospun nylon nanofibers as effective reinforcement to polyaniline membranes. *ACS Appl mater Interfaces* 2009, 1, 2502–2508.

14. Ozden, E., Menceloglu, Y. Z., Papila, M. Engineering chemistry of electrospun nanofibers and interfaces in nanocomposites for superior mechanical properties. *ACS Appl Mater Interfaces* 2010, 2, 1788–1793.

15. Lu, X., Wang, C., Wei, Y. One-dimensional composite nanomaterials: Synthesis by electrospinning and their applications. *Small* 2009, 5, 2349–2370.

16. Jo, J. H., Lee, E. J., Shin, D. S., Kim, H. E., Kim, H. W., Koh, Y. H., Jang, J. H. In vitro/in vivo biocompatibility and mechanical properties of bioactive glass nanofiber and poly(ε-caprolactone) composite materials. *J Biomed Mater Res Part B Appl Biomater* 2009, 91B, 213–220.

17. Boccaccini, A. R., Erol, M., Stark, W. J., Mohn, D., Hong, Z., Mano, J. F. Polymer/bioactive glass nanocomposites for biomedical applications: A review. *Compos Sci Technol* 2010, 70, 1764–1776.

18. Liu, Y., Sagi, S., Chandrasekar, R., Zhang, L., Hedin, N. E., Fong, H. Preparation and characterization of electrospun SiO_2 nanofibers. *J Nanoscience Nanotechnol* 2008, 8, 1528–1536.

19. Wen, S., Liu, L., Zhang, L., Chen, Q., Zhang, L., Fong, H. Hierarchical electrospun SiO_2 nanofibers containing SiO_2 nanoparticles with controllable surface-roughness and/or porosity. *Mater Lett* 2010, 64, 1517–1520.

20. Gao, Y., Sagi, S., Zhang, L., Liao, Y., Cowles, D. M., Sun, Y., Fong, H. Electrospun nano-scaled glass fiber reinforcement of Bis-GMA/TEGDMA dental composites. *J Appl Polym Sci* 2008, 110, 2063–2070.

21. Chen, Q., Zhang, L., Yoon, M.-K., Wu, X.-F., Arefin, R. H., Fong, H. Preparation and evaluation of nano-epoxy composite resins containing electrospun glass nanofibers. *J Appl Poly Sci* 2012, 124(1), 444–451.

22. Almeida, R. M. In *Hand Book of Sol–Gel Science and Technology*; Sakka, S., ed.; Kluwer Academic: Norwell, 2005; Vol. 2, pp. 65–90.

23. Xu, D., Sun, L., Li, H., Zhang, L., Guo, G., Zhao, X., Gui, L. Hydrolysis and silanization of the hydrosilicon surface of freshly prepared porous silicon by an amine catalytic reaction. *New J Chem* 2003, 27, 300–306.

24. Zhu, J., Imam, A., Crane, R. Processing a glass fiber reinforced vinyl ester composite with nanotube enhancement of interlaminar shear strength. *Compos Sci Technol* 2007, 67, 1509–1517.

25. Bekyarova, E., Thostenson, E. T., Yu, A. Multiscale carbon nanotube—carbon fiber reinforcement for advanced epoxy composites. *Langmuir* 2007, 23, 3970–3974.

26. Kulkarni, M., Carnahan, D., Kulkarni, K., Qian, D., Abot, J. L. Elastic response of a carbon nanotube fiber reinforced polymeric composites: A numerical and experimental study. *Composites: Part B* 2010, 41, 414–421.

27. Zhamu, A., Zhong, W. H., Stone, J. J. Experimental study on adhesion property of UHMWPE fiber/nano-epoxy by fiber bundle pull-out tests. *Compos Sci Technol* 2006, 66, 2736–2742.

28. Gaicia, E. J., Wardle, B. L., Hart, A. J. Joining prepreg composite interfaces with aligned carbon nanotubes. *Composites: Part A* 2008, 39, 1065–1070.
29. Gojny, F. H., Wichmann, M. H. G., Fiedler, B. Influence of nano-modification on the mechanical and electrical properties of conventional fiber-reinforced composites. *Composites: Part A* 2005, 36(11), 1525–1535.
30. Wichmann, M. H. G., Sumfleth, J., Gojny, F. H. Glass fiber-reinforced composites with enhanced mechanical and electrical properties-benefits and limitations of a nanoparticle modified matrix. *Eng Fract Mech* 2006, 73(16), 2346–2359.
31. Thostenson, E. T., Li, W. Z., Wang, D. Z., Ren, Z. F., Chou, T. W. Carbon nanotube/carbon fiber hybrid multiscale composites. *J Appl Phys* 2002, 91, 6034–6037.
32. Tsai, J. L., Wu, M. D. Organoclay effect on mechanical response of glass/epoxy nano-composites. *J Compos Mater* 2008, 42(6), 553–568.
33. Park, J. K., Do, I. H., Askeland, P. Electrodeposition of exfoliated graphite nanoparticles onto carbon fibers and properties of their epoxy composites. *Compos Sci Technol* 2008, 68, 1734–1741.
34. Allaoui, A., Bai, S., Cheng, H. M. Mechanical and electrical properties of a MWNT/epoxy composites. *Compos Sci Technol* 2002, 62, 1993–1998.
35. Kinloch, A. J., Masania, K., Taylor, A. C. The fracture of glass-fiber-reinforced epoxy composites using nanoparticle-modified matrices. *J Mater Sci* 2008, 43, 1151–1154.
36. Chen, Q., Zhang, L., Zhao, Y., Wu, X.-F., Fong, H. Hybrid multi-scale composites developed from glass microfiber fabrics and nano-epoxy resins containing electrospun glass nanofibers. *Compos Part B Eng* 2012, 43(2), 309–316.
37. Schrauwen, B., Bertens, P., Peijs, T. Influence of hybridization and test geometry on the impact response of glass-fiber-reinforced laminated composites. *Polym Polym Compos* 2002, 10(4), 259–272.

13 Templated Self-Assembly for Nanolithography and Nanofabrication
Overview and Selected Examples

Albert Hung

CONTENTS

13.1 INTRODUCTION

Miniaturization allows more function to be packed into the same amount of space. Since the invention of the integrated circuit more than 50 years ago, the need for more powerful devices for sensing, communication, and computation has thus driven the development of methods for fabricating smaller structures. The current "top-down" microfabrication techniques in which structures are carved out from larger substrates in multiple steps can produce features smaller than 100 nm, but the cost of doing so becomes prohibitive as sizes shrink. In addition, the vision of ubiquitous computing—the integration of information processing into a greater number of everyday objects and actions—might require the use of nontraditional materials to build sensors and circuits that are more flexible and durable. New fabrication methods that reduce costs or expand the range of materials and substrates that can be patterned are in demand now more than ever. Recent advances in nanotechnology further fuel this demand, but also offer promising solutions to these challenges. Unique, functional nanomaterials such as nanoparticles, carbon nanotubes, and graphene might enable revolutionary applications. However, controlling their placement on a substrate and incorporating them into the current manufacturing processes remains difficult.

Self-assembly is the spontaneous organization of rationally designed molecules or particles into well-defined aggregates, patterns, or lattices through noncovalent interactions with little manual intervention.[1] In this manner, nanoscale structures are constructed all at once from smaller subunits, potentially with control of chemical functionality and feature sizes <20 nm. Many examples of self-assembly naturally exist in biological systems, making them an important source of inspiration for designing new assembly strategies, especially the ones that proceed under mild conditions and are amenable to a wide variety of materials. However, assembly is often governed by relatively weak, reversible, and nondirectional interactions, resulting in higher defect rates or patterns limited to simple repeating motifs. Perfectly ordered assembly might be observed over small length scales, but entropic fluctuations increase the likelihood of defects over larger distances.

An attractive strategy for addressing these challenges is to integrate top-down lithographic techniques with "bottom-up" self-assembly, combining the strengths of each approach to fabricate both high-resolution and high-fidelity patterns.[2–4] Also called "templated assembly" or "directed assembly," this method uses a chemically or topographically patterned substrate to control alignment and defects. Confining assembly within or around a lithographically defined feature leads to locally enhanced order, and replication of the feature over the template surface too extends this short-range order to macroscopic length scales. Templated alignment is usually thermodynamically stable, driven by intermolecular and surface forces. This chapter presents a brief overview of directed self-assembly and selected examples focusing on the use of two-dimensional (2-D), prepatterned surfaces to guide the organization of colloidal particles, diblock copolymers, and deoxyribonucleic acid (DNA) nanostructures.

13.2 TEMPLATE DESIGN AND FABRICATION

Templated assembly is analogous in many ways to crystallographic epitaxy, in which the physical and chemical organization of atoms at a nominally flat surface of a crystalline substrate, through organized interatomic bonds made at the interface, subsequently determines the crystal structure and orientation of the material deposited on top. One of the earliest examples of templated assembly is the use of a rubbed glass substrate to align a liquid crystal film.[5] Unidirectional rubbing of the glass surface with paper or cloth generates parallel nanogrooves along which the rod-shaped molecules of the liquid crystal align. Since glass is an amorphous material, the surface topology rather than specific molecular bonds influences assembly. This idea can be expanded upon using more intricately fabricated templates or other assembling materials. Features such as posts and sidewalls can nucleate ordered assembly around them, and repeating the features in a regular array thus extends short-range order over a large area. Conversely, microscale channels or wells can be fabricated so that assembly occurs within a spatially confined volume, typically resulting in alignment commensurate with the shape of the confining space. Oriented assembly in templates with topographical features much larger than the nanostructure size is often called "graphoepitaxy" to distinguish this effect from molecular epitaxy.[6]

The chemical nature of the substrate surface is also an important variable that can be used to control assembly independent of surface topology. Deposition of molecules or nanoparticles from the solution can be guided by patterning the surface with clearly delineated areas that either favor or disfavor adsorption. Some commonly utilized interactions include hydrophobic forces, electrostatic forces, hydrogen bonding, thiol bonds, and biomolecular recognition. The exact character of the chemical interaction depends on the material of interest, but the strength and specificity of the bond are critical parameters to consider. A very strong bond may result in very fast but irreversible, disordered aggregation, whereas weaker bonds are prone to slower assembly and lower yields. Greater chemical selectivity and a reduction in nonspecific binding are always better in theory, but are usually more complicated to implement in practice. Another concern is whether chemically induced order at the substrate surface persists deeper into the bulk of a deposited thick film.

The mechanism and efficacy of chemical or physical templating critically depend on the size of the patterned features relative to the size of the assembled structures. For large structures, gravity and capillary forces tend to dominate assembly and make topological patterns more effective templates. Greater attention must be given to surface chemistry as sizes shrink because intermolecular forces become more important. It stands to reason that features of a size that matches the characteristic length scale of the nanostructures would be able to confine assembly and direct order and placement most precisely, thereby eliminating defects. But the sizes in question can be <10 nm, making them costly and difficult to fabricate. Since the goal of directed assembly is to avoid expensive, high-resolution lithography, using templates with larger features to control assembly such as by graphoepitaxy is more practical but also more defect prone. Complicating matters is the fact that some assembled materials possess multiple characteristic length scales that range over several orders of magnitude. Thus, an enduring question of templated assembly is how large can the lithographically patterned features be while still suppressing defects to an acceptable level.

Templates can be fabricated using any of a number of different lithographic techniques depending on the specifications of the pattern desired and the properties of the sample or substrate material. A full review of each process is beyond the scope of this chapter, but some common methods and relevant considerations bear mentioning. Photolithography and electron-beam (e-beam) lithography are the most mature and widely practiced methods.[7] Photolithography can quickly and inexpensively pattern large areas with 1 μm resolution, but smaller features require more expensive projection systems. Alternately, interference lithography can also achieve higher resolutions over large areas, <10 nm at best for extreme ultraviolet (EUV) interference lithography, but only for simple stripe or dot array patterns.[8,9] E-beam lithography routinely reaches sub-100 nm resolution, but is much slower than photolithography because each feature is drawn in series. Other direct-write methods such as dip-pen nanolithography (DPN)[10,11] expand the variety of materials and substrates that can be patterned, but suffer from the same drawback. This trade-off between resolution and time or cost for a given area to be patterned is a fundamental concern, but one that is alleviated to a degree if the fabricated template can be reused multiple times. Both nanoimprint lithography[12] and soft lithography[13,14] apply this idea, using the topographically patterned rigid template as a master mold to repeatedly emboss or cast negative relief copies in softer materials. The copies can be directly used as substrates for templated assembly or deposition, or as intermediaries for further lithography. Soft lithography in particular is useful for patterning materials that require mild conditions such as biomolecules on curved or flexible substrates.

13.3 SELECTED EXAMPLES OF TEMPLATED ASSEMBLY

13.3.1 COLLOIDS AND NANOPARTICLES

Colloids are broadly defined as micrometer- or nanometer-sized particles of any material dispersed in a matrix of another substance. Solid colloidal particles that are monodisperse (uniform in size and shape) and suspended in a liquid medium are of particular interest for their ability to assemble out of suspension into well-ordered, periodic arrangements analogous to atomic crystals as shown in Figure 13.1.[15,16] This section primarily focuses on isotropic spherical particles that typically assemble into a mixture of close-packed face-centered cubic (FCC) and hexagonally close-packed (HCP) lattices in the absence of a template,[17,18] although nonclose packed,[19] liquid crystalline,[20] and binary superlattice assemblies[17,21–23] are possible for anisotropic or mixed colloidal systems.

Colloidal crystals are potentially useful for a wide range of applications, leading to significant interest in finding ways to control assembly better and build complex structures.[4,24] 2-D colloidal monolayers can be used as masks for lithographic etching or deposition with a resolution determined by the particle diameter, offering an

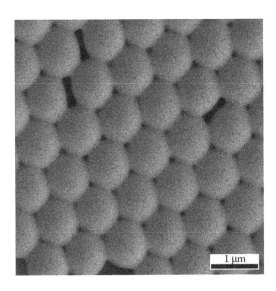

FIGURE 13.1 Scanning electron microscope (SEM) image of 1-μm-diameter polystyrene colloids assembled in a hexagonal monolayer. Hexagonal symmetry is characteristic of many simple self-assembling systems, including spherical or cylindrical block copolymers and surfactant micelles.

inexpensive route to generate nanostructure arrays with unique optical, electrical, or chemical properties.[25,26] Three-dimensional (3-D) ordered, macroporous materials and photonic crystals fabricated by colloidal assembly and templating are valued for their special optical properties, high surface area, and bicontinuous pore structure. Potential areas of application for these materials include catalysis, energy generation and storage, sensors, and optical computing devices.[27–29] Superlattices of sub-20 nm nanoparticles may also be useful as thermoelectric or photovoltaic materials due to increased interface area, quantum confinement, and possible electrical or magnetic coupling between particles.[23,30–32]

Colloidal assembly is mediated by interparticle interactions that tend to be weak, nondirectional, and dependent on particle size, chemistry, and deposition conditions. Gravity, capillary forces, and excluded volume interactions are especially important at longer length scales, which make topological patterns very effective at templating the assembly of colloids that are roughly 50 nm or larger.[33,34] Colloids are typically concentrated and deposited from liquid suspension, although they can also be organized onto topological templates by mechanically applying a dry powder.[35] Within large wells or channels, colloids assemble into HCP arrangements with number and orientation commensurate with the size and shape of the confining space.[36–41] As the confining features shrink, assemblies of discrete number and unique symmetry are obtained. If the channel or well width is incommensurate with a close-packed arrangement, the colloids can exhibit non-close-packed order, close-packed order with regular defects, or complete disorder depending on the properties of the particles and deposition method.[36–38] Over features with dimensions less than the particle diameter, colloids deposit with precise positioning as single-file lines or individual colloids. In this manner, arrays of thin grooves defined lithographically[42,43] or by wrinkling of a soft substrate[39,40] can align rows of colloids over large areas.

Using 2-D templates to direct 3-D colloidal assembly is of particular interest for photonic applications.[27,28] An array of large square pyramidal pits or ridges, fabricated by crystallographic etching of a (100) silicon substrate, can template the colloidal assembly of a bulk single crystal with FCC symmetry.[44,45] The geometry of the well itself templates epitaxial growth of the crystal along the (100) facet and eliminates the HCP stacking faults typically seen with unguided assembly. For obtaining different lattice structures, arrays of single-particle-sized wells are used to first direct the assembly of a monolayer of spheres with controlled spacing and symmetry.[38,46–52] These arrays then direct the assembly of subsequent layers of colloids deposited on top, serving as a seed layer for growing non-close-packed colloidal crystals. Body-centered cubic (BCC) or face-centered tetragonal (FCT) lattices can be grown from square arrays of wells that are spaced to match the (001) facet of the desired crystal.[46–48] Recently, templated assembly combined with tuning of interparticle electrostatic interactions showed progress toward assembling a diamond cubic lattice.[52]

Chemically patterned surfaces can also control the adsorption of colloidal particles from the solution.[53–56] Depending on the nature of the interaction at the surface, adsorption can be designed to be reversible or stimuli responsive. However, chemical patterns tend to be less successful than topographically patterned substrates at determining order in colloidal assembly, especially with large spheres where the contact area with the surface is comparatively small. With nanoscale particles, surface interactions have a greater impact on assembly, but the patterns need to be of higher resolution to eliminate defects. As one example, gold nanoparticles coated and linked together with DNA strands have been shown to assemble from suspension into bulk colloidal crystals of different symmetries depending on the size of the nanoparticles and the DNA sequences.[57–59] When patterns of DNA are printed by soft lithography onto a silicon substrate, the DNA-coated gold nanoparticles bind to these areas and assemble into similar ordered superlattices.[60] The colloidal crystal, now confined as thin-film patches of a defined area and placement, orients itself to present the most close-packed crystal facet at its top surface. Limited control over the in-plane alignment of the crystal lattice is also possible by printing small patterns with defined shapes.

13.3.2 BLOCK COPOLYMERS

Block copolymers are polymer chains that consist of two or more chain segments or "blocks" of different chemistry covalently linked together.[61–63] Diblock copolymers consisting of two blocks joined linearly are the simplest example and the focus of this section. More complicated architectures including triblock, comb-like, branched, and radial star-like copolymers are also possible, but their assembly is harder to predict. Although covalently attached, the different blocks prefer not to mix with each other; so, they microphase separate into ordered arrays of spherical, cylindrical, or lamellar nanostructures that are 10–200 nm in size depending on the lengths of each block in the polymer chain.[64,65] Ordered assembly results from balancing interface energy with chain conformational entropy and requires the copolymer to be monodisperse. By preferential degradation or mineralization of one of the assembled blocks, copolymer films can be used as inexpensive tools for fabricating nanoscale patterns with high feature density.[66,67] However, the nanostructure array is only coherent over distances shorter than a few micrometers, theoretically because long-wavelength thermal phonons give rise to defects such as grain boundaries and disclinations that disrupt order.[68] Defect density and long-range order also depend on the degree of chemical incompatibility between the blocks.[69] Low incompatibility allows for diffuse interfaces and weak phase separation, whereas high incompatibility can inhibit molecular diffusion and freeze defects in place, leading to high defect densities in both cases. Unlike colloidal systems, neat copolymer films can be thermally annealed to remove defects, but only to a limited extent.

Directed assembly of block copolymers on either chemical or topographical patterns extends and guides order over macroscopic areas.[2,70,71] The template features may vary from molecular to multidomain in scale due to the hierarchical nature of copolymer assembly. At the smallest length scale, an organic crystal substrate can template the long-range assembly of a semicrystalline copolymer by molecular epitaxy and directional solidification.[72] Molecular interactions directly affect the local organization of copolymer chains, influencing film morphology and assembly. At a slightly larger scale, single crystal sapphire surfaces with one-dimensional (1-D) faceting can induce ordered assembly over large areas.[73] Epitaxy on single-crystal substrates does not require lithography, but the range of patterns and polymers that can be templated in this way is limited. Noncrystalline surface chemistries also affect assembly and may be tuned to be neutral or favor adsorption of specific blocks. Cylindrical and lamellar phases tend to align parallel to flat surfaces that favor the adsorption of one block over the other.[61] Perpendicular alignment is possible if the surface has equal affinity for both blocks and the film thickness does not match the characteristic nanostructure periodicity, but in-plane orientation remains random.[74–76]

Complex order and orientation of copolymer films can be realized by graphoepitaxy on lithographically defined templates. Within large topographical features, block copolymer spheres assemble into close-packed layers against both vertical and horizontal surfaces, nucleating the growth of large, ordered domains with orientation commensurate with the topological pattern.[77,78] These templated domains span up to several micrometers, many times larger than ordered domains in unpatterned films and orders of magnitude larger than the individual nanostructures. As the features shrink, the number of nanostructures packed within is dictated by the size of the confining space.[79–82] Unlike colloids, the copolymer assembly can strain itself to fill space without introducing packing defects when the feature size is incommensurate with an integer number of nanostructures.

Fully controlled alignment can be obtained by patterning the substrate with a heterogeneous chemical motif that matches the shape and pitch of the assembled structure, leading to domain-level epitaxy. For example, a pattern of nanoscale stripes with alternating affinity for each block orient a lamellar copolymer perpendicular to the surface and parallel to the stripes, replicating the pattern.[83,84] The closer the stripes match the chemistry and thickness of the copolymer lamellae, the fewer defects there are. This method can be extended to template complex patterns[85] and is useful for developing a better understanding of the assembly process. However, fabricating templates with such high resolution and feature density is impractical for real applications. Chemical[86] or topographical[87,88] patterns with larger pitch and lower feature density are easier to manufacture and are still capable of guiding assembly. Films of a block copolymer with a larger domain size can also be used to template the assembly of small-domain copolymers.[89,90] Copolymer assembly on these periodic patterns results in multiplication of feature density and alignment of the hexagonal or striped array. Varying the template array spacing and symmetry can result in more complex 2-D[91] or even 3-D[92] assembly.

13.3.3 DNA Nanostructures

As the understanding of and the ability to manipulate biomaterials and biochemical interactions has grown, so too has the interest in adapting bioinspired strategies for assembly and fabrication. Biomolecules are capable of strong and selective molecular binding, potentially enabling complex, directed assembly at the single-molecule level. DNA, in particular, has attracted significant attention because its structure and assembly are well known and relatively simple, being composed of only four distinct nucleic acids. In contrast, proteins are composed of 20 different amino acids for which the rules of assembly and target binding have yet to be fully elucidated. Single-strand DNA (ssDNA) is a linear sequence of nucleic acids that hybridizes with a complementary sequence to make double-stranded DNA (dsDNA) with the familiar double-helix structure. Since DNA assembly is highly predictable, synthetic DNA nanostructures for both biological and nonbiological applications can be built from rationally designed ssDNA sequences.[3,93–97] DNA assemblies can be 100 nm or larger, but an individual double helix is 2 nm wide. One idea is to position these nanostructures precisely on a lithographically patterned substrate, and then use them to direct the assembly of smaller nanoparticles and nanowires. In this way, arrays of sub-20 nm components can be constructed without the need for high-resolution lithography.

Of particular relevance to the scope of this chapter are 2-D DNA objects and lattices. DNA "tiles" are the simplest of these constructs, typically composed of at least three different DNA strands hybridized together in a branched or interlocked subunit.[94,96] Extending from each tile are ssDNA "sticky" strands that allow them to link with each other to build larger discrete objects, 1-D ribbons or tubes, or 2-D lattices. Additional elements can be built into the tiles so that they bind nanomaterials including nanoparticles,[98–102] fullerenes,[103] and proteins[104–107] in a regular array on the assembled DNA lattice. DNA tiles can direct the ordered assembly of nanomaterials with sub-20 nm resolution over small areas with ease, but controlling the placement and orientation of the tiles over large areas remains a challenge.[108] Tiles are too small to be positioned and aligned individually, and 2-D unbounded lattices and mats are difficult to orient because they lack consistency in shape. However, 1-D assemblies might be aligned by applied forces or graphoepitaxy. For example, DNA tubules decorated with quantum dots can be deposited in an aligned macroscopic array by drying front combing over a topologically patterned substrate.[109]

Larger DNA assemblies of finite shape and size are good candidates for templated assembly on lithographically patterned substrates. DNA origami[110] and modular canvases[111] are the two types of assemblies that are particularly useful for this purpose. DNA origami consists of a loop of plasmid ssDNA, typically about 7000 bases long and derived from the M13 bacteriophage, which is folded into a desired shape by the addition of around 200 short, complementary strands or "staples" that link together different sections of the plasmid scaffold. The origami itself is on the order of 100 nm in size, a length scale accessible to commercial top-down lithography methods. A modular DNA canvas is a newer, alternative design to DNA origami that forgoes the plasmid scaffold and is purely composed of staples (or "single-strand tiles," [SSTs]) linked to each other and arrayed like bricks in a brick wall. Unlike origami, which requires an entirely unique set of staples for each desired shape, the modular canvas uses only one set of tiles and can assemble different shapes by omitting specific tiles.

Each staple strand in an origami or canvas structure is unique and can be individually modified to present different chemical or biomolecular binding groups to capture specific nanostructures. In this way, the origami construct can be treated as a molecular breadboard on which smaller components can be precisely positioned with 6 nm resolution. Since the origami is larger and more complex than DNA tiles, the number of possible patterns and applications that can be generated is much greater. DNA origami has been used as a platform to build multiplexed sensors for detecting RNA targets,[112] to assemble nanoparticles[113] or carbon nanotubes[114] into plasmonic

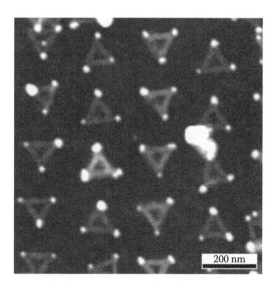

FIGURE 13.2   Atomic force microscope (AFM) height image of triangular DNA origami assembled with DNA-coated 5 nm gold nanoparticles at each corner and templated with controlled placement and orientation on an e-beam-patterned silicon substrate following a previously reported procedure.[119]

lenses or field-effect transistors, and even to construct a nanomechanical assembly line.[115] Alternately, the DNA origami itself can be used for lithography to mask the chemical etching of a silica substrate.[116]

To build devices, the origami needs to be assembled with controlled placement and orientation on a substrate. Origami has been trapped between gold nanoelectrodes by dielectrophoresis and gold–thiol bonds,[117] but the templated assembly of origami on chemically patterned surfaces has generally proven to be more effective.[118–120] DNA preferentially adsorb to hydrophilic or charged surfaces via electrostatic interactions that are strongly dependent on the salt concentration of the solution.[121] Surface topology can aid in templating the assembly, but is not the primary driving force since the origami is only 2 nm thick. Using e-beam lithography to pattern hydrophilic regions that match the size and shape of the origami, single DNA nanostructures can be positioned with precise placement and orientation.[120] An attractive surface interaction that is still weak enough to allow the origami to rotate into position is critical for obtaining oriented assembly. Within larger hydrophilic regions, the origami geometrically tile to a limited extent to fill the area. The templated origami can subsequently be used as templates themselves for assembling gold nanoparticles, generating a large-area array of sub-20 nm components as shown in Figure 13.2.[119]

13.4   OUTLOOK

The fields of self-assembly and nanotechnology have rapidly advanced in recent years, and their parallel growth is not entirely coincidental. The rational design of self-assembly requires an understanding of the chemical and physical characteristics and behaviors of matter at the nanoscale. Similarly, many functional nanomaterials either self-assemble or are made by assembly. However, a better fundamental understanding of the phenomena that drive assembly is needed to increase yields, decrease defects, and make such tools commercially viable. Significant knowledge gaps persist regarding the roles of kinetics and surface interactions. Self-assembly most often focuses on thermodynamic factors that drive assembly and determine the most stable state. But diffusion and activation energies decide how long the process will take and what actually happens along the way. Understanding kinetics would not only help optimize and accelerate assembly, but could also expand the possibility of designing and assembling kinetically trapped structures with unique properties. The majority of the interactions that regulate assembly operate between surfaces and interfaces. As sizes decrease, the significance of surface forces grows, but the task of characterizing and controlling the structure of surfaces and interfaces with sufficient detail at the molecular level becomes increasingly difficult. Concerns about kinetics and surfaces are particularly relevant to the growing arena of bioinspired assembly where biomolecular stability and activity are very sensitive to changes in temperature and chemical structure.

Templated assembly is a compelling idea because the strengths and weaknesses of top-down lithography and bottom-up assembly complement each other so well. Simple and versatile, templating can be theoretically adapted for any material or desired pattern to obtain a stable, ordered configuration, although challenges remain before its promise can be fully realized. Among the many lithographic techniques developed, no one method is perfect and the trade-off between resolution and time/cost for a given area is an ever-present concern. Newer methods such as nanoimprint or soft lithography have the potential to shift this balance favorably, but must prove themselves to be scalable. The integration of top-down and bottom-up strategies can also reveal unforeseen and underappreciated roadblocks. For example, the use of DNA or other biomolecules requires mild templating conditions and the use of aqueous buffer solutions that are foreign to conventional lithographic processes may need to be handled differently.[119] Selection of the proper lithographic technique also depends on the minimum template resolution needed to sufficiently control defects. This minimum resolution requirement may be thought of as part of a larger question: How strong an influence can the template exert on assembly? The primary effect of the template is felt at surfaces and interfaces. More research is needed to elucidate the physical nature of that effect, its role in alignment and defect control, and how its influence is translated deeper into the bulk of the assembly.

Further studies seek to expand the ways in which assembly can be controlled. Most investigations of templated self-assembly have focused on defect elimination and the organization of periodic arrays. The ability to deterministically assemble defects or nonperiodic arrays would greatly expand the amount of spatial information that could be encoded into the template as well as the range of possible applications. Extending directed assembly from 2-D into 3-D is also an area of significant interest. Both colloidal crystal lattices[122] and anodic aluminum oxide pore arrays[123] have already been explored as templates for 3-D block copolymer assembly. Applied forces such as mechanical shear or electric fields can be harnessed to dictate orientation, although the alignment is typically uniaxial and must be frozen in place lest it fades away when the force is removed. However, innovative methods that combine applied forces with topological and chemical patterns may be more effective at directing assembly.

While sub-20 nm lithography is a primary goal of computer chip manufacturers, it is not the sole focus of templated assembly. A myriad of functional nanomaterials with unique properties have been developed in recent years, and controlling the placement and orientation of nanostructures is potentially useful not only for lithography but also medicine, structural composites, energy, and other fields. Novel applications and unique system designs may demand different requirements than state-of-the-art integrated circuits, perhaps favoring the adoption of new materials and eschewing resolution and performance for increased durability, flexibility, and affordability. This convergence of new techniques, new materials, and new applications means that less traditional lithographic methods can be viable options despite the remaining challenges. Templated assembly is a powerful and adaptable strategy with the potential to significantly advance commercial nanomanufacturing in the near future.

ACKNOWLEDGMENTS

The author gratefully acknowledges financial support from the Joint School of Nanoscience and Nanoengineering, North Carolina State A&T University, and the State of North Carolina.

REFERENCES

1. G. M. Whitesides and B. Grzybowski, *Science*, 2002, 295, 2418–2421.
2. J. Y. Cheng, C. A. Ross, H. I. Smith, and E. L. Thomas, *Advanced Materials*, 2006, 18, 2505–2521.
3. A. M. Hung, H. Noh, and J. N. Cha, *Nanoscale*, 2010, 2, 2530–2537.
4. D. Wang and H. Mohwald, *Journal of Materials Chemistry*, 2004, 14, 459–468.
5. P. J. Collins, *Liquid Crystals*, 1st edn., Princeton University Press, Princeton, NJ, 1990.
6. J. W. Matthews, *Epitaxial Growth*, Academic Press, New York, NY, 1975.
7. M. J. Madou, *Fundamentals of Microfabrication: The Science of Miniaturization*, 2nd edn., CRC Press, New York, NY, 2002.
8. B. Paivanranta, A. Langner, E. Kirk, C. David, and Y. Ekinci, *Nanotechnology*, 2011, 22, 375302.
9. G. Tallents, E. Wagenaars, and G. Pert, *Nature Photonics*, 2010, 4, 809–811.
10. K. Salaita, Y. Wang, and C. A. Mirkin, *Nature Nanotechnology*, 2007, 2, 145–155.
11. R. D. Piner, J. Zhu, F. Xu, S. H. Hong, and C. A. Mirkin, *Science*, 1999, 283, 661–663.
12. L. J. Guo, *Advanced Materials*, 2007, 19, 495–513.
13. Y. N. Xia and G. M. Whitesides, *Angewandte Chemie-International Edition*, 1998, 37, 551–575.
14. D. J. Lipomi, R. V. Martinez, L. Cademartiri, and G. M. Whitesides, in *Polymer Science: A Comprehensive Reference*, eds. K. Matyjaszewski and M. Moller, Elsevier B. V., Amsterdam, 2012, vol. 7, pp. 211–231.
15. E. D. Shchukin, A. V. Pertsov, E. A. Amelina, and A. S. Zelenev, eds., *Colloid and Surface Chemistry*, Elsevier Science, New York, NY, 2001.
16. T. Cosgrove, ed., *Colloid Science: Principles, Methods, and Applications*, Blackwell Publishing, Ames, IA, 2005.

17. M. D. Eldridge, P. A. Madden, and D. Frenkel, *Nature*, 1993, 365, 35–37.
18. N. D. Denkov, O. D. Velev, P. A. Kralchevsky, I. B. Ivanov, H. Yoshimura, and K. Nagayama, *Langmuir*, 1992, 8, 3183–3190.
19. A. M. Kalsin, M. Fialkowski, M. Paszewski, S. K. Smoukov, K. J. M. Bishop, and B. A. Grzybowski, *Science*, 2006, 312, 420–424.
20. L. S. Li, J. Walda, L. Manna, and A. P. Alivisatos, *Nano Letters*, 2002, 2, 557–560.
21. C. J. Kiely, J. Fink, M. Brust, D. Bethell, and D. J. Schiffrin, *Nature*, 1998, 396, 444–446.
22. E. V. Shevchenko, D. V. Talapin, N. A. Kotov, S. O'Brien, and C. B. Murray, *Nature*, 2006, 439, 55–59.
23. E. V. Shevchenko, D. V. Talapin, C. B. Murray, and S. O'Brien, *Journal of the American Chemical Society*, 2006, 128, 3620–3637.
24. Y. Xia, B. Gates, Y. Yin, and Y. Lu, *Advanced Materials*, 2000, 12, 693–713.
25. J. Zhang, Y. Li, X. Zhang, and B. Yang, *Advanced Materials*, 2010, 22, 4249–4269.
26. C. H. Chang, L. Tian, W. R. Hesse, H. Gao, H. J. Choi, J. G. Kim, M. Siddiqui, and G. Barbastathis, *Nano Letters*, 2011, 11, 2533–2537.
27. J. D. Joannopoulos, R. D. Meade, and J. N. Winn, *Photonic Crystals: Molding the Flow of Light*, Princeton University Press, Princeton, NJ, 1995.
28. C. Lopez, *Advanced Materials*, 2003, 15, 1679–1704.
29. A. Stein, B. E. Wilson, and S. G. Rudisill, *Chemical Society Reviews*, 2013, 42, 2763–2803.
30. A. Courty, A. Mermet, P. A. Albouy, E. Duval, and M. P. Pileni, *Nature Materials*, 2005, 4, 395–398.
31. C. R. Kagan, C. B. Murray, and M. G. Bawendi, *Physical Review B*, 1996, 54, 8633–8643.
32. H. Zeng, J. Li, J. P. Liu, Z. L. Wang, and S. H. Sun, *Nature*, 2002, 420, 395–398.
33. N. V. Dziomkina and G. J. Vancso, *Soft Matter*, 2005, 1, 265–279.
34. A. vanBlaaderen, R. Ruel, and P. Wiltzius, *Nature*, 1997, 385, 321–324.
35. N. N. Khanh and K. B. Yoon, *Journal of the American Chemical Society*, 2009, 131, 14228–14230.
36. E. Kumacheva, R. K. Golding, M. Allard, and E. H. Sargent, *Advanced Materials*, 2002, 14, 221–224.
37. Y. N. Xia, Y. Yin, Y. Lu, and J. McLellan, *Advanced Functional Materials*, 2003, 13, 907–918.
38. Y. Yin, Y. Lu, B. Gates, and Y. Xia, *Journal of the American Chemical Society*, 2001, 123, 8718–8729.
39. C. Lu, H. Mohwald, and A. Fery, *Soft Matter*, 2007, 3, 1530–1536.
40. D. C. Hyun, G. D. Moon, E. C. Cho, and U. Jeong, *Advanced Functional Materials*, 2009, 19, 2155–2162.
41. E. Kim, Y. N. Xia, and G. M. Whitesides, *Advanced Materials*, 1996, 8, 245–247.
42. M. Allard, E. H. Sargent, P. C. Lewis, and E. Kumacheva, *Advanced Materials*, 2004, 16, 1360–1364.
43. Y. H. Ye, S. Badilescu, V. V. Truong, P. Rochon, and A. Natansohn, *Applied Physics Letters*, 2001, 79, 872–874.
44. Y. Yin, Z. Y. Li, and Y. Xia, *Langmuir*, 2003, 19, 622–631.
45. Y. D. Yin and Y. N. Xia, *Advanced Materials*, 2002, 14, 605–608.
46. N. V. Dziomkina, M. A. Hempenius, and G. J. Vancso, *Advanced Materials*, 2005, 17, 237–240.
47. C. Jin, M. A. McLachlan, D. W. McComb, R. M. de la Rue, and N. P. Johnson, *Nano Letters*, 2005, 5, 2646–2650.
48. J. P. Hoogenboom, C. Retif, E. de Bres, M. van de Boer, A. K. van Langen-Suurling, J. Romijn, and A. van Blaaderen, *Nano Letters*, 2004, 4, 205–208.
49. J. Hur and Y. Y. Won, *Soft Matter*, 2008, 4, 1261–1269.
50. H. K. Choi, S. H. Im, and O. O. Park, *Langmuir*, 2009, 25, 12011–12014.

51. M. E. Abdelsalam, P. N. Bartlett, J. J. Baumberg, and S. Coyle, *Advanced Materials*, 2004, 16, 90–93.
52. V. Sharma, D. Xia, C. C. Wong, W. C. Carter, and Y. M. Chiang, *Journal of Materials Research*, 2011, 26, 247–253.
53. J. Aizenberg, P. V. Braun, and P. Wiltzius, *Physical Review Letters*, 2000, 84, 2997–3000.
54. H. P. Zheng, M. F. Rubner, and P. T. Hammond, *Langmuir*, 2002, 18, 4505–4510.
55. M. H. S. Shyr, D. P. Wernette, P. Wiltzius, Y. Lu, and P. V. Braun, *Journal of the American Chemical Society*, 2008, 130, 8234–8240.
56. P. A. Maury, D. N. Reinhoudt, and J. Huskens, *Current Opinion in Colloid and Interface Science*, 2008, 13, 74–80.
57. D. Nykypanchuk, M. M. Maye, D. van der Lelie, and O. Gang, *Nature*, 2008, 451, 549–552.
58. S. Y. Park, A. K. R. Lytton-Jean, B. Lee, S. Weigand, G. C. Schatz, and C. A. Mirkin, *Nature*, 2008, 451, 553–556.
59. R. J. Macfarlane, B. Lee, M. R. Jones, N. Harris, G. C. Schatz, and C. A. Mirkin, *Science*, 2011, 334, 204–208.
60. H. Noh, A. M. Hung, C. Choi, J. H. Lee, J. Y. Kim, S. Jin, and J. N. Cha, *Acs Nano*, 2009, 3, 2376–2382.
61. M. J. Fasolka and A. M. Mayes, *Annual Review of Materials Research*, 2001, 31, 323–355.
62. N. Hadjichristidis, S. Pispas, and G. A. Floudas, *Block Copolymers: Synthetic Strategies, Physical Properties, and Applications*, Wiley Interscience, Hoboken, NJ, 2003.
63. M. Lazzari, G. Liu, and S. Lecommandoux, eds., *Block Copolymers in Nanoscience*, Wiley-VCH, Weinheim, 2006.
64. A. K. Khandpur, S. Forster, F. S. Bates, I. W. Hamley, A. J. Ryan, W. Bras, K. Almdal, and K. Mortensen, *Macromolecules*, 1995, 28, 8796–8806.
65. L. Leibler, *Macromolecules*, 1980, 13, 1602–1617.
66. I. W. Hamley, *Nanotechnology*, 2003, 14, R39–R54.
67. F. H. Schacher, P. A. Rupar, and I. Manners, *Angewandte Chemie International Edition*, 2012, 51, 7898–7921.
68. N. D. Mermin, *Physical Review*, 1968, 176, 250–254.
69. R. A. Segalman, A. Hexemer, R. C. Hayward, and E. J. Kramer, *Macromolecules*, 2003, 36, 3272–3288.
70. B. H. Kim, J. Y. Kim, and S. O. Kim, *Soft Matter*, 2013, 9, 2780–2786.
71. A. P. Marencic and R. A. Register, *Annual Review of Chemical and Biomolecular Engineering*, 2010, 1, 277–297.
72. C. Park, C. De Rosa, and E. L. Thomas, *Macromolecules*, 2001, 34, 2602–2606.
73. S. Park, D. H. Lee, J. Xu, B. Kim, S. W. Hong, U. Jeong, T. Xu, and T. P. Russell, *Science*, 2009, 323, 1030–1033.
74. E. Huang, T. P. Russell, C. Harrison, P. M. Chaikin, R. A. Register, C. J. Hawker, and J. Mays, *Macromolecules*, 1998, 31, 7641–7650.
75. G. J. Kellogg, D. G. Walton, A. M. Mayes, P. Lambooy, P. D. Gallagher, and S. K. Satija, *Physical Review Letters*, 1996, 76, 2503–2506.
76. E. Han, K. O. Stuen, Y. H. La, P. F. Nealey, and P. Gopalan, *Macromolecules*, 2008, 41, 9090–9097.
77. R. A. Segalman, A. Hexemer, and E. J. Kramer, *Macromolecules*, 2003, 36, 6831–6839.
78. R. A. Segalman, H. Yokoyama, and E. J. Kramer, *Advanced Materials*, 2001, 13, 1152–1155.
79. J. Y. Cheng, A. M. Mayes, and C. A. Ross, *Nature Materials*, 2004, 3, 823–828.
80. N. Koneripalli, N. Singh, R. Levicky, F. S. Bates, P. D. Gallagher, and S. K. Satija, *Macromolecules*, 1995, 28, 2897–2904.
81. D. Sundrani, S. B. Darling, and S. J. Sibener, *Nano Letters*, 2004, 4, 273–276.

82. Y. H. La, E. W. Edwards, S. M. Park, and P. F. Nealey, *Nano Letters*, 2005, 5, 1379–1384.
83. S. O. Kim, H. H. Solak, M. P. Stoykovich, N. J. Ferrier, J. J. de Pablo, and P. F. Nealey, *Nature*, 2003, 424, 411–414.
84. X. M. Yang, R. D. Peters, P. F. Nealey, H. H. Solak, and F. Cerrina, *Macromolecules*, 2000, 33, 9575–9582.
85. M. P. Stoykovich, M. Muller, S. O. Kim, H. H. Solak, E. W. Edwards, J. J. De Pablo, and P. F. Nealey, *Science*, 2005, 308, 1442–1446.
86. R. Ruiz, H. Kang, F. A. Detcheverry, E. Dobisz, D. S. Kercher, T. R. Albrecht, J. J. De Pablo, and P. F. Nealey, *Science*, 2008, 321, 936–939.
87. I. Bita, J. K. W. Yang, Y. S. Jung, C. A. Ross, and E. L. Thomas, *Science*, 2008, 321, 939–943.
88. J. Y. Cheng, C. T. Rettner, D. P. Sanders, H. C. Kim, and W. D. Hinsberg, *Advanced Materials*, 2008, 20, 3155–3158.
89. Y. S. Jung, J. B. Chang, E. Verploegen, K. K. Berggren, and C. A. Ross, *Nano Letters*, 2010, 10, 1000–1005.
90. J. G. Son, A. F. Hannon, K. W. Gotrik, A. Alexander-Katz, and C. A. Ross, *Advanced Materials*, 2011, 23, 634–639.
91. J. K. W. Yang, Y. S. Jung, J. B. Chang, R. A. Mickiewicz, A. Alexander-Katz, C. A. Ross, and K. K. Berggren, *Nature Nanotechnology*, 2010, 5, 256–260.
92. A. Tavakkoli K. G., K. W. Gotrik, A. F. Hannon, A. Alexander-Katz, C. A. Ross, and K. K. Berggren, *Science*, 2012, 336, 1294–1298.
93. H. A. Becerril and A. T. Woolley, *Chemical Society Reviews*, 2009, 38, 329–337.
94. C. Lin, Y. Liu, and H. Yan, *Biochemistry*, 2009, 48, 1663–1674.
95. C. M. Niemeyer and U. Simon, *European Journal of Inorganic Chemistry*, 2005, 3641–3655.
96. N. C. Seeman, *Annual Review of Biochemistry*, 2010, 79, 65–87.
97. A. V. Pinheiro, D. Han, W. M. Shih, and H. Yan, *Nature Nanotechnology*, 2011, 6, 763–772.
98. J. D. Le, Y. Pinto, N. C. Seeman, K. Musier-Forsyth, T. A. Taton, and R. A. Kiehl, *Nano Letters*, 2004, 4, 2343–2347.
99. J. Sharma, R. Chhabra, Y. Liu, Y. G. Ke, and H. Yan, *Angewandte Chemie-International Edition*, 2006, 45, 730–735.
100. J. Sharma, Y. G. Ke, C. X. Lin, R. Chhabra, Q. B. Wang, J. Nangreave, Y. Liu, and H. Yan, *Angewandte Chemie-International Edition*, 2008, 47, 5157–5159.
101. J. P. Zhang, Y. Liu, Y. G. Ke, and H. Yan, *Nano Letters*, 2006, 6, 248–251.
102. J. W. Zheng, P. E. Constantinou, C. Micheel, A. P. Alivisatos, R. A. Kiehl, and N. C. Seeman, *Nano Letters*, 2006, 6, 1502–1504.
103. C. Song, Y. Q. Chen, S. J. Xiao, L. Ba, Z. Z. Gu, Y. Pan, and X. Z. You, *Chemistry of Materials*, 2005, 17, 6521–6524.
104. R. Chhabra, J. Sharma, Y. G. Ke, Y. Liu, S. Rinker, S. Lindsay, and H. Yan, *Journal of the American Chemical Society*, 2007, 129, 10304–10305.
105. Y. He, Y. Tian, A. E. Ribbe, and C. D. Mao, *Journal of the American Chemical Society*, 2006, 128, 12664–12665.
106. S. H. Park, P. Yin, Y. Liu, J. H. Reif, T. H. LaBean, and H. Yan, *Nano Letters*, 2005, 5, 729–733.
107. B. A. R. Williams, K. Lund, Y. Liu, H. Yan, and J. C. Chaput, *Angewandte Chemie-International Edition*, 2007, 46, 3051–3054.
108. K. Sarveswaran, W. C. Hu, P. W. Huber, G. H. Bernstein, and M. Lieberman, *Langmuir*, 2006, 22, 11279–11283.
109. C. X. Lin, Y. G. Ke, Y. Liu, M. Mertig, J. Gu, and H. Yan, *Angewandte Chemie-International Edition*, 2007, 46, 6089–6092.
110. P. W. K. Rothemund, *Nature*, 2006, 440, 297–302.

111. B. Wei, M. Dai, and P. Yin, *Nature*, 2012, 485, 623–626.
112. Y. G. Ke, S. Lindsay, Y. Chang, Y. Liu, and H. Yan, *Science*, 2008, 319, 180–183.
113. B. Q. Ding, Z. T. Deng, H. Yan, S. Cabrini, R. N. Zuckermann, and J. Bokor, *Journal of the American Chemical Society*, 2010, 132, 3248–3249.
114. H. T. Maune, S. P. Han, R. D. Barish, M. Bockrath, W. A. Goddard, P. W. K. Rothemund, and E. Winfree, *Nature Nanotechnology*, 2010, 5, 61–66.
115. H. Z. Gu, J. Chao, S. J. Xiao, and N. C. Seeman, *Nature*, 2010, 465, U202–U286.
116. S. P. Surwade, S. Zhao, and H. Liu, *Journal of the American Chemical Society*, 2011, 133, 11868–11871.
117. A. Kuzyk, B. Yurke, J. J. Toppari, V. Linko, and P. Torma, *Small*, 2008, 4, 447–450.
118. A. E. Gerdon, S. S. Oh, K. Hsieh, Y. Ke, H. Yan, and H. T. Soh, *Small*, 2009, 5, 1942–1946.
119. A. M. Hung, C. M. Micheel, L. D. Bozano, L. W. Osterbur, G. M. Wallraff, and J. N. Cha, *Nature Nanotechnology*, 2010, 5, 121–126.
120. R. J. Kershner, L. D. Bozano, C. M. Micheel, A. M. Hung, A. R. Fornof, J. N. Cha, C. T. Rettner et al. *Nature Nanotechnology*, 2009, 4, 557–561.
121. D. Pastre, L. Hamon, F. Landousy, I. Sorel, M. O. David, A. Zozime, E. Le Cam, and O. Pietrement, *Langmuir*, 2006, 22, 6651–6660.
122. A. C. Arsenault, D. A. Rider, N. Tetreault, J. I. L. Chen, N. Coombs, G. A. Ozin, and I. Manners, *Journal of the American Chemical Society*, 2005, 127, 9954–9955.
123. Y. Y. Wu, G. S. Cheng, K. Katsov, S. W. Sides, J. F. Wang, J. Tang, G. H. Fredrickson, M. Moskovits, and G. D. Stucky, *Nature Materials*, 2004, 3, 816–822.

Section VI

Nanosafety

14 Toward Understanding Toxicity of Engineered Nanomaterials

Komal Garde, Karshak Kosaraju, Soodeh B. Ravari, and Shyam Aravamudhan

CONTENTS

14.1 INTRODUCTION AND MOTIVATION

There is an accelerating and nonuniform progress of innovations and discoveries leading to many emerging technologies. Ramifications of these technologies, budding mostly as a result of nanotechnology, are enabling the introduction of new and more efficient products in the market. The use of these products could lead to exposure of engineered nanomaterials (ENs) to the environment. Figure 14.1 shows the pathways of exposure to EN, affected organs, and associated diseases (Table 14.1) from epidemiological *in vivo* and *in vitro* studies. Since ancient times, humans have

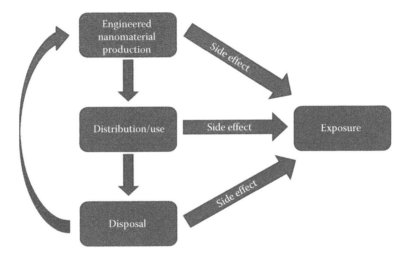

FIGURE 14.1   Exposure from NP production cycle.

been exposed to nanoparticles (NPs) from nature and other anthropogenic sources, but there is a heightened concern in the current times with the development of nanotechnology. The effect of these products containing nanomaterials on humans and the environment are not completely understood. Therefore, it becomes extremely important to analyze the risks posed by EN to the environment, health, and safety (EHS) to optimize design and/or control their production, distribution/use, storage and disposal, and hence control toxicity.[1]

The ENs on contact with biological systems can be absorbed via dermal, gastrointestinal (GI) tract, respiratory, intravenous, and subcutaneous routes. Once absorbed, they can be internalized and biodistributed. Biodistribution will help to predict the plausible target system and the effects that ENs might have on the system as a whole. To understand EN toxicity, a systematic and focused approach is required to understand the many variables involved in their effects. Toxic effects can occur at different

TABLE 14.1
Diseases Associated with Various Systems on Exposure to Nanomaterials

Exposed Systems	Associated Diseases
Skin	Dermatitis and autoimmune diseases
Nervous system	Neurological diseases: Alzheimer's disease and Parkinson's disease
Respiratory system	Bronchitis, asthma, emphysema, and cancer
Circulatory system	Artheriosclerosis, vasoconstriction, arrhythmia, thrombus, high BP
Lymphatic system	Podoconiosis, and Kaposi's sarcoma
Digestive system	Crohn's disease and colon cancer
Other organs	Diseases of unknown origin in kidneys and liver

Source: Adapted from Buzea, C., Pacheco, I. I., Robbie, K., *Biointerphases* 2007, 2(4), MR17.

biological levels, that is, organism, organ, cells, or nuclear level. In addition, there are different types of toxic effects, namely acute, chronic, cyto- and geno-toxicities, reproductive and developmental toxicities.

Depending on the NP application, specific tests related to understanding toxicity of EN *in vitro* and *in vivo* need to be performed. For example, genotoxicity and cytotoxicity assays can be performed in cases where an EN is used for diagnostic purposes as it enters the human body and thereby enters the cell affecting the structural and functional components of the cell.

14.2 ENs AND THEIR APPLICATIONS

Materials that are manufactured with at least one dimension between approximately 1 and 100 nm are classified as ENs.[3] They can be broadly classified into carbon-based (nanotubes, fullerenes, and carbon black), mineral-based (metal and metal oxide), and organic composites/hybrids (polymers, dendrimers, surfactant coatings, quantum dots, and doped meta/metal oxides).[4,5] These ENs and their applications are aiding to revolutionize many sectors namely health care, information technology, energy, environmental science, homeland security, food safety, transportation, and many others. The incorporation of ENs in the existing products induces many desirable effects on their properties. Everyday products that use ENs include polymer composites, fabrics, thin-film coatings, cosmetics, automotive, and household products to name a few. The use of nanoscale transistors,[6] magnetic random access memory, organic light-emitting diodes,[7] and flash memory chips in electronics and information technology applications has helped manage and store larger amounts of information. In the sustainable energy arena, scientists have developed nanostructured solar cells, efficient fuel production processes, nano-bioengineered enzymes, and designed thin-film solar electric panels.[8] In the environmental sector, products such as nanofabric paper towels that absorb 20 times their weight in oil, air filters having nanopores that allow the finest mechanical filtration, and nanosensors to detect and filter out chemical and biological agents have been developed. Potential applications in medical and health sectors include quantum dots for biological imaging and medical diagnostics,[9] gold nanorods to detect Alzheimer's disease, multifunctional therapeutics for targeting and treating cancer using the same EN, microfluidic chip-based nanolabs for monitoring and manipulating individual cells, and the use of nanofibers to cure spinal injuries along with regenerative medicine.[10] Apart from maintaining smarter, efficient, and greener vehicles, cementitious materials are being engineered with nanoscale sensors for structural monitoring.[11] Along with this tremendous potential to revolutionize the world, it is evident that the use of EN at a larger scale could have potential impacts on EHS. Therefore, understanding the EN's physicochemical characteristics, along with their fate and behavior is important.

14.3 DEPENDENCE OF TOXICITY ON EN CHARACTERISTICS

To understand EN toxicity, a thorough characterization of properties and any change/variation in the properties that can lead to toxicity is important. From the knowledge of toxicological properties of EN, it is understood that the most important parameter determining the adverse effects of ENs are dose, dimension, and durability. But recent studies show that there is a different correlation between the properties of EN and its toxicological profiles. This leads to uncertainties in the dependence of toxicity on the properties of EN, such as particle size distribution, shape, size, agglomeration, chemical composition, purity, solubility, surface properties, physical properties (density or crystallinity), bulk powder properties, and last but not the least, concentration-dependent toxicity. The effects of these properties on toxicity are discussed later in this chapter. Some of the properties mentioned above are discussed in the following section. In addition, the cellular assays that are used in the following section are also discussed later.

14.3.1 CHEMICAL COMPOSITION-DEPENDENT TOXICITY

Chemical composition is defined as the arrangement, type, and ratio of atoms in molecules. The toxicity of nanomaterials is very much dependent on the chemical composition and different chemical compositions exhibit different levels of toxicity. This is mainly attributed to the different means of cellular interaction with different compositions of NPs. The difference in atomic and molecular arrangement leads to different levels of cytotoxicity. A comparison study done by Zhang et al.[12] showed that graphene and single-walled carbon nanotubes (SWCNTs) exhibit different levels of cytotoxicity on PC12 cells. They performed lactate dehydrogenase (LDH) release assays that indicate that the cytotoxicity was higher in the case of PC12 cells exposed to SWCNTs than that of cells exposed to graphene. Even though, both graphene and SWCNTs are mainly composed of carbon, the difference in atomic arrangement is attributed to different toxicities.

In Section 14.5, we will look in detail at the difference in cytotoxicity of ENs of interest (gold, silica, and zinc oxide). The difference in cytotoxicities can be attributed to different elemental composition and atomic arrangement.

14.3.2 SIZE-DEPENDENT TOXICITY

In the past few decades, toxicological studies have demonstrated that smaller particles (<100 nm) have the potential to be more toxic compared to larger counterparts. *In vitro* and *in vivo* studies conducted by exposure to EN have shown that smaller particles have greater toxicological effects. This is mainly attributed to the surface area, which is higher in the case of smaller-sized particles. A recent study on size-dependent toxicity comparison of 20 and 100 nm silica NPs on the cutaneous tissue concluded that the cytotoxicity of 20 nm silica NPs was higher than that of 100 nm NPs.[13] This is highly attributed to the larger surface area offered by the 20 nm NP than that of the 100 nm NP.

14.3.3 SURFACE-AREA-DEPENDENT TOXICITY

Smaller NPs have a higher surface area and particle number per unit mass when compared to their larger counterparts. When particles of the same mass, chemical composition, and crystalline nature are compared, toxicity was found to be greater for NPs than their larger counterparts. This led to the understanding that the adverse effects imparted may be dependent on the surface area of the EN leading to change in the regulations based on dose and exposure limits. The larger surface area of EN leads to increased reactivity and, in turn, leads to increased adverse effects.[14] The higher surface area of NPs leads to a dose-dependent increase in production of reactive oxygen species.

14.3.4 CONCENTRATION-DEPENDENT TOXICITY

Concentration can be described as the abundance of a constituent per total volume of the mixture. Concentration of EN is one of the reasons affecting agglomeration of particles. Even though it depends on the solvent or the medium, it can be concluded that higher concentrations of EN would promote agglomeration. Most aggregates formed as a result are observed to be larger than 100 nm, a size that seems to be the threshold for EN to exhibit adverse effects. Even though there are various other factors to be considered, it can be concluded that at higher concentrations EN tends to form agglomerates, which in turn reduce the toxicity of EN when compared to a lower concentration of the same composition. Figure 14.2 shows a comparison of

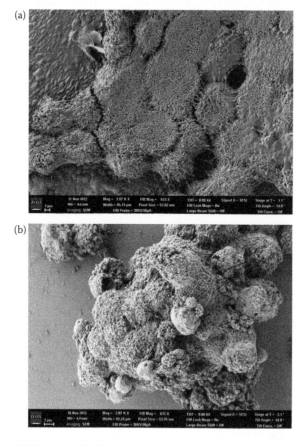

FIGURE 14.2   SEM image of PC12 cells: (a) Healthy cells and (b) cells exposed to silica NPs at 75 µg/mL.

SEM (scanning electron microscopy) images of PC12 cells cultured under humidified conditions and those exposed to silica NPs at a concentration of 75 µg/mL. It can be observed that the cell morphology changed in the case of cells exposed to the NPs (Figure 14.2b) when compared to that of healthy PC12 (Figure 14.2a).

Figure 14.3 shows the MTT [3-(4,5-dimethylthiazol-2-yl)-2,5-diphenyltetrazolium bromide] and LDH assay of PC12 cells exposed to silica NPs, after 24-h exposure, with concentrations varying from 15 to 100 µg/mL at 2-, 4-, 7-, and 14-day time points. It can be observed from the MTT and LDH assay that with increasing concentration of EN, the cell viability increases, which indirectly means that the cytotoxicity decreases. This could be due to increases in agglomeration of silica NPs as the concentration increases.

14.3.5 PARTICLE CHEMISTRY OR CRYSTALLINITY-DEPENDENT TOXICITY

Particle chemistry is also an important factor to consider in understanding the toxicity of EN. Depending on the chemistries, EN can show different cellular uptake, subcellular localization, and an ability to produce reactive oxygen species. NPs can change the crystal structure after interaction with water or liquids. For example, it is reported that zinc sulfide (ZnS) NPs (3 nm across containing around 700 atoms) rearrange their crystal structure in the presence of water and become more ordered, closer to the structure of a bulk piece of solid ZnS. NPs often exhibit unexpected crystal structures due to surface effects.[2] This will contribute to different types of interactions with cells leading to various levels of toxicity depending on the arrangement of molecules.

14.3.6 ASPECT RATIO-DEPENDENT TOXICITY

It was observed that particles with higher aspect ratios exhibit higher toxicity when compared to the particles of the same kind with lower aspect ratios. For example, SWCNTs with a higher aspect ratio were observed to create more pulmonary toxicity when compared to similar doses of spherical amorphous carbon or silica particles.[2,11]

14.3.7 SURFACE FUNCTIONALIZATION-DEPENDENT TOXICITY

Particle surface morphology could play an important role in toxicity of EN as it comes in contact with cells and other biological materials. For example, the cytotoxicity of C60 molecules correlates with their surface functionality.[15] Quantum dots of CdSe can be rendered nontoxic when coated with functional groups.[16] Along with controlling toxicity, surface functionalization also changes the various properties of EN such as solubility, which can be used to a greater advantage.

Depending on composition, EN released into the environment can be a source of contamination. Surface chemistry is governed by the functionality and hence, solubility, charge, and adsorption/desorption characteristics should be taken into account as these change the way EN interacts with biomolecules. The reactivity of the EN surface controls the potential to generate reactive oxygen species and in turn its damaging potential. Size and size distribution are another important criteria as they influence the particle settling velocity, thus affecting mobility, potential transport,

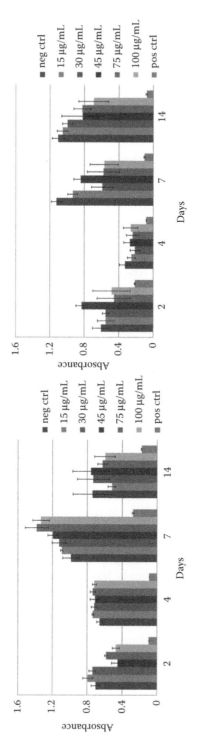

FIGURE 14.3 MTT assay data for PC12s exposed to silica NPs.

and bioavailability in the environment. The potential to transport across membranes and cause damage to cell organelles is also determined by EN size. A crystalline structure has more reactive sites as compared to amorphous materials having more toxicity; so, morphology is another important aspect. Concentration and purity are other factors that affect toxicity. With so many properties and their dependence, understanding toxicity is a challenge!

14.3.8 DEPENDENCE OF TOXICITY ON CELL TYPE

Along with the properties of the EN, the type of cell that the EN interacts with will also determine the toxicity. Even though one cell type can exhibit toxicity to EN, it is not necessary that another cell type should exhibit the same toxicity profile when exposed to the same EN. For example, when 3T3 and PC12 cells were exposed to gold nanorods, the cells did not exhibit the same levels of toxicity. This was concluded from the MTT assay done on 3T3 (Figure 14.4a) and PC12 (Figure 14.4b) cells on exposure to the same concentrations of gold nanorods, which were predispersed in cetrimonium bromide (CTAB), for 48 h and subsequent incubation for different periods of time (2, 4, 7, and 14 days). On the basis of comparison of the MTT assays, PC12 cells exhibit more cytotoxicity when compared to 3T3 cells.

Figure 14.5 shows fluorescence images of 3T3. All fluorescent images were recorded on cells stained by Actin-Green and Nuc-Blue, which stain the cell membrane and nucleus, respectively (Figure 14.5a). We see a huge reduction in cell density and some damage to the cytoskeleton in the case of cells exposed to gold nanorods (Figure 14.5b).

14.3.8.1 Toxicity Studies on Gold, Silica, and Zinc Oxide NPs

To understand better in terms of EN composition, which is related to characteristics and cell-type dependence on toxicity, we will take a closer look at the toxicity of selected nanomaterials. Gold NPs have interesting physicochemical properties and have been in existence since Faraday attempted to make stable aqueous dispersions of gold NPs. Gold has numerous applications because thiol and amine groups strongly bind to the gold surfaces, thus enabling modifications of the surface of gold with proteins and amino acids leading to the main applications including cell imaging, biodiagnostics, drug/DNA delivery, immunostaining, and biosensing.[17] The toxicity of gold NPs also varies depending on the physicochemical properties of the particle. Spherical particles of gold are found to be less toxic than its rod counterparts. Gold dispersed in CTAB is found to be toxic as CTAB adds to the toxicity. To overcome this toxicity, CTAB can be replaced by PEG (polyethylene glycol) that reduces the zeta potential; hence, it is more stable and less cytotoxic.[18] Marsich et al.[19] observed that the toxicity of NPs having the same surface area is dependent on the composition. They also demonstrated that the uptake of NPs is not only dependent on the NP but also on the cell type, C17.2; these cells showed the highest cellular uptake followed by HUVEC and PC12 cells, possibly owing to the smaller cell size. Organ biodistribution of gold was found to be gender dependent.[20] Among aspartic acid, trisodium citrate dehydrate, and bovine serum albumin, the latter was found to be the most biocompatible; showing surface modification caused by the capping material

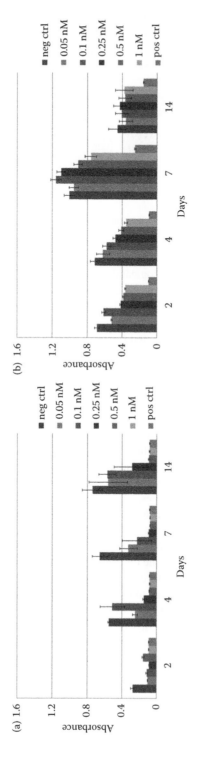

FIGURE 14.4 MTT assay for (a) 3T3s and (b) PC12 cells exposed to gold nanorods.

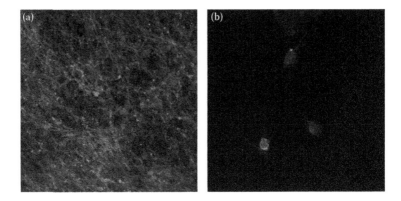

FIGURE 14.5 (**See color insert.**) Fluorescence image of 3T3 cells: (a) Healthy cells and (b) cells exposed to gold nanorods at 0.5 nM.

has an effect on toxicity.[21] Morphological damages were seen on PC12 and NIH 3T3 cells that were exposed to gold nanorods at 0.5 nM as seen in Figures 14.4 and 14.5. It was observed from both the MTT and LDH assay data that as the concentration of gold increases, the viability of the cells decreases for all time points. The negative control has the healthiest cells and the positive control (treated with H_2O_2) has the least viability. SEM images reveal cell membrane damage and complete change in morphology in PC12 (Figure 14.6b) and 3T3 cells (Figure 14.6c) exposed to gold nanorods when compared to unexposed cells (Figure 14.6a) that revealed a reduction in cell density and some damage to the cytoskeleton.

On comparing MTT assay data of PC12 and 3T3 cells, 3T3 cells are found to be more susceptible to the gold nanorods as compared to PC12 cells. The same conclusion can be made when the SEM images of PC12 and 3T3 cells exposed to gold nanorods are compared. Table 14.2 summarizes the influence of selected characteristics of gold NPs on their toxicity profiles on exposure to various cell types, along with their possible mechanism routes.

Silica is produced on a large scale in industries to be used as additives to drugs, cosmetics, varnishes, printer toners, and food products. From cytotoxic studies on NPs ranging from 20 to 50 nm, the smallest NPs were observed to have the highest toxicity. This can be explained by the increasing specific surface area of the nanomaterial of decreasing sizes, increasing cell–particle interaction, and greater penetration into the cells. This penetration leads to excessive reactive oxygen species (ROS) generation and reduces glutathione (GSH) levels.[11] Polystyrene NPs (30 nm) and mesoporous (10 nm) NPs were taken up via calcium-dependent and cholesterol-dependent mechanisms, respectively.[22] Thermally hydrocarbonized silicon (THCPSi) of size range 1–10 μm was found to be more toxic than 10–25 μm (THCPSi) and NPs with an average size of 142 nm (PSi), showing microsized particles to be more toxic than nanosized particles.[23] Colloidal silica (40 m²/g) was highly toxic at 100 μg/mL while mesoporous silica (1150 m²/g) did not have any effect on cell viability up to 100 μg/mL.[24] Sun et al.[25] demonstrated that silica NPs of 43 nm increased ROS levels in HepG2 up to 100 μg/mL, but it decreased at 200 μg/mL. When NIH 3T3 cells were

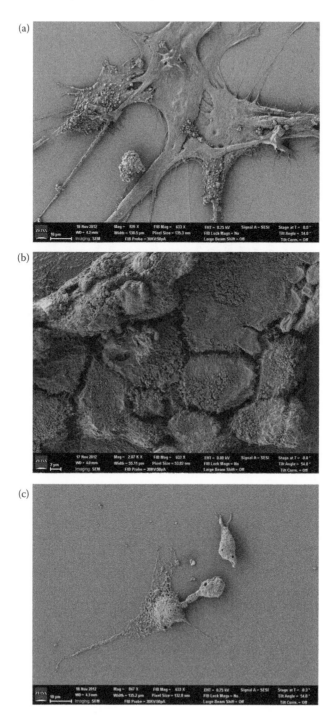

FIGURE 14.6 SEM image of (a) healthy 3T3 cells, (b) PC12 cells, and (c) 3T3 cells after exposure to gold nanorods at 0.5 nM.

TABLE 14.2
Toxicity Dependence of Various Physicochemical Properties of Gold NPs

Size	Surface Area	Zeta Potential (mV)	Animal/Cell	Method	Result	Mechanism
Length 5/ diameter 4 Nm	1.25 cm^2/mL	—	HepG2 and MG63	DCFH-DA	ROS increased but it's level was dominant in Ac–Au	Due to apoptotic mechanism[19]
4 nm	—	27.6 ± 5.6	C17.2, UVECs, and PC12	(ICP)-MS	Cellular uptake decreased (C17.2 highest, closely followed by HUVEC and PC12)	Due to size of the cells (lower uptake results from smaller cell size)
				LDH	Toxicity increased (decrease in cell viability)	Immediate exposure of the cells to NPs and not by secondary effects
5.06 nm 4.12 nm 4.3 nm	3.64 × 10^8 nm^2/cm^3, 1.68 × 10^7 nm^2/cm^3, and 1.9 × 10^6 nm^2/cm^3	—	Rat	Pulmonary function testing	Decreased the function at high surface area in the lung	Dose-dependent manner for both male and female rats[20]
15–20 nm (GNPA, GNPB, and GNPC)	—	—	MRC-5 (mice)	WST-1/LDH	Cell viability decreased negligibly with increasing concentration (even with different surfactant)	Due to the interference of NPs with membrane function[21]
Length (65 ± 5)/width (11 ± 1) nm	—	−0.5 ± 0.4 (neutral surfaces) +41 ± 1 (cationic surfaces)	HeLa (mice)	MTT	Surface modification reduces zeta potential Cytotoxicity decreased dramatically (90% viability)	PEG modification (neural) reduces surface charge[18]

imaged (Figure 14.7a) after exposure to silica NPs (Figure 14.7b), at a concentration of 75 μg/mL, only minimal morphological damage was observed (Figure 14.7c). This is possibly due to aggregation of silica.

The MTT assay on 3T3 cells shows cytotoxicity immediately after exposure (2 days), which seems to be an acute response, but subsequent incubation decreases toxic response for all silica concentrations and for all different time points. Table 14.3 summarizes the toxicity dependence of silica NPs on their characteristics and cellular environment along with predicted mechanisms.

Zinc oxide NPs are used in many applications including sunscreen products, textile, paintings, industrial coatings, and antibacterial agents. Even though zinc oxide NPs are believed to be nontoxic, literature suggests that they can exert negative cellular responses.[28] The toxicity is not only associated with dose and concentration but also with other physicochemical properties. Heng et al.[29] demonstrated that the toxicity associated with zinc oxide spheres is slightly more than the zinc oxide sheets. Kao et al.[30] showed that smaller-sized NPs with a larger surface area cause higher LDH release, indicating less membrane integrity and hence, are more toxic. The smaller the size, the greater the surface area and toxicity. Prach et al.[31] demonstrated that the positively charged NPs enhance cytotoxicity. The toxicity of the ZnO NP (20 nm) was found to be higher at lower concentrations (5–6 μg/mL) and steeply declined at concentrations between 6 and 10 μg/mL.[32] Drastic morphological abnormalities were observed on PC12 and NIH 3T3 cells at 25 μg/mL as seen in Figure 14.8a and b. Fluorescence images of 3T3 cells treated with zinc oxide NPs show that the cytoskeleton is completely lost after exposure (Figure 14.8c).

In the case of PC12 cells, we observe that zinc oxide is slightly less toxic to the cells as indicated in MTT data (Figure 14.9a). Concentrations ranging from 1 to 25 μg/mL show gradual dose-dependent responses. While 3T3 cells are seen to be more susceptible with a steep dose-dependent response.

Table 14.4 summarizes the dependence of toxicity of zinc oxide NPs on their characteristics and cellular environment along with predicted mechanisms.

14.4 CYTO- AND GENOTOXICITY OF ENs

14.4.1 ENDOCYTOSIS AND INTRACELLULAR TRANSPORT OF NPS

The most common way by which the EN exhibits toxicity on cells is by entering the cells. This happens through the process of endocytosis and so, it is really important to understand the mechanisms of endocytosis. Many attempts have been made to understand the cellular uptake of nanomaterials. The plasma membrane plays an important role in maintaining mechanical, chemical, osmotic, and electrical equilibrium of cells relative to the outside membrane. It also controls the translocation of the molecules in and out of the cytoplasm. Active or passive transport is usually used by the cell membrane to transport materials across the membrane, but since the NP size is bigger, cellular uptake by endocytosis of these NPs is most likely. Endocytosis can take place by phagocytosis (cellular eating), pinocytosis (cell drinking), and clathrin-dependent/clathrin-independent receptor-mediated endocytosis.[34]

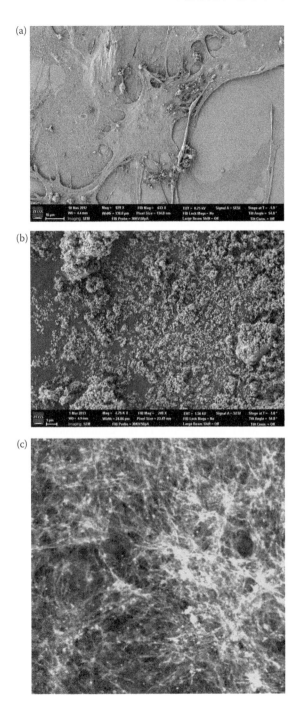

FIGURE 14.7 (**See color insert.**) SEM image of (a) 3T3 cells exposed to silica NPs at 75 μg/mL, (b) SEM image of silica NPs, and (c) fluorescence image of 3T3 cells exposed to silica NPs.

TABLE 14.3

Toxicity Dependence of Various Physicochemical Properties of Silica NPs

Size	Surface Area	Zeta Potential (mV)	Animal/Cell	Method	Result	Mechanism
14.23 ± 2.16 nm	640 m²/g	−28	HepG2	MTT/NRU	Viability decreased as concentration increased, but no toxicity at <10 µg/mL	Dose-dependent manner[26]
20 and 50 nm			PC12	MTT/	NP is toxic at concentrations above 20 µg/mL. Smallest NP shows highest toxicity	Larger specific surface area induces interactions with cells and smaller size facilitates NP penetration into cells
Polystyrene (30 nm) and Mesoporous silica (10 nm)	—	—	RBL-2H3	Cellular viability	Both NPs showed dose-/time-dependent response	Polystyrene via calcium-dependent (ionic interaction) and mesoporous silica via cholesterol[27]
THC (Psi) (97, 142, and 188) nm, (1–10), and (10–25) µm	202 m²/g	−30	Caco-2 and RAW 264.7	ROS assay using DCF-DA	Oxidative stress in RAW 264.7 was higher than Caco-2. But all sizes did not induce toxicity	Due to interaction between particles and extracellular membranes. ROS level was low[28]
Colloidal silica and MPS NPs (100 nm)	Colloidal (40 m²/g) and MPS (1150 m²/g)	—	J774A.1	MTT	Colloidal silica was highly toxic at 100 µg/mL while MPS did not affect cell viability up to 100 µg/mL	Due to surface area effect (increase in surface area causes decrease in toxicity)[24]

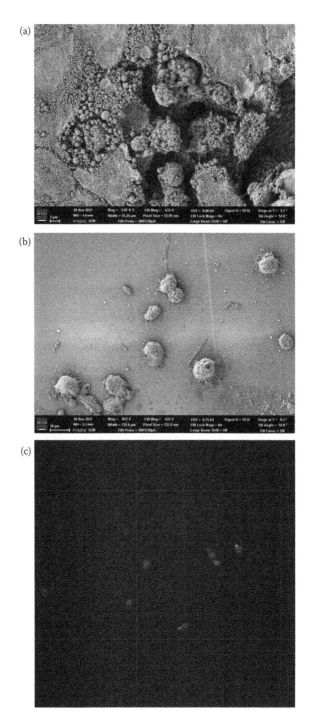

FIGURE 14.8 SEM image of (a) PC12 cells and (b) 3T3 cells exposed to zinc oxide NPs at 25 μg/mL, along with (c) fluorescence image of exposed 3T3 cells.

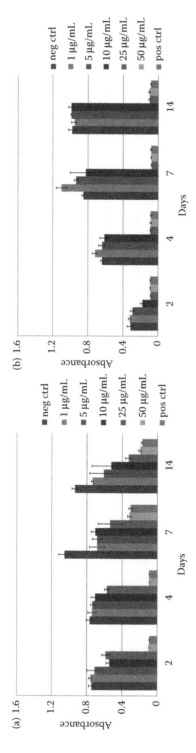

FIGURE 14.9   MTT data for (a) PC12 cells and (b) 3T3 cells exposed to zinc oxide NPs.

TABLE 14.4
Toxicity Dependence of Various Physicochemical Properties of Zinc Oxide NPs

Size	Surface Area (m²g)	Zeta Potential (mV)	Animal/Cell	Method	Result	Mechanism
20*20 nm (sphere) and 325*15 nm (sheet)	35.3, 28.6	−36 ± 0.2, −31 ± 0.3	RAW-264.7 (mouse) and BEAS-2B (human)	WST-8 assay	Sphere is more toxic than the sheet for both cell lines, especially at a low concentration RAW-264.7 is more sensitive to sphere NPs	Due to higher cellular association of spherical versus sheet No significant effect of shape[29]
			BEAS-2B	DCFH-DA	Sphere NP induces more ROS (at high concentration)	Attributed to the difference in cytotoxicity of the sheet versus sphere
			RAW-264.7	ELISA	Both NPs increased TNF-a Production at low concentration (spherical shape produces more)	Due to higher metabolic activity
20 nm and micro size	50	−30.9	Rat	ALT and AST activity	Both activities level increased No toxicity from microsize Nanosize causes toxicity at lower dose	Inverse dose-dependent manner at nanosize[33]
90 nm (<100) and 22 nm (<50)	15–25, 10.8	−9.8, −9.4	H1355 (lung)	LDH	Smaller-sized NPs cause higher LDH release	Size-dependent manner[30]
20 nm	47.47	—	BEAS-2B	MTS	Cytotoxic at low concentration	There is a relationship between concentration and reduction in cell viability[32]

All these mechanisms are activated differently; so, it is important to understand which mechanism is responsible for NP uptake.[34]

14.4.2 MECHANISMS OF CYTOTOXICITY

The prediction of toxicological consequences requires an understanding of the mechanisms of toxicity.[2] There is no particular design that explains the toxicity, but numerous approaches have been proposed for action of the NPs depending on the physicochemical characteristic properties of nanomaterials (as discussed in Tables 14.2 through 14.4). The most accepted mechanisms include ROS production (for TiO_2, ZnO, SiO_2, Fe_3O_4, CNTs, fullerenes, and Al_2O_3), dissolution and release of toxic cations (ZnO, Ag, CdSe, Fe_3O_4, Co/Ni ferrite NPs, and magnetic metallic NPs), lysosomal damage (ZnO), NP-mediated cell membrane disruption leading to cell death (necrosis/apoptosis) (for TiO_2, SiO_2, and MoO_3), protein fibrillation (TiO_2 and CeO_2), disruption of membrane integrity and transport processes (Ag), disruption of protein conformation (Au NPs and nanorods), inflammation (ZnO), and proinflammatory effects (SWCNT, MWCNT, and Al_2O_3).[35] Most often, mechanisms of cytotoxicity are neglected as it is an expensive and time-consuming process, riddled with uncertainty.

14.4.3 CYTOTOXICITY ASSAYS

There are various methods that have been developed and used for measuring cellular cytotoxicity. Some of the most common methods are discussed below.

Trypan blue exclusion assay: In this assay, the treated cells are trypsinized and treated with trypan blue, a diazo dye that is taken up by dead cells, but excluded by live cells. Unstained cells thus reflect the total number of viable cells in a given culture. This method is advantageous because it gives the actual number of cells in comparison to the control.

Microculture tetrazolium assay (MTA): These are metabolic assays that do not provide direct information about the total cell count, but measure viability of a cell population relative to control/untreated cells. Cells whose viability is to be assessed are treated with soluble yellow tetrazolium salts such as MTS [3-(4,5-dimethylthiazol-2-yl)-5-(3-carboxymethoxyphenyl)-2-(4-sulfophenyl)-2H-tetrazolium], or MTT at 37°C for 2–4 h. During this time, viable cells with active respiratory mitochondrial activity reduce MTT or MTS into an insoluble purple formazan product through mitochondrial dehydrogenases, which are then dissolved in DMSO (dimethyl sulfoxide) or detergent and quantified on a visible light spectrophotometer. This method has some shortcomings, including inefficient processing of tetrazolium salts and protocol lengthening due to dissolution of the dye in DMSO. Modification to this protocol can be made by using XTT [2,3-bis(2-methoxy-4-nitro-5 sulfophenyl)-5-[(phenylamino)carbonyl]-2H-tetrazolium hydroxide] that is metabolized to a water-soluble product, thereby eliminating the solubilization step.

Clonogenic assay or colony-forming efficiency: This assay allows assessment of survival and proliferation over extended periods of time. Cells are plated with very low density and observed for 10 days to 3 weeks and then stained with crystal violet or nuclear stains and quantified according to numbers and/or size. This assay is used for tracheal epithelial cells, A549 cells, NHBE, and also keratinocyte cells.

Lactate dehydrogenase assay: LDH is a soluble cytosolic enzyme that serves as an indicator of lytic cell death as it is released into the extracellular medium following cellular membrane damage resulting from apoptosis or necrosis. This test is used to study the membrane integrity.

TdT dUTP nick end labeling and apostain assays: TdT dUTP nick end labeling (TUNEL) and apostain techniques are the most commonly used immunohistochemical techniques for *in vitro* visualization of apoptotic cells. Apoptosis is a programmed cell death where a cell is degraded so that it can be engulfed and recycled. This is characterized by cell membrane blebbing, mitochondrial DNA damage, cytoplasmic and nuclear shrinkage, fragmentation, and chromatin fragmentation. So, the DNA fragmentation that takes place as a result of the activity of endonucleases can be detected using TUNEL assay, where the ends of DNA are labeled and analyzed under light or fluorescence microscope depending on the dye used. TUNEL assay detects fragmented DNA, which is a characteristic feature of both necrosis and apoptosis. Hence, apostain used to label condensed DNA is considered to be a precise marker of apoptosis. Other techniques to measure cellular cytotoxicity include flow cytometry with PI (propidium iodide), 7AAD (7-aminoactinomycin D), and/or annexin V, lipid peroxidation, cytochrome c release from mitochondria, and caspase activation.[36] Even though performing cytotoxic assays is straightforward, making sure that the ENs do not interfere with the assay is a challenge. Usually, more than one assay needs to be performed to obtain reliable results.

14.4.4 GENOTOXICITY MEASUREMENTS

When a cell encounters EN, it might impair the genetic content of the cell. This has the potential to instigate and promote carcinogenesis or impact fertility; hence, nanogenotoxicity is essential to be understood for safety of EN. There are direct and indirect mechanisms that can promote DNA damage once they gain entry inside the body. A number of genotoxic assays can be performed to check for gene mutations and chromosome damage. Ames test, chromosome aberration test, comet assay, cytokinesis-blocked micronucleus assay, HPRT forward mutation assay, γ-H2AX staining, and 8-hydroxydeoxyguanosine DNA adducts are the most commonly used assays to assess genotoxic responses, such as chromosomal fragmentation, DNA strand breakages, point mutations, oxidative DNA adducts, and alterations in gene-expression profile. These assays are usually labor intensive and require a skilled worker. Hence, most of the current studies involve just the study of cytotoxicity. Often, one assay is not sufficient to draw a conclusion as each assay will measure one specific endpoint; hence, the response is dependent on the mechanism of action. For example, the micronucleus assay quantifies clastogenic and aneugenic events resulting in chromosome fragmentation or loss of the entire chromosomes. On the other hand, the comet assay measures unrepaired DNA strand breaks and alkali-labile sites in DNA. The micronucleus assay measures lower levels of damage than the comet assay because the former detects damage that persists after cell division, but the comet assay detects breakages that are repairable as well. Hence, a micronucleus assay is a more demanding assay. In addition, it is important to study the long-term fate of the nanomaterials inside the cells.

14.5 SUMMARY

In summary, it is important to have a firm understanding of the toxic potential of nanomaterials before incorporating them into consumer products. A complete characterization of ENs physiochemical properties is also necessary. Short-term toxicity is widely studied, but the long-term fate of EN needs to be understood. Most of the current studies have focused on cytotoxicity as it is relatively straightforward and fast, but in the future the genotoxic potential of the nanomaterial needs to be measured as well. Moreover, currently, there is a diverse assortment of nanomaterials and experimental conditions available with no reliable and reproducible methods, models, and standardized ENs. In the future, a more ordered and systematic approach (with computational and experimental methodology) is necessary to both assess the biological responses and to address fundamental mechanistic questions. There will be a point in the near future, when we will know how changing a specific parameter affects the toxicity of the NP and then we will be able to design tailored NPs. We should not merely be the observers of toxicity but the choreographers to minimize and control the toxic effects.

REFERENCES

1. Nel, A. X., Madler, T., Li, L. N., Toxic potential of materials at the nanolevel. *Science* 2006, 311, 622–627.
2. Buzea, C., Pacheco, I. I., Robbie, K., Nanomaterials and nanoparticles: Sources and toxicity. *Biointerphases* 2007, 2(4), MR17.
3. Eastlake, A., Hodson, L., Geraci, C., Crawford, C., A critical evaluation of material safety data sheets (MSDSs) for engineered nanomaterials. *Journal of Chemical Health and Safety* 2012, 19(5), 1–8.
4. Tervonen, T., Linkov, I., Figueira, J. R., Steevens, J., Chappell, M., Merad, M., Risk-based classification system of nanomaterials. *Journal of Nanoparticle Research* 2008, 11(4), 757–766.
5. Stone, V., Nowack, B., Baun, A., van den Brink, N., Kammer, F., Dusinska, M., Handy, R. et al. Nanomaterials for environmental studies: Classification, reference material issues, and strategies for physico-chemical characterisation. *The Science of the Total Environment* 2010, 408(7), 1745–1754.
6. Kettle, J., Whitelegg, S., Song, A. M., Wedge, D. C., Kotacka, L., Kolarik, V., Madec, M. B., Yeates, S. G., Turner, M. L., Fabrication of planar organic nanotransistors using low temperature thermal nanoimprint lithography for chemical sensor applications. *Nanotechnology* 2010, 21(7), 075301.
7. Yamamoto, H., Wilkinson, J., Long, J. P., Bussman, K., Christodoulides, J. A., Kafafi, Z. H., Nanoscale organic light-emitting diodes. *Nano Letters* 2005, 5(12), 2485–2488.
8. Serrano, E., Rus, G., García-Martínez, J., Nanotechnology for sustainable energy. *Renewable and Sustainable Energy Reviews* 2009, 13(9), 2373–2384.
9. Fernandez-Fernandez, A., Manchanda, R., McGoron, A., Theranostic applications of nanomaterials in cancer: Drug delivery, image-guided therapy, and multifunctional platforms. *Applied Biochemistry and Biotechnology* 2011, 165(7–8), 1628–1651.

10. Kubinová, Š., Syková, E., Nanotechnologies in regenerative medicine. *Minimally Invasive Therapy and Allied Technologies* 2010, 19(3), 144–156.

11. Wang, Z. L., Self-powered nanosensors and nanosystems. *Advanced Materials* 2012, 24(2), 280–285.

12. Zhang, Y., Ali, S. F., Dervishi, E., Xu, Y., Li, Z., Casciano, D., Biris, A. S., Cytotoxicity effects of graphene and single-wall carbon nanotubes in neural phaeochromocytoma-derived PC12 cells. *ACS Nano* 2010, 4(6), 3181–3186.

13. Park, Y.-H., Bae, H., Jang, Y., Jeong, S., Lee, H., Ryu, W.-I. et al. Effect of the size and surface charge of silica nanoparticles on cutaneous toxicity. *Molecular and Cellular Toxicology* 2013, 9(1), 67–74.

14. Passagne, I., Morille, M., Rousset, M., Pujalté, I., L'Azou, B., Implication of oxidative stress in size-dependent toxicity of silica nanoparticles in kidney cells. *Toxicology* 2012, 299(2–3), 112–124.

15. Trpkovic, A., Todorovic-Markovic, B., Trajkovic, V., Toxicity of pristine versus functionalized fullerenes: Mechanisms of cell damage and the role of oxidative stress. *Archives of Toxicology* 2012, 86(12), 1809–1827.

16. Liu, W., Choi, H. S., Zimmer, J. P., Tanaka, E., Frangioni, J. V., Bawendi, M., Compact cysteine-coated CdSe(ZnCdS) quantum dots for *in vivo* applications. *Journal of the American Chemical Society* 2007, 129(47), 14530–14531.

17. Zeng, S., Yong, K.-T., Roy, I., Dinh, X.-Q., Yu, X., Luan, F., A review on functionalized gold nanoparticles for biosensing applications. *Plasmonics* 2011, 6(3), 491–506.

18. Niidome, T., Yamagata, M., Okamoto, Y., Akiyama, Y., Takahashi, H., Kawano, T., Katayama, Y., Niidome, Y., PEG-modified gold nanorods with a stealth character for *in vivo* applications. *Journal of Controlled Release : Official Journal of the Controlled Release Society* 2006, 114(3), 343–347.

19. Marsich, E., Travan, A., Donati, I., Di Luca, A., Benincasa, M., Crosera, M., Paoletti, S., Biological response of hydrogels embedding gold nanoparticles. *Colloids Surface B Biointerfaces* 2011, 83(2), 331–339.

20. Sung, J. H., Ji, J. H., Park, J. D., Song, M. Y., Song, K. S., Ryu, H. R., Yoon, J. U. et al. Subchronic inhalation toxicity of gold nanoparticles. *Part Fibre Toxicology* 2011, 8, 16.

21. Das, S., Debnath, N., Mitra, S., Datta, A., Goswami, A., Comparative analysis of stability and toxicity profile of three differently capped gold nanoparticles for biomedical usage. *Biometals : An International Journal on the Role of Metal Ions in Biology, Biochemistry, and Medicine* 2012, 25(5), 1009–1022.

22. Ekkapongpisit, M., Giovia, A., Nicotra, G., Ozzano, M., Caputo, G., Isidoro, C., Labeling and exocytosis of secretory compartments in RBL mastocytes by polystyrene and mesoporous silica nanoparticles. *International Journal of Nanomedicine* 2012, 7, 1829–1840.

23. Bimbo, L. M., Sarparanta, M., Santos, H. A., Airaksinen, A. J., Mäkilä, E., Laaksonen, T., Peltonen, L., Lehto, V.-P., Hirvonen, J., Salonen, J., Biocompatibility of thermally hydrocarbonized porous silicon nanoparticles and their biodistribution in rats. *ACS Nano* 2010, 4(6), 3023–3032.

24. Lee, S. Y., Kim, S. H., The comparative effects of mesoporous silica nanoparticles and colloidal silica on inflammation and apoptosis. *Biomaterials* 2011, 32, 9434–9443.

25. Sun, L., Li, Y., Liu, X., Jin, M., Zhang, L., Du, Z., Guo, C., Huang, P., Sun, Z., Cytotoxicity and mitochondrial damage caused by silica nanoparticles. *Toxicology in Vitro* 2011, 25(8), 1619–1629.

26. Ahmad, J., Ahamed, M., Akhtar, M. J., Alrokayan, S. A., Siddiqui, M. A., Musarrat, J., Al-Khedhairy, A. A., Apoptosis induction by silica nanoparticles mediated through reactive oxygen species in human liver cell line HepG2. *Toxicology and Applied Pharmacology* 2012, 259, 160–168.

27. Ekkapongpisit, M., Giovia, A., Nicotra, G., Ozzano, M., Caputo, G., Isidoro, C., Labeling and exocytosis of secretory compartments in RBL mastocytes by polystyrene

and mesoporous silica nanoparticles. *International Journal of Nanomedicine* 2012, 7, 1829–1840.

28. Luis, M., Bimbo, T. L., Mirkka, S., Leena, P., Helder, A. S., Vesa-Pekka, L., Anu, J. A., Jouni, H., Ermei, M., Jarno, S., Biocompatibility of thermally hydrocarbonized porous silicon nanoparticles and their biodistribution in rats. *American Chemical Society* 2010, 6, 3023–3032.

29. Heng, B. C., Zhao, X., Tan, E. C., Khamis, N., Assodani, A., Xiong, S., Ruedl, C., Ng, K. W., Loo, J. S., Evaluation of the cytotoxic and inflammatory potential of differentially shaped zinc oxide nanoparticles. *Archives of Toxicology* 2011, 85, 1517–1528.

30. Kao, Y. Y., Chen, Y. C., Cheng, T. J., Chiung, Y. M., Liu, P. S., Zinc oxide nanoparticles interfere with zinc ion homeostasis to cause cytotoxicity. *Toxicology Science* 2012, 125, 462–472.

31. Prach, M. S., Proudfoot, V. L., Zinc oxide nanoparticles and monocytes: Impact of size, charge and solubility on activation status. *Toxicology Application Pharmacology* 2013, 266, 19–26.

32. Huang, C. C., Aronstam, R. S., Chen, D. R., Huang, Y. W., Oxidative stress, calcium homeostasis, and altered gene expression in human lung epithelial cells exposed to ZnO nanoparticles. *Toxicology in Vitro: An International Journal Published in Association with BIBRA* 2010, 24(1), 45–55.

33. Pasupuleti, S., Alapati, S., Ganapathy, S., Anumolu, G., Pully, N. R., Prakhya, B. M., Toxicity of zinc oxide nanoparticles through oral route. *Toxicology and Industrial Health* 2012, 28(8), 675–686.

34. Iversen, T.-G., Skotland, T., Sandvig, K., Endocytosis and intracellular transport of nanoparticles: Present knowledge and need for future studies. *Nano Today* 2011, 6, 176–185.

35. Andre, E., Nel, L. M., Fred, K., Darrell, V., Vince, C., Tian, X., Eric, M. V., Hoek, T. A. M., Understanding biophysicochemical interactions at the nano–bio interface. *Nature Materials* 2009, 8, 543–557.

36. Hillegass, J. M., Shukla, A., Lathrop, S. A., MacPherson, M. B., Fukagawa, N. K., Mossman, B. T., Assessing nanotoxicity in cells *in vitro*. *Wiley Interdisciplinary Reviews. Nanomedicine and Nanobiotechnology* 2010, 2, 219–231.

15 The Safety of Nanomaterials

What We Know and What We Need to Know

Joseph L. Graves Jr.

CONTENTS

Advances in nanomanufacturing have outpaced studies of nanosafety. Despite this, there has been an explosion of nanoproducts on the market, including medical devices, pharmaceuticals, and cosmetics. Existing studies of nanotoxicity show general trends for increased toxicity as particulate sizes decrease. This increase may be associated with increased surface area, ability to pass through cellular barriers, interaction with subcellular structures, increased ability to activate neutrophils, and the stimulation of the release of inflammatory mediators. Chronic inflammation is of particular concern because it has been associated with a number of complex diseases including atherosclerosis and cancer. Few studies have examined the impact of nanomaterials (NMs) on reproductive fitness. This is particularly disturbing because it has been shown that nanoparticles (NPs) and fibers can interact with the spindle apparatus of dividing cells. In addition, existing studies have tended to examine nanotoxicity within one generation, in young animals, and have been biased toward mammalian systems. Given the rate at which nanomanufacturing and research is accelerating, there will be increased amounts of NMs entering waste streams, and therefore eventually entering the environment. Thus, it is imperative that nanosafety now include studies that examine both intergenerational and age-associated effects of nanoexposure across broad taxa. These will be best carried out in model organisms with rapid generation times and well-known genetics (such as bacteria (*E. coli*), algae, yeast, flowering plants (*Arabidopsis*), roundworm, fruit fly, and mouse).

15.1 INTRODUCTION: *SILENT SPRING* FOR NANO?

Rachel L. Carson's *Silent Spring* first published in 1962 is hailed as beginning the "ecology movement." The book begins with a fable. In it, life in a prosperous farming community suddenly turns horribly wrong:

> Then a strange blight crept over the area and everything began to change. Some evil spell had settled on the community: mysterious maladies swept the flocks of chickens; the cattle and sheep sickened and died. Everywhere was a shadow of death. The farmers spoke of much illness among their families. In the town the doctors had become more and more puzzled by new kinds of sickness appearing among their patients. There several sudden and unexplained deaths, not only among adults but even among children, who would be stricken suddenly at play and die within a few hours.
>
> **Rachel Carson**
> Silent Spring, *p. 2, 1962*

Carson's tale ends with the eventual extinction of the human species, hence, the title of the book. She saw this coming from the massive disruption of ecosystems that she predicted would result from the unregulated use of organochloride pesticides such as DDT (dichlorodiphenyltrichloroethane). Even before Carson, evolutionary biologists warned of the danger of the increasing use of these pesticides. This warning was based upon the power of natural selection to produce resistance. Any novel toxic compound will show high efficacy in its initial applications. However, if genetic variants exist within populations that confer resistance to the compound, individuals with them will exhibit differential reproductive success, such that within a few generations their progeny will dominate the species. Thus, for the compound to continue its utility, greater and greater amounts will have to be used. Unfortunately this practice has an unanticipated consequence: that the application of greater amounts of poison in the ecosystem will impact other desirable species, including working its way into the food chain that humans ingest. There is considerable evidence that organochloride and organophosphate pesticides accumulate in human populations as a result of occupational or residential exposure. Despite their occurrence in considerably low levels in humans, their biological effects are hazardous since they interact with a plethora of enzymes, proteins, receptors, and transcription factors (Androutsopoulos et al. 2013). In addition, these compounds have considerable staying power in ecosystems. For example, despite the fact that the use of DDT and hexachlorocyclohexanes (HCH) were banned in China in the 1980s; it was estimated that in Shanghai the current daily intake via ingestion, inhalation, and dermal contact in dust was 79.4 and 4.9 ng/day, respectively, in children and 131.1 and 8.0 ng/day, respectively, in adults (Yu et al. 2012).

15.2 NANOPARTICLE DEFINITION

While organochlorides are still a problem, they are last generation toxins. They are manufactured in large scale and their effectiveness as toxins against insects and their side effects on other species does not rely on the compound's characteristics at different size scales. NMs are those who are characterized by a change in behavior when the substance is encountered on the nanoscale. Based on size units, a NP is anything in the range of 1–1000 nm (which is the edge of the microparticle range, Keck and Muller 2013). However, the Food and Drug Administration (FDA) defines a NP as one having any dimension from 1 to 100 nm (Hubbs et al. 2011; Keck and Muller 2013). NPs may result from natural processes (such as fires) or be industrial by-products (such as those produced in diesel exhaust), or can be specifically engineered for their nanoscale properties. It is the latter two categories that will be the primary concern of this review. This is because the last decade has seen a massive revolution in the production of engineered NMs (National Research Council 2006). This has been brought about by improved techniques for synthesis, rapid advances in chemistry and physics on the nanoscale, and improved understanding of intracellular structures at the molecular scale. In the United States, investments in nanotechnology had reached $1.1 billion by 2006. The estimated market value of nanotechnology products for 2009 was $254 billion, and growth is anticipated to reach $2.5 trillion by 2015 (Aitken et al. 2006; Hubbs et al. 2011). These new, engineered NPs include particulates that have never been studied and other particulates that have been previously only studied as components of mixtures (Hubbs et al. 2011). To provide the reader with the pace of expansion in nano-related research, at the time of the Hubbs et al. (2011) review, an Entrez Pubmed search for the terms, "nanotechnology," "nanomedicine," "nanotoxicology," and "nano" and "pathology" returned >20,000, 1417, 77, and 430 references, respectively. At the time of this writing, the same search returned 38,817, 3890, 528, and 853 references. This means that in 2011, application-based research publications were at a ratio of 42.24 to 1 compared to safety research. In 2013, this ratio has improved some (30.92 to 1), but is still way out of line considering the unknown but highly likely risks associated with many of the classes of engineered NMs. One explanation for why more safety studies have not yet been conducted on nanoparticulates is that, with the exception of mineral fibers most nonpharmaceutical particulates are regulated by their chemical composition, not by their shape and size. In the workplace, airborne particulates with a chemical composition not specifically noted in regulations, are regulated by the Occupational Safety and Health Administration (OSHA 2006, Hubbs et al. 2011). These particulates are classified by OSHA as "particulates not otherwise regulated" or PNOR. OSHA has set a limit of exposure (PEL) on PNORs for an 8-h time-weight average as 15 mg/m^3. However, the fraction of those PNOR in the respirable range is less than that for total PNOR, so another PEL has been set for particulates with an aerodynamic diameter of 5 μm or less at 5 mg/m^3 (OSHA 2006; Hubbs et al. 2011). This means that the current regulations permit the commercial production and use of most NP's without additional safety testing, using standards developed for larger respirable particulates of the same chemical composition (Murashov et al. 2009; Hubbs et al. 2011). A major concern of the current state of affairs is that a large number of studies have shown that the toxicity of particulates increases as the surface area-to-mass ratio increases, which of course occurs when the size of the particulate decreases (reviewed in Stern and McNeil 2008; Hubbs et al. 2011; Liu et al. 2012; Keck and Muller 2013).

Ironically, the chemical properties that make NMs so useful in technology, for example high surface-area-to-mass ratio, large quantum effects, deformability, durability, tendency to aggregate, optical sensitivity, hydrophobicity, high reactivity, rapid dissolution (for soluble particles), electrical conductivity, and great tensile strength, are the very same reasons why these materials may be highly toxic to biological systems. For example, these characteristics will help bond them to other pollutants in the environment such as heavy metals (cadmium) and can help transport these materials into cells (Maynard and Aitken 2007, Musee 2011). Carbon nanotubes (single-walled, SWCNT and multiple-walled, MWCNT) were first synthesized in the early 1990s. Since then a variety of nanoscale products have been developed, including those composed of fullerene carbon, minerals, metals, or light-emitting structures such as quantum dots. In addition, a variety of different elements have been included such as nanotubes and sheets made of boron, zinc oxide nanowires, ultrathin zinc oxide nanorods, gold nanotubes, and nanorods. As the chemistry improves, there seems to be little limit to the sorts of nanostructures that can be developed. The problem with all of this from the safety perspective is that we have no studies showing the long-term effects of exposure to NMs. For example, not even one cohort of humans has completed its life cycle since the development of nanoengineered products. However, we already have several examples of the toxic effect of high-level acute exposure to nanoparticulates; such as in the case of elevated levels of carbon nanotubes (CNTs) found in the lungs of first responders who are suffering lung disease resulting from dust inhalation during the attack on the World Trade Center in 2001 (Wu et al. 2010). Other examples of toxic impacts on humans are not so clear, but highly suggestive, such as the relationship between nanoparticulates in dust inhaled by highway workers and its impact on their sperm motility and subsequent studies, validating a role for CNTs in impairing male reproductive function in rats (Rosa et al. 2003; Yoshida et al. 2010). Also, nanoparticulates derived from combustion are a significant part of air pollution, and there are several examples of the impact of air pollution on mortality in humans (e.g., Dockery et al. 1993).

15.3 NANOPARTICLE EXPOSURE

The physical characteristics of NMs mentioned above may also aid in their translocation through soil, air, and water (Musee 2011). For example, NMs are known to move through soil and aquifers with high velocity (Lecoanet et al. 2004). NMs have been shown to impact the fate, transformation, and transportation of chemical compounds in the environment (Gao et al. 2008; Musee 2011). An example of this is NMs increasing the toxicity of polycyclic aromatic hydrocarbons (PAHs, Yang et al. 2006). These kinds of issues led Musee (2011) to point out that at present we are completely unprepared to handle significant amounts of NMs in our current waste treatment stream. Musee presented the following general points: (1) nanowaste sources are increasing at exponential rates (consider the personal products and cosmetics sector). (2) We have no data at present that supports the idea that current waste stream treatment methodologies can effectively remove or neutralize NMs (Westerhoff et al. (2008) showed that existing methods removed 0–40% of NMs depending upon the material). (3) NMs challenge the current regulatory frameworks (e.g., since little is known on how to clean up NM's spills at the research scale, let alone the industrial scale; how are we to develop appropriate regulations for such instances?). (4) To effectively remediate NM waste streams, we must be able to quantify the NM volumes within them (this is necessary to develop regulations and guidelines to effectively handle industrial scale NM waste; global production of NMs is expected to grow exponentially, considering that globally 10^4–10^5 tons annually of NMs used in structural applications by 2020, 10^3 tons skin care products, information communications technology industries >10^3 tons, biotechnology 10 tons, and environmental industry 10^3–10^4, source Royal Society and Royal Academy of Engineering Report, 2004). (5) Little has been done to study the ecotoxicity of NMs, particularly issues such as bioaccumulation in food chains. And (6) standardization of NM classification is lacking (this will be crucial in developing standardized waste treatment protocols and regulations).

Currently, no internationally agreed upon system of classification exists for NMs or their waste streams (Musee 2011, Keck and Muller 2013). Keck and Muller suggest that the most important variables to consider in such a classification scheme would be size and biodegradability, since these determine the NMs interaction with cells, and hence their biocompatibility. Their suggested classification system is presented in Table 15.1A and Musee's proposed classification of NM waste streams is presented in Table 15.1B. This distinction is important because there is evidence that the toxic risk of NMs changes at different portions of its life cycle (production, use, disposal). For example, SWCNTs, MWCNTs, and fullerenes used in automobile parts have a high hazard during their use, but are a medium hazard during waste disposal (Musee 2011).

While the invention of new nanotechnologies and products has now become exponential we have few studies documenting the impact of exposure to NMs. Those

---

TABLE 15.1A

Nanotoxicological Classification Systems

Class I: Size above 100 nm	Class II: Size above 100 nm
Biodegradable	Nonbiodegradable
Class III: Size below 100 nm	Class IV: Size below 100 nm
Biodegradable	Nonbiodegradable

Source: After Keck, C.M. and Muller, R.H., Nanotoxicological classification system (NCS)—A guide for risk–benefit assessment of nanoparticle drug delivery systems, *Eur. J. Pharmaceut. Biopharmaceut.* 2013, http://dx.doi.org/10.1016/j.ejpb.2013.01.001.

Note: Class I least dangerous to Class IV most dangerous; persistence of materials in individual organisms and the ecosystem is based on their biodegradability.

TABLE 15.1B

Classification of Nanowaste Streams

NM Class	Description	Comments	Examples
Class I	NM hazard: nontoxic, Exposure: low to high.	Concerns might arise from large amounts of waste, with threshold limit for toxicity.	Display backplanes of TV screens, solar panels, memory chips.
Class II	NM hazard: harmful or toxic, Exposure: low to medium.	Need to determine acute or chronic effects of such materials, to generate most effective way to handle in waste stream.	Display backplanes of TV screens, solar panels, polishing agents, paints, coatings.
Class III	NM hazard: Toxic to very toxic, Exposure: low to medium.	Will require protocols for handling hazard material during entire waste stream.	Food packaging, food additives, wastewater containing personal care products, pesticides.
Class IV	NM hazard: Toxic to very toxic, Exposure: medium to high.	Waste stream should be disposed of only in specialized hazardous waste sites.	Paints and coatings, personal care products, pesticides.
Class V	NM hazard: Very toxic to extremely toxic, Exposure: medium to high.	Waste stream should be disposed of only in specialized hazardous waste sites. Poor management of such streams could cause severe nanopollution.	Pesticides, sunscreen lotions, food and beverages containing fullerenes in colloidal suspensions.

Source: After Musee, N., Nanowastes and the environment: Potential new waste management paradigm, *Environ. Int.* 37: 112–128, 2011.

that exist are examples of acute exposure. For example, I have already mentioned the WTC first responders study. A similar case occurred in China, when seven workers in a print factory were exposed to a polyacrylic ester paste that contained NPs. The workers developed an unusual and progressive lung disease of which two of them died (Song et al. 2009). In this sense, NMs are not tremendously different from other toxic chemicals used in industrial processes or produced by combustion. High, acute doses of such materials will cause illness and possibly death. What are desperately needed are studies of chronic, long-term NP exposure. This will be particularly important for workers and researchers who will be exposed over long periods to NPs (such as graduate students, postdoctoral researchers, and faculty members; as well as employees in the growing private nanoproduct sector). We already know the lifespan of chemists (particularly those working in organic chemistry) are shorter than the general population (Olin 1978). In that study the causes of death among the 93 chemists belonging to a cohort of 857 men, who graduated from the Schools of Chemical Engineering in Sweden between 1930 and 1950 were analyzed. The chemists showed a significant increase of cancer, particularly of leukemias and malignant lymphomas, as well as urogenital tumors. The study indicated that chemists who continued with laboratory work for at least a few years after graduation and specifically worked with organic compounds displayed an increased frequency of death from cancer. These results may not be immediately relevant to today's chemists, given that better safety protocols are in place; however, they do raise a cause for alarm, given that we are still unclear what safety protocols will be most effective for NMs. The principal routes of exposure to NPs are skin, gastrointestinal tract, and respiratory tract (Aitken et al. 2006; Hubbs et al. 2011). Much of the current NP exposure is due to consumer products (such as makeups and sunblock). Also many NMs are being investigated for use in medical imaging and/or therapeutics, with intravascular and other forms of parenteral exposure as major exposure routes (Hubbs et al. 2011). It will be important to recognize going forward that the sort of exposures to NMs experienced by medical researchers, workers, and patients will be different.

15.4 DEMONSTRATING/ASSAYING NANOPARTICLES IN TISSUE

The fact that people are being exposed to NMs, and the high amount of NPs present in even a microgram of NM is not proof positive that these NMs are translocated throughout their bodies, or in the end causing tissue damage. For example, a 50 μg lung burden of well-dispersed MWCNTs in a mouse lung could easily distribute 10 billion nanotubes in that tissue (Hubbs et al. 2011). For a mouse this would be 10,000 or more nanotubes per alveolus (the mouse has about 4 million alveoli). The visibility of NMs in tissue may be compromised by a variety of factors, including lack of adequate dispersion, limitations of visualization due to narrow depth of field, lack of contrast between the biological material and the NM, and the small fraction of the section's area covered by the NPs (Hubbs et al. 2011). In experimental work, this can be overcome by labeling the NPs, particularly when using the TEM (transmission electron microscope) and FESEM (field emission scanning electron microscope). At the JSNN we can image NMs in biological tissue via our Zeiss helium ion (HIM) microscope. The HIM is optimized for imaging biological samples. It operates without a huge charge build-up, and therefore it is superior to FESEM in imaging NMs in biological tissue (see Chapter 5, this volume). In addition to imaging the NMs it must be quantified (He et al. 2012). Quantifying is a key to determining the bio-disruption potential of the NM, especially when considering realistic exposure levels, long-term deposition, and ability to cross natural barriers within the tissue. Evidence that NMs do end up in biologic tissue comes from the study of ultrafine particles (Oberdorster et al. 2005b; Oberdorster et al. 2007; Bonner 2010; Hubbs et al. 2011). For example, Bonner (2010) was able to show MWCNTs incorporated into the alveolar macrophages of C57BL6 mouse 24 h after a 6-h exposure to 30 mg/m^3 of MWCNT.

15.5 MECHANISMS OF NANOPARTICLE DAMAGE

A variety of NMs cause a variety of damages to specific cells and biological tissues. For example, engineered CNTs have been shown to cause pulmonary inflammation (Warheit et al. 2004; Shvedova et al. 2005, 2008; Hubbs et al. 2011; Win-Shwe and Fujimaki 2011). The Shvedova et al. and Warheit et al. studies were *in vivo* utilizing model organisms (mouse and rat, with acute exposure to CNTs). Some early responses to these studies questioned the legitimacy of their results, claiming that they were artifacts associated with the way the CNTs were administered to the animals (intratracheally or aspiration). Subsequent studies have demonstrated that the inflammation engendered by SWCNTs in biological tissue is real (Hubbs et al. 2011; Castranova et al. 2013; Liu et al. 2012). The Shvedova et al. studies also demonstrated fibrosis in the lungs of SWCNT exposed mice. These rodent studies, in addition to the observation of SWCNT's in the lungs of exposed first responders with interstitial/parenchymal lung disease at the WTC (Wu et al. 2010), is solid evidence of the damage that CNTs can produce in lung tissue. Not surprisingly, MWCNTs have also been implicated as a causal factor in pulmonary inflammation (Porter et al. 2010). The Porter study showed that in mice exposed to MWCNTs by aspiration showed granulomatous inflammation and very early fibrosis, at 7 days pass exposure. Finally, this study showed that the NMs penetrated through the pleura to reach the mesothelium, which is the classic target for fiber-induced carcinogenesis. Furthermore, the similarity between physical shape and biological persistence of MWCNT's and amphibole asbestos fibers has caused several research groups to hypothesize that MWCNT's like asbestos could cause mesothelioma (Takagi et al. 2008; Donaldson et al. 2010). Indeed, the Takagi study showed that 87.5% of p53 heterozygous mice developed mesothelioma after an intraperitoneal injection of 3 mg exposure of MWCNTs/mouse. The p53 mouse strain used in this experiment was chosen because it was already known to be sensitive to asbestos. This result is very strong evidence that MWCNTs have the potential to provoke cancer (at least within p53-deficient individuals).

With nonfibrous NPs the data on respirable particles tells us about how particle number, mass, surface area, surface properties, and composition impact particle-induced tissue alterations (Hubbs et al. 2011; Win-Shwe and Fujimaki 2011; Manzetti and Andersen 2012). For soluble particles, their toxicity is usually determined by the toxicity of their components. In addition, for metals and some other elements that may have low solubility but high cytotoxicity, the composition may have major effects on toxicity (Hubbs et al. 2011; Tchounwou et al. 2012). For example, nickel (which is usually insoluble), undergoes dissolution in the acidic environment of cytoplasmic vacuoles, releasing nickel ion (Costa et al. 2005). In this case, nickel causes its harm by mutagenesis, which may result in cancer. In the case of NMs such as these, composition mainly determines the histopathologies they cause, not their shape. For example, beryllium particles persist in the lung after exposure. If this occurs in an individual with a major histocompatibility complex (MHC) class II protein that presents beryllium to the immune system, then chronic beryllium disease is likely to result (Snyder et al. 2008; McCleskey et al. 2009).

For poorly soluble particles of low acute toxicity, histopathologic alterations in the exposed lung generally develop with time and are usually characterized by dose-dependent chronic inflammation, fibrosis, alveolar epithelial hypertrophy and hyperplasia, and with some species and particles, lung tumors (Porter et al. 2001; Roller 2009; Hubbs et al. 2011). In this case, composition matters since cessation of exposure causes regression of histopathology (as in TiO_2 exposed, but not in crystalline exposed silica rats, Baggs et al. 1997; Porter et al. 2004). The characteristics of the particle surface can also influence toxicity. For example, some particles have a nonhomogeneous composition that affects the particle surface and impacts cytotoxicity. Other NPs may generate more hydroxyl radicals from their surface than larger but still respirable particulates (Donaldson et al. 1996; Hubbs et al. 2011). Finally, although surface composition influences NP toxicity, there are general trends for NPs that cannot be explained by their surface chemistry and composition (Hubbs et al. 2011). Chief among these differences is, when compared with larger particles on a per mass basis, NPs cause more pulmonary inflammation (Renwick et al. 2004). This again can result from the greater surface area per mass shown by NPs (Sager et al. 2008).

The bulk of past NP tissue effect studies were conducted on lung tissue. However, recent work has begun to show that NPs can be incorporated into gastrointestinal cells (Gaiser et al. 2009). Previous studies had indicated that fine and ultrafine particles might be correlated with the development of Crohn's disease in humans (Lomer et al. 2002). A variety of experiments have now demonstrated the uptake of NPs in the gastrointestinal tract and a variety of pathologic effects. For example, intragastric administration of nanoclay to rats during 28 days led to reductions in the relative weight of the liver, the activity of its conjugating enzymes, the antagonistic activity of bifidoflora, and the hyperproduction of colonic yeast microflora (Smirnova et al. 2012). Zhang et al. (2005) showed that nanoselenium was less toxic when administered to mice than non-NP selenium (as measured by growth rates, oxidative stress, and liver injury). Wang et al. (2006) showed that zinc NPs by gastrointestinal administration had severe impacts on mice, including the development of lethargy, anorexia, vomiting, and diarrhea compared to controls. In addition, the zinc NP exposed mice had a 22% reduction of weight gain. Similar symptoms were seen in mice exposed to metallic copper NPs (Chen et al. 2006). The LD50 for mice exposed to NP copper compared to microcopper particles was 413 versus > 5000 mg/kg body weight. Wang et al. (2008) evaluated the possible effect of CdSe quantum dots (QD) via ingestion by evaluating the impact of NP Cd on enterocyte-like Caco-2 cells (human brush border expressing cells). Cells were incubated in Cd^{2+} (2–200 nmol/mL) containing medium for 24 h. The 200 nmol/mL concentration led to a 62% drop in the viability of these cells. However, cytotoxicity was strongly influenced by the QD coating and treatment (acid treatment vs. dialysis). Finally, Kalive et al. (2012) examined the impact of hematite NPs on human intestinal epithelial cells (Caco-2 cells). The hematite NPs were evaluated at three sizes (17, 53, and 100 nm). Their results indicated that the 17 nm NPs were the most toxic. The mechanism of toxicity was the disruption of epithelial structures (microvilli) and disruption of the cell–cell junctions (which lead to a reduction in the transepithelial electrical resistance). They also showed that the NPs were impacting the expression of genes in this cell line, particularly those related to cell junction maintenance. In conclusion, new studies of NP impact on gastrointestinal cells strongly suggests that NPs cause significant toxicity to GI cells and that this is dependent on both NP composition and size (Pattan and Kaul 2012). In addition, to disruption of the host gastrointestinal tract, the Smirnova et al. (2012) study demonstrates the potential for NP disruption of the microbiome. This of course could have accounted for the diarrhea and vomiting observed in the mouse studies. Given the importance of the microbiome to the overall health of animals (humans), and the fact that NPs have multiple well-documented impacts on bacteria and other microorganisms, GI tract NM effects must be considered a very serious concern. In conclusion, a series of studies have now shown the impact of various NMs and NPs on a variety of tissue types in metazoans (particularly mammals). These include NPs and NMs of various types (magnetite, hematite, TiO_2, gold NPs, ZnO, Fe_2O_3, selenium, silver NPs, fullerenes, SWCNTs, and MWCNTs, among others, Pattan and Kaul 2012).

15.6   SUBCELLULAR IMPACTS OF NPs

Clearly, the effect of NPs on tissues must be mediated via impacts on subcellular and molecular components of the organism. Particles that primarily enter the body via inhalation such as nanotubes, nanofibers, and nanowires have very high-aspect ratios (the ratio of length to width, Oberdorster et al. 2005a; Takagi et al. 2008; Porter et al. 2010; Hubbs et al. 2011; Castranova et al. 2013; Liu et al. 2012). Many of these materials have aspect ratios higher than asbestos, which has long-established carcinogenicity. Some asbestos fibers have diameters less than 100 nm and are thus natural NPs (Hubbs et al. 2011). The generation of free radicals due to incomplete phagocytosis has been proposed as a mechanism explaining the carcinogenicity of asbestos fibers (Kamp et al. 1992). Nanofibers may also interfere with the function of cellular organelles and structures. Ma et al. (2012) demonstrated that SWCNT's altered cytochrome c (*Cyt c*) function in human epithelial cells *in vitro*. They found that mitochondrial membrane potential and mitochondrial oxygen uptake were greatly decreased in cells treated with SWCNTs. The SWCNT's deoxidized *Cyt c* in a pH-dependent manner. Finally, electron transfer was also disturbed by SWCNTs.

Nanofibers can cause multinucleated giant cells to form via inhibition of cytokinesis, as well as via cell fusion (Jensen and Watson 1999; Asakura et al. 2010). Indeed, some nanofibers can enter cells without killing them and cause chromosomal missegregation. For example, Shvedova et al. (2008) observed a dividing macrophage with SWCNT nanoropes bridging anaphase chromosomes from opposite sides, with additional nanoropes located at the spindle poles. This may be a key mechanism in the initiation of mesothelioma by asbestos fibers of nano-dimension (Yegles et al. 1995). In studies of macrophages, MWCNTs have been shown to extend from the cytoplasmic margin and penetrate subcellular structures such as nuclei (Cheng et al. 2009). Porter et al. (2010) and Shedova et al. (2008) reported macrophages without nuclei, multinucleate epithelial cells, as well as mutations within specific genes (K-ras) following exposure to CNTs. Lindberg et al. (2013) showed that CNTs (both SWCNT and MWCNT) caused DNA damage in bronchial epithelial and mesothelial cells *in vitro*. This damage was both dose and cell-type dependent. This result makes the study of the mutagenic impact of NMs even more complicated because the effects will always be influenced by the characteristics of the NM (composition, shape, size, etc.) in question, as well as the genetic features of the organism, tissues, and cell types. Finally, there is evidence that NMs may also impact gene expression via interaction with the histones (Toyooka et al. 2012). Indeed, NPs and NMs do exist within cells as "naked" surfaces, but are spontaneously coated by proteins. This coating may actually aid in the dispersion of NPs/NMs throughout an organism's body and possibly throughout the ecosystem (Landsiedel et al. 2012). Obviously, mutagenesis of nanofibers and other NMs as well as their interaction with subcellular organelles will be a key issue with regards to determining their systemic effects.

15.6.1 Systemic Effects of NMs

There is abundant evidence for the association of levels of inhaled particulates (which include nanoparticles, NPs) with elevation of cardiovascular mortality and morbidity. Since particulates contain particles of various sizes, it is difficult to assess just how much of that elevation is resulting from combustion produced NP's (Hubbs et al. 2011). However, research on the pulmonary effects of engineered NP's demonstrates a variety of cardiovascular impacts. Li et al. (2007) exposed mice to purified SWCNTs at 10–40 mg per mouse by pharyngeal aspiration. They monitored oxidative stress in aortic and cardiac tissues at 1, 7, and 28 days postexposure. The exposure caused rapid but transitory pulmonary inflammation. Du et al. (2013) investigated the cardiovascular toxicity of different sizes and different dosages of silica NPs in Wistar rats. To determine the effect of particle size they examined NPs (30, 60, and 90 nm) and one silica particle outside the "nanoparticle" range (fine particle, 600 nm). These were administered at three doses of 2, 5, and 10 (mg/kg) by weight. Rats had intratracheal instillation for a total of 16 times and the concentration of silica (Si) in hearts and serum was measured. They also measured hematology parameters, inflammatory reaction, oxidative stress, endothelial dysfunction, and the level of myocardial enzymes in serum. They demonstrated that intratracheal-instilled silica NPs passed through the alveolar–capillary barrier into systemic circulation. The concentration of Si in the heart and serum depended on the particles size and dosage. The levels of reactive oxygen species (ROS) at 5 and 10 mg/kg by weight of the three silica NPs were higher than the fine silica particles. Blood levels of inflammation-related high-sensitivity C-reactive protein and cytokines such as interleukin-1beta (IL-1β), interleukin-6 (IL-6), and tumor necrosis factor-alpha were increased after exposure to three silica NPs at 10 mg/kg by weight. Again, the levels of IL-1β and IL-6 at 10 mg/kg by weight of silica NPs (30 nm) were higher than the fine silica particles. Significant decrease in superoxide dismutase, glutathione peroxidase and significant increase in malondialdehyde were observed at 10 mg/kg by weight of the three silica NPs. A significant decrease in nitric oxide (NO) production was induced which coincided with the reduction of nitric oxide synthase (NOS) activity and the excessive generation of ROS in rats. The levels of intercellular adhesion molecule-1 and vascular cell adhesion molecule-1 elevated significantly after exposure to three silica NPs at 10 mg/kg by weight. This is considered as early steps of endothelial dysfunction. They concluded that cardiovascular toxicity of silica NPs was related to the particles size and dosage. The mechanism they proposed to explain these results was that oxidative stress is involved in the inflammatory reaction and endothelial dysfunction, all of which could aggravate cardiovascular toxicology. Finally, they propose that endothelial NO/NOS system disorder caused by NPs could be one of the mechanisms for endothelial dysfunction. The evidence to date suggests that a wide variety of NP's can contribute to both pulmonary and cardiovascular disease, including SWCNTs, MWCNTs, and TiO$_2$. Du et al. (2013), as in earlier studies (Oberdorster et al. 2002) suggest that the small size of NP's aids in their ability to cross membrane barriers and thus be translocated throughout the body after inhalation. This means that NPs could interact with a variety of cell and tissue types to cause systemic impacts (Hubbs et al. 2011). This includes impacting blood cytokine levels (which in turn would impact systemic inflammation), activating peripheral neutrophils (Nurkiewicz et al. 2008), which in turn generate reactive oxygen species (ROS) at vessel walls. It is also suggested that NP deposition in the lungs can activate neuronal input to the brain, altering microvascular function in the tissue (Nurkiewicz et al. 2009). Furthermore it is possible that NPs can enter circulation (Elder et al. 2006; Kao et al. 2012) and cross the blood–brain barrier (Takenaka et al. 2001; Burch 2002; Sharma and Sharma 2007; Hubbs et al. 2011; Win-Shwe and Fujimaki 2011; Landsiedel et al. 2012). Indeed, many of the proposed nanomedicine applications of engineered NMs are predicated on the fact that they can deliver drugs to brain tissue via this mechanism (Kulkarni and Feng 2013).

15.7 EVOLUTIONARY TOXICOLOGY AND NMs

Nothing in biology makes sense, save in the light of evolution. Therefore, since toxicology attempts to understand how various substances impact the morbidity and mortality of biological organisms, it too should be guided by evolutionary principles. This is crucially important for understanding how we should construct an effective research program to gage the safety of NMs. It is now clear that there are multiple and quite serious hazards to the invention, production, and application of NMs. However, at the same time, NMs offer unprecedented opportunities to create new technologies in a variety of crucial applications, including ICT, medicine, environment, transportation, food science, and so on. The current rate of expansion of nanotechnology applications is exponential and at the same time, the attention to nanosafety issues has not kept pace. This immediately brings into question the legitimacy of our current nanosafety research program and exactly how we can develop a proactive and comprehensive safety research agenda that is guided by best principles. Currently, I will argue that we do not have such a program and it is imperative for us to develop one as rapidly as possible.

One of the core claims of evolutionary approaches toward health is that much of the modern disease burden results from the fact that humans live in environment that has little in common with our ancestral one. We are currently observing unprecedented increases in the prevalence of complex diseases in Western societies, including cancer, heart disease, stroke, and in addition behavioral diseases such as depression, obsessive–compulsive disorder, Parkinson's disease, and bipolar mania. The question can be immediately asked, how much of this is due to chronic low-level exposure to novel toxic compounds in the modern environment? For example, heavy metal exposure in humans has been increasing rapidly due to the exponential use of these materials in industry (Bradl 2002). While these materials are naturally occurring in the Earth's crust, and some are absolutely essential to the metabolism (cobalt, Co; copper, Cu; chromium, Cr; iron, Fe; magnesium, Mg; manganese, Mn; molybdenum, Mo; nickel, Ni; selenium, Si; and zinc, Zn), the vast majority of human toxic exposure comes from anthropogenic activities (such as mining, metal refining, fossil fuel burning, nuclear power stations, and paper-processing plants). For example, Saunders et al. (2012) measured fingernail metal levels as determinants of auditory performance in 59 subjects residing in the gold mining community of Bonanza, Nicaragua. Their auditory testing revealed widespread hearing loss in the cohort. Nail metal concentrations (mercury, lead, aluminum, manganese, and arsenic) far exceeded reference levels. This study could not evaluate how much of their exposure was due to NPs; however, we have already seen that nanoscale particles of heavy metals are generally more toxic than microscale versions of those same elements.

Most importantly, this kind of harm cannot be observed using the standard toxicologic protocols. These studies generally utilize either model organisms, or cultured cell lines exposed to an acute high dosage of the suspected toxin. For example, *Journal of Toxicology and Industrial Health* (2013) published 10 research studies. Of these, 4 were whole rat, 1 whole mouse, 2 human cultured cell lines, 1 plant, 1 alga, and 1 microbiome community. The rodent studies were all acute exposure and used only young animals. The rat studies used the Wistar rat, which is claimed as an "outbred" strain. However, in reality this strain was created by crossing former inbred strains, to produce a strain with greater genetic variability than its inbred progenitors. The mouse was an inbred albino strain that was prone to diabetes. Clearly, these stocks suffer from serious deficiencies of genetic structure, which actually make their utility for drawing inferences in humans (who are normally outbred) compromised. The error here is using biological materials that were designed for other purposes, to study toxicology. Cell line studies also have limited utility to study toxicity as well. Generally, cells are exposed to acute high dosages of a suspected toxin and then observed for some sort of damage. If the damage occurs, then one has some idea that a material may be toxic. For example, Kalive et al. (2012) examined a human intestinal cell (Caco-2) culture for response to 17, 53, and 100 nm hematite particles. The cells were exposed to NPs at 10 ppm for 48 h and cell viability was examined after 5, 14, and 28 days. This experiment demonstrated that the smaller NPs tended to kill more cells than the larger ones. However, the relevance to this sort of experiment to chronic whole organism exposures can be seriously questioned. This is for two reasons: first cell cultures are composed of individual cells that must be undergoing some amount of Darwinian adaptation to living as an individual cell. Thus, we expect that the gene expression patterns of cells in culture will not be equivalent to those in an intact whole organism; second, the environment of the cells in culture (sterile, for example), is not like that of an epithelial cell in an intact organism (e.g., with its complex microbiome).

These sorts of standard protocols should be jettisoned if we are to recover reliable information concerning the toxicity of NM. Clearly, we cannot wait for observable cohort studies of human beings exposed to various NMs (that would be like closing the barn door, after the horse has gone). Therefore, it is crucial that we design evolutionarily informed studies of NM toxicity in model organisms (e.g., fruit fly, round worm). These will have greater potential to be applicable to more "real" world exposures of NMs that are likely to be at low concentrations, but also occur over an individual's entire life span. For example, *Journal of Toxicology and Industrial Health* (2013) exposed *Drosophila melanogaster* (fruit fly) to different levels of deuterium oxide (D_2O) dissolved in the water used to make its initial food allocation and showed that at the highest percentage used (22.5%) a reduction in life span. However, in that same experiment, a lesser amount of D_2O, 15.0% actually increased life span, without causing a reduction of fecundity. This is an excellent example of how the use of model organisms will allow us to gather information concerning dosage and exposure of potentially toxic materials on longevity, fecundity, development, responses to environmental stress, and behavior. Model organisms are more useful to understanding physiological mechanisms in humans than suspected. For example, it is relatively easy to extrapolate genomic in *Drosophila* to human applications. This is possible due to the use of genomic "orthology," the ability to identify the genes that share common ancestry among *Drosophila* and humans. For example, the C2H2 zinc-finger superfamily (ZNF) has over 600 members in humans, comprising 1–2% of all human genes. C2H2 ZNF primarily encodes DNA- and chromatin-binding transcription factors, including developmental genes such as *Krox-20, snail, Gli, Kruppel,* and *hunchback.* Of these 600, at least 39 families are orthologous between the invertebrates and vertebrates, and thus thought to have originated in the common Metazoan ancestor (Knight and Shimeld 2001). It is also clear that these results are an underestimate of the power of orthology-based approaches to the functional characterization of genetic loci. Korcsmaros et al. (2011) have shown that orthology alignment and functional genomics allows determination of shared functional pathways between species and simultaneously enables the identification of novel signaling pathways. These orthology methodologies increase the usefulness of model species such as *Drosophila,* helping to identify new signaling mechanisms as well as novel genomic responses to toxins. Indeed, bioinformatics tools of considerable sophistication already exist which allow researchers to use orthology to map pathways between model organisms and humans. Finally, the model organism approach would allow us to more rapidly assess a specific NMs impact across generations.

Since the exponential growth in the use of NMs will mean a broader exposure of all organisms in the environment, the sooner we study more taxa, the more likely we are to head off potential ecological catastrophes (like those envisioned by Carson). For example, honeybee colonies in the Northern Hemisphere are now undergoing massive collapses. Parasites and pathogens are considered as principal actors, in particular the ectoparasitic mite *Varroa destructor,* associated viruses, and the microsporidian *Nosema ceranae* (Dainat et al. 2012). However, bees that are exposed to the commonly used pesticide imidacloprid are far more susceptible to the gut parasite, *Nosema* spp. Nosema infections increased significantly in the bees from pesticide-treated hives when compared to bees from control hives (Pettis et al. 2012). NMs have tremendous potential to create these sorts of unintended damages to the human environment. Consider that heavy metal exposure in bacteria has been associated with the co-evolution of antibiotic resistance (Seiler and Berendonk 2012). Thus, any heavy metal-based NP has the potential after entering the waste stream to accelerate antibiotic resistance in bacteria. In this vein, we should be conducting experiments now to evaluate the potential for heavy metal (e.g., silver and titanium) NPs to have this unintended impact, since they are proposed for widespread use against bacteria and protozoa purposes (Allahverdiyev et al. 2011). Heavy metal NPs are potentially quite toxic to humans, and just like antibiotics bacteria will undoubtedly evolve resistance to them. Resistance has the potential to create a NP thread mill where greater concentrations of NPs will be required to achieve the initial successes of the treatments. Thus, a crucial set of experiments will determine how quickly bacteria evolve resistance to proposed heavy metal NPs, what correlated traits will emerge, and what will be the associated fitness costs. Experimental evolution is precisely the tool required to examine such questions (Garland and Rose 2009). Clearly, if single strains of bacteria can evolve resistance to NP treatment, so too will the entry of NMs in significant concentrations in the broader ecosystem impact microbiomes. For example, Claire et al. (2013) examined the impact of the pesticide chlorpyrifos on the intestinal microbiome. They utilized an intestinal tract simulator (SHIME) and the intestinal tract of experimental rats. As predicted, chronic low-level exposure to this pesticide induced dysbiosis of the microbiome; with some strains drastically increasing in frequency and other strains decreasing in frequency. We already have emerging evidence that NPs may have similar dysbiotic impacts on microbiomes. For example, Garcia et al. (2012) examined the impact of cerium dioxide (CeO_2), titanium dioxide (TiO_2), silver (Ag), and gold (Au) NPs on the activity of the most important components of the waste water treatment community using respiration tests and biogas analysis. Not surprisingly, differences were found in the impact of each NP, with CeO_2 causing the greatest inhibition in biogas production (100%) as well as a strong inhibitory action of other biomasses; Ag NPs had an intermediate inhibitory action (33–50%) and a slight inactivation of other biomasses, while Au and TiO_2 caused only slight inhibition. Merrifield et al. (2013) has demonstrated that copper (Cu) and silver (Ag) NPs disrupt the intestinal microbiome of zebrafish (*Danio rerio*). The fish were fed on diets containing Cu and Ag NPs at a concentration of 500 mg kg^{-1} food. In particular, some beneficial strains (e.g., *Cetobacterium somerae*) were suppressed to nondetectable amounts. These sorts of alterations in the microbiome are immediate causes for concern. Alterations in the human microbiome have been associated with colorectal cancer, inflammatory bowel disease, and rheumatic disease (de Wouters et al. 2012; Gallimore and Godkin 2013; Yeoh et al. 2013). In addition to diseases within specific organisms, large-scale NP pollution could have the potential to disrupt terrestrial microbiomes associated with agriculture, causing potentially damaging reduction of agricultural production.

15.8 CONCLUSION: CARSON REVISITED

This chapter has reviewed a variety of potential dangers which we face accompanying the exponential increase in research, production, and distribution of NMs. The potential for NMs to improve the human condition are tremendous. Applications are planned in information technology, medicine, transportation, consumer products, and a wide variety of other areas. For example, Quick et al. (2008) demonstrated that a carboxyfullerene could mimic Cu–Zn superoxide dismutase and significantly increase both the age-related cognition and the life span of mice. El-Safty et al. (2013) has demonstrated how nanosphere sensors can be used to facilitate the removal of toxic heavy metals from the blood. Clearly, such a technology will have wide applications, including the possibility of helping to detoxify drinking water supplies; the lack of drinkable water is one of the most significant problems facing the world today. The production of even greater computing power via the use of NMs could revolutionize all scientific endeavors. However, the implication of such technologies is not without a cost. The danger that NMs/NPs present to living systems is real and is a clarion call for an immediate and aggressive research program addressing how to best implement this technology while protecting human life and the biosphere as a whole. Less some future species inherits a world without *Homo sapiens* in it.

REFERENCES

Aitken, R.J., Chaudhry, M.Q., Boxall, A.B.A., and Hull, M., Manufacture and use of nanomaterials: Current status in the UK and global trends, *Occup. Med.* (Lond.) 56: 300–306, 2006.

Allahverdiyev A.M., Abamor E.S., Bagirova M., and Rafailovich M., Antimicrobial effects of TiO(2) and Ag(2)O nanoparticles against drug-resistant bacteria and Leishmania parasites, *Future Microbiol.* 6(8): 933–940, 2011.

Androutsopoulos, V.P., Hernandez, A.F., Liesivuori, J., and Tsatsakisa, A.M., A mechanistic overview of health associated effects of low levels of organochlorine and organophosphorous pesticides, *Toxicology* 307: 89–94, 2013.

Asakura M., Sasaki T., Sugiyama T. et al., Genotoxicity and cytotoxicity of multi-wall carbon nanotubes in cultured Chinese hamster lung cells in comparison with chrysotile A fibers, *J. Occup. Health.* 52(3):155–166, 2010.

Baggs, R.B., Ferin, J., and Oberdorster, G., Regression of pulmonary lesions produced by inhaled titanium dioxide in rats, *Vet. Pathol.* 34: 592–597, 1997.

Bonner, J., Nanoparticles as a potential cause of pleural and interstitial lung disease, *Proc. Am. Thorac. Society* 7: 138–141, 2010.

Bradl, H., *Heavy Metals in the Environment: Origin, Interaction, and Remediation,* Vol. 6 (London, UK: Academic Press), 2002.

Burch, W.M., Passage of inhaled particles into the blood circulation in humans, *Circulation* 106: e141–e142, 2002.

Carson, R., *Silent Spring* (New York, NY: Houghton Mifflin and Co.), 1962.

Castranova, V., Schulte, P.A., and Zumwalde, R.D., Occupational nanosafety considerations for carbon nanotubes and carbon nanofibers, *Accounts of Chemical Research* 46(3): 605–872, 2013.

Chen, Z., Meng, H., Xing, G. et al., Acute toxicological effects of copper nanoparticles *in vivo*, *Toxicology Letters* 163(2): 109–120, 2006.

Cheng, C., Muller, K.H., Koziol, K.K. et al., Toxicity and imaging of multi-walled carbon nanotubes in human macrophage cells, *Biomaterials* 30: 4152–4160, 2009.

Claire, J., Gay-Quéheillard, J., Léké, A. et al., Impact of chronic exposure to low doses of chlorpyrifos on the intestinal microbiota in the Simulator of the Human Intestinal Microbial Ecosystem (SHIME®) and in the rat, *Environ. Sci. Pollut. Res.,* 20(5): 2726–2734, 2013.

Costa, M., Davidson, T.L., Chen, H. et al., Nickel carcinogenesis: Epigenetics and hypoxia signaling, *Mutat. Res.* 592: 79–88, 2005.

Dainat, B., Evans, J.D., Chen, Y.P., Gauthier, L., and Neumann, P., Predictive markers of honey bee colony collapse, *PLoS One* 7(2): e32151, 2012.

de Wouters, T, Doré, J., and Lepage, P., Does our food (environment) change our gut microbiome ('in-vironment'): A potential role for inflammatory bowel disease?, *Dig. Dis.* 30 Suppl 3: 33–39, 2012. doi: 10.1159/000342595.

Dockery, D.W., Pope, C.A., Xu, X. et al., An association between air pollution and mortality in six U.S. cities, *N. Engl. J. Med.* 329: 1753–1759, 1993.

Donaldson, K., Beswick, P.H., and Gilmour, P.S., Free radical activity associated with the surface of particles: A unifying factor in determining biological activity? *Toxicol. Lett.* 88: 293–298, 1996.

Donaldson, K., Murphy, F.A., Duffin, R., and Poland, C.A., Asbestos, carbon nanotubes and the pleural mesothelium, a review of the hypothesis regarding the role of long fiber retention in the parietal pleura, inflammation and mesothelioma, *Part. Fibre. Toxicol.* 7: 5, 2010.

Du, Z., Zhao D., Jing L. et al., Cardiovascular toxicity of different sizes amorphous silica nanoparticles in rats after intratracheal instillation, *Cardiovasc. Toxicol.*, 13(3): 194–207, 2013.

El-Safty, S.A., Abdellatef, S., Ismael M., and Shahat, A. Optical nanosphere sensor based on shell-by-shell fabrication for removal of toxic metals from human blood, *Adv. Healthcare Mater.* 2(6): 854–862, 2013.

Elder, A., Gelein, R., Silva, V. et al., Translocation of inhaled ultrafine manganese oxide particles to the central nervous system, *Environ. Health Perspect.* 114(8): 1172–1178, 2006.

Gaiser, B.K, Fernandes, T.F., Jepson, M. et al., Assessing exposure, uptake, and toxicity of silver and cerium dioxide particles nanoparticles from contaminated environments, *Environ. Health* 8(Suppl. 1), S2, 2009.

Gallimore, A.M. and Godkin, A., Epithelial barriers, microbiota, and colorectal cancer, *N. Engl. J. Med.* 368(3): 282–284, 2013.

Gao, J., Bonzong, J-CJ, Britton, G. et al., Nanowastes and the environment: Using mercury as an example of a pollutant to assess environmental fate of chemicals absorbed onto manufactured nanomaterials, *Environ. Toxicol. Chem.* 27(4): 808–810, 2008.

Garland, T. and M.R. Rose, Eds., *Experimental Evolution* (University of California Press, Berkeley), 2009.

He, X., Ma, Y., Li, M. et al., Quantifying and imaging engineered nanomaterials *in vivo*: Challenges and techniques, *Small* 9(9–10): 1482–1491, 2012.

Hubbs, A.F., Mercer, R.R., Benkovic, S.A. et al., Nanotoxicology—A pathologist's perspective, *Toxicol. Pathol.* 39: 301–324, 2011.

Jensen, C.G. and Watson, M., Inhibition of cytokinesis by asbestos and synthetic fibers, *Cell Biol. Int.* 23: 829–840, 1999.

Journal of Toxicology and Industrial Health, 29(1), 2013, Sage Journals. http://tih.sagepub.com/content/29/1.toc

Kalive, M, Zhang, W., Chen, Y., and Capco, D.G., Human intestinal epithelial cells exhibit a cellular response indicating a potential toxicity upon exposure to hematite nanoparticles, *Cell Biol. Toxicol.* 28: 343–368, 2012.

Kamp, D.W., Graceffia, P., Pryor, W.A. et al., The role of free radicals in abestos-induced diseases, *Free Radic. Biol. Med.* 12: 293–315, 1992.

Kao, Y., Cheng, T., Yang, D. et al., Demonstration of an olfactory bulb-brain translocation pathway for ZnO nanoparticles in rodent cells *in vitro* and *in vivo*, *J. Mol. Neurosci.* 48: 464–471, 2012.

Keck, C.M. and Muller, R.H., Nanotoxicological classification system (NCS)—A guide for risk–benefit assessment of nanoparticle drug delivery systems, *Eur. J. Pharmaceut. Biopharmaceut.* 84(3): 445–448, 2013.

Knight, R.D. and Shimeld, S.M., Identification of conserved C2H2 zinc-finger gene families in the Bilateria, *Genome Biol.* 2(5): RESEARCH0016, 2001.

Korcsmaros, T., Szalay, M.S., Rovo, P. et al., Signalogs: Orthology based-identification of novel signaling pathway components in three metazoans, *Plos One* 6(5): e19240, 2011.

Kulkarni, S.A. and Feng, S.S., Effects of particle size and surface modification on cellular uptake and biodistribution of polymeric nanoparticles for drug delivery, *Pharm Res.* 30(10): 2512–2522, 2013.

Landsiedel, R., Fabian, E., Ma-Hock, L. et al., Toxico-/biokinetics of nanomaterials, *Arch Toxicol.* 86: 1021–60, 2012.

Lecoanet, H.F., Bottero, J-Y, and Wiesner, M.R., Laboratory assessment of the mobility of nanomaterials in porous media, *Environ. Sci. Technol.* 38(19): 5164–5169, 2004.

Lindberg, H., Falck, G., Singh, R. et al., Genotoxicity of short single-wall and multi-wall carbon nanotubes in human bronchial epithelial and mesothelial cells in vitro, *Toxicology* 313(1): 24–37, 2013.

Liu, Y., Zhao, Y., Sun, B. et al., Understanding the toxicity of carbon nanotubes, *Acc. Chem. Res.* 46(3): 702–713, 2012.

Lomer, M.C., Thompson, R.P., and Powell, J.J., Fine and ultrafine particles of the diet: Influence on the muscosal immune response and association with Crohn's disease, *Proc. Nutr. Soc.* 61: 123–130, 2002.

Ma, X., Zhang, L., Wang, L. et al., Single-walled carbon nanotubes alter cytochrome c electron transfer and modulate mitochondrial function, *ACSNANO* 6(12): 10486–10496, 2012.

Manzetti, S. and Andersen, O, Toxicological studies of nanomaterials used in energy harvesting consumer electronics, *Renew. Sustain. Energy Rev.* 16: 2102–2110, 2012.

Maynard, A.D. and Aitken, R.J., Assessing exposure to airborne nanomaterials: Current abilities and future consequences, *Nanotoxicology* 1(1): 26–41, 2007.

McCleskey, T.M., Buchner, V., Field, R.W., and Scott, B.L., Recent advances in understanding the biomolecular basis of chronic beryllium disease: A review, *Rev. Environ. Health* 24: 75–115, 2009.

Murashov, V., Engel, S., Savolainen, K. et al., Occupational safety and health in nanotechnology and Organization for Economic Cooperation and Development, *J. Nanopart. Res.* 11: 1587–1591, 2009.

Musee, N., Nanowastes and the environment: Potential new waste management paradigm, *Environ. Int.* 37: 112–128, 2011.

National Research Council, *A Matter of Size: Triennial Review of the National Nanotechnology Initiative,* (Washington, DC: National Academies Press), 2006.

Nurkiewicz, T.R., Porter, D.W., Hubbs, A.F. et al., Nanoparticle inhalation, augments particle-dependent systemic microvascular dysfunction, *Part. Fibre Toxicol.* 5: 1, 2008.

Nurkiewicz, T.R., Donlin, M., Hubbs, A. et al., Mechanistic links between the lung and the systematic microcirculation alter nanoparticle exposure, *The Toxicologist* 108: A1353, 2009.

Oberdorster, G., Sharp, Z., Atudorei, V. et al., Extrapulmonary translocation of ultrafine carbon particles following whole-body inhalation exposure of rats, *J. Toxicol. Environ. Health* 65: 1531–1543, 2002.

Oberdorster, G., Maynard, A, Donaldson, K. et al., Principles for characterizing the potential human health effects from exposure to nanomaterials: Elements of a screening strategy, *Part. Fiber Toxicol.* 2: 8, 2005a.

Oberdorster, G., Oberdorster, E., Oberdorster, J., Nanotoxicology: An emerging discipline evolving from studies of ultrafine particles, *Environ. Health Perspectives* 113: 823–839, 2005b.

Oberdorster, G., Stone, V., and Donaldson, K., Toxicology of nanoparticles: A historical perspective, *Nanotoxicology* 1: 2–25, 2007.

Olin, G.R., The hazards of a chemical laboratory environment—A study of the mortality in two cohorts of Swedish chemists, *Am. Ind. Hyg. Assoc. J.* 39(7): 557–562, 1978.

OSHA, 29 CFR—Occupational Safety and Health Regulations (OSHA Standards). In *Toxic and Hazardous Substances*, Vol. *Standards—29* CFR. Occupational, Safety, and Health Administration, Washington, DC. http://www.osha.gov/pls/oshaweb/owadisp.show_document?p_table=STANDARDS&p_id=9992.

Pattan, G. and Kaul, G., Health hazards associated with nanomaterials, *Toxicol. Ind. Health*, 2012, (published online 25 September 2012). DOI: 10.1177/0748233712459900.

Pettis, J.S., van Engelsdorp, D., Johnson, J., and Dively, G., Pesticide exposure in honey bees results in increased levels of the gut pathogen *Nosema*, *Naturwissenschaften* 99(2): 153–158, 2012.

Porter, D.W., Ramsey, D., Hubbs, A.F. et al., Time course of pulmonary response of rats to inhalation of crystalline silica: Histological results and biochemical indices of damage, lipidosis, and fibrosis, *J. Environ. Pathol. Oncol.* 20(Suppl. 2), 1–14, 2001.

Porter, D.W, Hubbs, A.F., Mercer, R.R. et al., Mouse pulmonary dose- and time course-responses induced by exposure to multi-walled carbon nanotubes, *Toxicology* 269: 136–147, 2010.

Porter, D.W., Hubbs, A.F., Mercer, R.R. et al., Progression of lung inflammation and damage in rats after cessation of silica inhalation, *Toxicol. Sci.* 79: 370–380, 2004.

Quick, K.L., Ali, S.S, Arch, R. et al., A carboxyfullerene SOD mimetic improves cognition and extends the lifespan of mice, *Neurobiol. Aging* 29: 117–128, 2008.

Renwick, L.C., Brown, D., Clouter, A., and Donaldson, K., Increased inflammation and altered macrophage chemotactic responses caused by two ultrafine particle types, *Occup. Environ. Med.* 61: 442–447, 2004.

Roller, M., Carcinogencity of inhaled nanoparticles, *Inhal. Toxicol.* 21(Suppl. 1): 144–157, 2009.

Rosa, M., Zarrilli, S., Paesano, L. et al., Traffic pollutants affect fertility in men, *Hum. Reprod.* 18: 1055–1061, 2003.

Royal Society and Royal Academy of Engineering Report on Nanotechnology. Nanoscience, and nanotechnologies: Opportunities and uncertainties. *Roy. Soc. Roy. Acad. Eng.*, 2004. Available from: The Royal Academy of Engineering's; website: www.raeng.org.uk

Sager, T.M., Kommineni, C., and Castranova, V., Pulmonary response to intratracheal instillation of ultrafine versus fine titanium oxide: Role of particle surface area, *Part. Fibre Toxicol.* 5: 17, 2008.

Seiler, C. and Berendonk, T.U., Heavy metal driven co-selection of antibiotic resistance in soil and water bodies impacted by agriculture and aquaculture, *Front. Microbiol.* 3: 399, 2012.

Sharma, H.S. and Sharma, A., Nanoparticles aggravate heat stress induced cognitive deficits, blood–brain barrier disruption, edema formation, and brain pathology, *Prog. Brain Res.* 162: 245–273, 2007.

Shvedova, A.A., Kisin, E.R., Mercer, R. et al., Unusual inflammatory and fibrogenic pulmonary responses to single-walled carbon nanotubes in mice, *Am. J. Physiol. Lung Cell Mol. Physiol.* 289: L698–L708, 2005.

Shvedova, A.A., Kisin, E.R., Murray, A.R. et al., C57BL/6 mice: Inflammation, fibrosis, oxidative stress, and mutagenesis, *Am. J. Physiol. Lung Cell Mol. Physiol.* 295(4): L552–L565, 2008.

Smirnova, V.V., Tananova, O.N., Shumakova, A.A. et al., Toxicological and sanitary characterization of bentonite nanoclay, *Gig Sanit.* (3): 76–8, 2012.

Snyder, J.A., Demchuk, E., McCanlies, E.C. et al., Impact of negatively charged patches on the surface of MHC class II antigen-presenting proteins on risk of chronic beryllium disease, *J. R. Soc. Interface* 5: 749–758, 2008.

Song, Y., Li, X, and Du, X., Exposure to nanoparticles is related to pleural effusion, pulmonary fibrosis, and granuloma, *Eur. Respir. J.* 34: 559–567, 2009.

Stern, S.T. and McNeil, S.E., Nanotechnology safety concerns revisited, *Toxicol. Sci.* 101(1): 4–21, 2008.

Takagi A., Hirose A., Nishimura T. et al., Induction of mesothelioma in p53 +/− mouse by intraperitoneal application of multi-wall carbon nanotube, *J. Toxicol. Sci.* 33(1): 105–116, 2008.

Takenaka, S., Karg, E., Roth, C. et al., Pulmonary and systemic distribution of inhaled ultrafine silver particles in rats, *Environ. Health Perspect.* 109: 547–551, 2001.

Tchounwou, P.B., Yedjou, C.G., Patlolla, A.K., and Sutton, D., Heavy metal toxicity in the environment, *EXS* 101: 133–164, 2012. doi: 10.1007/978-3-7643-8340-4_6.

Toyooka, T., Amano, T., and Ibuki, Y., Titanium dioxide particles phosphorylate histone H2AX independent of ROS production, *Mut. Res.* 742: 84–91, 2012.

Wang, B., Feng, W.Y., Wang, T.C. et al., Acute toxicity of nano- and micro-scale zinc powder in healthy adult mice, *Toxicol. Lett.* 161: 115–113, 2006.

Wang, L., Nagesha, D.K., Selvarasah, S. et al., Toxicity of CdSe nanoparticles in Caco-2 cell cultures, *J. Nanobiotechnol.* 6: 11, 2008.

Warheit, D.B., Laurence, B.R., Reed, K.L., et al., Comparative pulmonary toxicity assessment of single-walled carbon nanotubes in rats, *Toxicol. Sci.* 77: 117–125, 2004.

Westerhoff, P., Zhang, Y., Crittenden, J., Chen, Y., Properties of commercial nanoparticles that affect their removal during water treatment, In: Grassian, V.H. Ed., *Nanoscience and*

Nanotechnology: Environmental and Health Impacts (NJ: John Wiley and Sons), pp. 71–90, 2008.

Win-Shwe, T. and Fujimaki, H., Nanoparticles and neurotoxicity, *Int. J. Mol. Sci.* 12: 6267–6280, 2011.

Wu, M., Gordon, R.E, Herbert, R. et al., Case report: Lung disease in World Trade Center responders exposed to dust and smoke: Carbon nanotubes found in the lungs of World Trade Center patients and dust samples, *Environ Health Perspect.* 118: 499–504, 2010.

Yang, K., Zhu, L., and Xing, B., Adsorption of polycyclic aromatic hydrocarbons by carbon nanomaterials, *Environ. Sci. Technol.* 40(6): 1855–1861, 2006.

Yegles, M., Janson, X., Dong, H. et al., Role of fiber characteristics on cytotoxicity and induction of anaphase/teleophase aberrations in rat pleural mesothelial cells *in vitro*: Correlations with *in vivo* animal findings, *Carcinogenesis* 16: 2751–2758, 1995.

Yeoh, N., Burton, J.P., Suppiah, P., Reid, G., and Stebbings, S., The role of the microbiome in rheumatic diseases, *Curr. Rheumatol. Rep.* 15(3): 314, 2013.

Yoshida, S., Hiyoshi, K, Oshio, S. et al., Effects of fetal exposure to carbon nanoparticles on reproductive function in male offspring, *Fertility Sterility* 93(5): 1695–1699, 2010.

Yu, Y., Li, C., Zhang, X. et al., Route-specific daily uptake of organochlorine pesticides in food, dust, and air by Shanghai residents, China, *Environ. Int.* 50(2012): 31–37, 2012.

Zhang, J., Wang, H., Yan, X., and Zhang, L., Comparison of short-term toxicity between nano-Se and selenite in mice, *Life Sci.* 76(10): 1099–1109, 2005.

Index

A

A-EGNFs, *see* APTES-treated
        EGNFs (A-EGNFs)
Abraxane®, 124
Action potential durations (APDs), 64
   cryogenic system, 36
   and RI, 64
Action potentials (APs), *see* Neural signals
AFM, *see* Atomic force microscopy (AFM)
ALD, *see* Atomic layer deposition (ALD)
Alumina nanoparticulate hybrid
        composites, 171
   atomistic analysis, 173
   BDA, 173
   EPON™ 9554 epoxy model, 172–173
   epoxy–alumina and silane–alumina
        interfaces, 175
   interface binding energy, 175
   material chemistry structures, 172
   mode-I fracture toughness, 176
   molecular layer configurations, 173
   surface interaction energy, 174–176
7-aminoactinomycin D (7AAD), 262
   and APTES, 203
   composite resins, 209
   data for PC12s, 249, 251, 259
   FT-IR spectra, 206
   GPTMS-treated fibers, 203, 204, 205
   and LDH assay data, 252
   silane-coupling agents, 207
3-aminopropyl triethoxysilane (APTES), 203
Antioxidants, 148–149
APDs, *see* Action potential durations (APDs)
Apoptosis, 262
Apostain assays, 262; *see also* Lactate
        dehydrogenase assay (LDH)
APTES-treated EGNFs (A-EGNFs), 206, 207
   amine functional groups on, 210
   epoxy composite resins, 208
   reinforcement effect, 210
   silanized EGNFs with, 221
APTES, *see* 3-aminopropyl triethoxysilane
        (APTES)
Arthritis
   activation and inflammatory, 153
   MC degranulation, 152
   MCs in, 152–154
   mouse models of, 149
Aspect ratio-dependent toxicity, 248

Asthma
   Cromolyn stabilizer, 150
   efficacy of FD, 151
   in vitro studies, 151
   MC and PBB in, 149
   novel fullerene derivative, 151, 152
   role of basophils in, 150
Atherosclerosis, 156, 267
Atherosclerotic plaque
   ATCA, 157
   metallo-fullerenes MRI, 156–157
Atherosclerotic plaque-targeting contrast
        agent (ATCA), 157
Atom by atom concept, 2, 5
Atomic force microscopy (AFM), 1, 19–20,
        70, 233; *see also* Helium ion
        microscopy (HIM)
   HS-AFM technique, 74
   intermittent contact mode, 73
   mechanisms of operation, 71–74
   noncontact mode imaging, 73, 74
   SCM, 74–75
   set point maintenance, 72
   SMM, 75, 76
   SPMs, 70
   tip–sample force regimes, 71
Atomic layer deposition (ALD), 14
AuNP, *see* Gold nanoparticle (AuNP)

B

Backscattered electrons (BSE), 84, 89
BAL, *see* Bronchoalveolar lavage (BAL)
BCC, *see* Body-centered cubic (BCC)
BDA, *see* Butan-diamine (BDA)
Beam equivalent pressure (BEP), 36
Bio-nano-electro-mechanical systems
        (bio-NEMS), 132
Biochips, *see* Bio-nano-electro-mechanical
        systems (bio-NEMS)
Biological paradigms, 7
   mechanical and, 6, 7
   of molecular assembly, 2
Biomaterial-based tubular nerve conduits, 53–54
Biomimetic nanotechnology, 1, 2
Biosensor, 134, 135
   molecular, 134
   multifunctional biosensor–chip, 137
   nano-FET, 141
   nanoscale, 135

Printed and bound by CPI Group (UK) Ltd, Croydon, CR0 4YY

18/10/2024

01776262-0011